WEATHERIZING YOUR HOME

WEATHERIZING YOUR HOME

George R. Drake

Reston Publishing Company, Inc., Reston, Virginia
A Prentice-Hall Company

Library of Congress Cataloging in Publication Data

Drake, George R.
  Weatherizing your home.

  Includes index.
    1. Dwellings—Insulation. 2. Dwellings—
Energy conservation. I. Title.
TH1715.D7   643'.7   78-17165
ISBN 0-8359-8592-X

10 9 8 7 6 5 4 3 2

to NATALIE

CONTENTS

PREFACE

My home, as were millions of others, was built before the energy crisis of the early 1970s. Builders and owners didn't know there would one day be a fuel crisis. So they built and maintained homes with little regard to future energy crises or rising costs. No insulation was installed. Entryways were not built. Double glazing and storm equipment were nonexistent. Heat energy from the sun wasn't considered. Weather stripping was too much trouble to install or add. Caulking crumbled over the years and was not replaced. Shutters were added, but as a decorative touch, not to close at sundown to keep heat in and wind out. Homeowners just shoved the thermostat a bit higher.

But today we stare at the monthly bills and wonder how we will ever be able to keep on paying them. Not only that: if we use up a tank of fuel, we might not be able to get it refilled! So we keep the thermostat lower and we're cold! In the summer, we keep the thermostat higher to reduce the cost of electricity for air-conditioning and we're too hot. We need to *weatherize* our homes.

Weatherization is the "buttoning up" of your home to prevent heat escapement in the winter and heat infiltration in the summer. Weatherization is the efficient use of heating and cooling equipment so that less fuel oil, gas, electricity, wood, coal or solar energy are used to produce heated rooms in the winter and cooled rooms in the summer. Weatherization is the use of supplemental equipment to heat, cool, humidify, dehumidify, ventilate and circulate air to decrease expenses of the primary fuel or to supplement the primary heater/air-conditioner because of its efficiency or inadequacy. Weatherization is the reclaiming of heated or cooled air normally lost; it is reusing the conditioned air for supplementary heating or cooling. And finally, weatherization is the effective use of nature to aid in heating and cooling our homes. Weatherization is not limited to the homeowner; weatherization techniques are applicable for use by apartment renters, condominium owners, and new home builders as well.

You and your family derive three major benefits from weatherizing. First, you spend a little money to save a lot of money in the future. Your initial expenditures are recouped in a matter of one to several years. Once you've spent the dollars for weatherizing, you need not spend them again; nearly all weatherizing installations are permanent and require very little, if any, recurring maintenance. You prevent your home from losing heat in the winter and from gaining heat in the summer. You make effective use of the heating and cooling equipment you have in your home. Therefore, you use less energy and save money. You prevent your money from "going up the chimney in smoke."

The second major benefit derived from weatherizing your home is family comfort. Gone are the cold drafts across the floor, the uneven heating of some rooms, the cold outer walls, the drafty windows, the frost on the inside of the windows, inefficient fireplaces, the dry air of winter, the humid air of summer, and the hot breathless nights of summer. After weatherizing, your family will feel more comfortable and their health will probably be somewhat better too.

Finally, by weatherizing your home, you'll use less oil, gas, coal, wood, or electricity. By using less, you will be contributing along with countless others to the saving of this country's fossil fuels. It has been said that even if oil and gas become more plentiful, the price will increase by an estimated 50 to 100 percent by the 1980s.

This book contains detailed information and step-by-step instructions to enable you to weatherize your home or apartment yourself, saving you the labor costs of a contractor. Most procedures are very easy; the novice do-it-yourselfer can accomplish the work like a professional. The book is presented in the order that you should proceed in weatherizing your home—from replacing broken glass, caulking, and weatherstripping, to solar heating. More than 205 illustrations show you how; 39 do-it-yourself procedures guide you step-by-step and include lists of materials and tools needed for the job; 42 quick reference charts provide you with tabulated data on your part of the country, material choices, where to weatherize, and other pertinent supplemental information. A glossary and a comprehensive index help you to readily use this book.

Scan the table of contents and some of the do-it-yourself procedures, illustrations, and quick reference charts. You should realize that by utilizing this information, you and your family can weatherize your house or apartment easily yourselves and save hundreds of dollars in contractors' fees and fuel and power costs.

Saving yourself money, increasing your living comfort, and helping prevent a fossil fuel depletion are three excellent reasons for weatherizing your home. Get started today.

GEORGE R. DRAKE

ACKNOWLEDGMENTS

Nearly every homeowner across this country as well as throughout the world is caught in the flurry of finding better and less expensive ways of home heating and cooling. Mankind's very inquisitive nature has aided this rush of enthusiasm by causing people to think, study, create, design, invent, implement and to contrive ways to reduce heating and cooling costs and increase comfort at the same time. The government, community agencies, universities, utilities, private industry, inventors, small businessmen and the do-it-yourselfer have joined together in a common cause to fight the increased costs of fuel, to prevent shortages, and to find alternate sources of energy to supplement and perhaps one day replace fossil fuels.

I would like to express a sincere *thank you* to the many persons in various companies, associations, agencies, and universities who generously provided information, photographs, and illustrations for this book. Each of the following contributors welcomes inquiries from you:

Amana Refrigeration, Inc., Amana, IA 52203

American Ventilation Association, P.O. Box 7464, Houston, TX 77008

Andersen Corporation, Bayport, MN 55003

Ashley, Martin Ind., P.O. Box 730, Sheffield, AL 35660

Association of Home Appliance Manufacturers, 20 North Wacker Drive, Chicago, IL 60606

Broan Manufacturing Co., Inc., Hartford, WI 53027

California Redwood Association, 617 Montgomery Street, San Francisco, CA 94111

Certain-Teed Corp., P.O. Box 860, Valley Forge, PA 19482

Community Services Administration, Washington, DC 20506

Consumer Information, Public Documents Distribution Center, Pueblo, CO 81009

Dolin Metal Products, Inc., 475 President Street, Brooklyn, NY 11215

The Dow Chemical Company, P.O. Box 2166, Midland, MI 48640

Dyna Corporation, 2540 Industry Way, Lynwood, CA 90262

Energy Research and Development Administration (ERDA), 20 Massachusetts Ave., MW, Washington, DC 20545

Federal Energy Administration, Twelfth Street and Pennsylvania Avenue, NW, Washington, DC 20461

Heatilator Fireplace, Div. of Vega Industries, Inc., 4096 W. Saunders Street, Mt. Pleasant, IA 52641

Isothermics, Inc., P.O. Box 86, Augusta, NJ 07822

Leigh Products Inc., 1870 Lee, Coopersville, MI 49404

Lennox Industries Inc., P.O. Box 250, Marshalltown, IA 50158

Leslie-Locke, A Questor Corp., Ohio Street, Lodi, OH 44254

Macklanburg-Duncan Co., P.O. Box 25188, Oklahoma City, OK 73125

The Majestic Co.—An American Standard Co., Huntington, IN 46750

Martin Industries, P.O. Box 1527, Huntsville, AL 35807

National Climatic Center, NOAA, Federal Building, Asheville, NC 28801

National Forrest Products Association, 1619 Massachusetts Avenue, NW, Washington, DC 20036

National Mineral Wool Insulation Association, Inc., 382 Springfield Avenue, Summit, NJ 07901

National Science Foundation, 1800 G Street NW, Washington, DC 20550

National Solar Heating and Cooling Information Center, P.O. Box 1607, Rockville, MD 20850

Pease Company, Ever-Strait Division, 7100 Dixie Highway, Fairfield, OH 45023

Plaskolite, Inc., 1770 Joyce Avenue, Columbus, OH 43216

Rapperswill Corp., 305 E. 40th Street, New York, NY 10016

M. H. Rhodes, Inc., 99 Thompson Road, Avon, CT 06001

Ridgway Steel Fabricators Inc., P.O. Box 382, Ridgway, PA 15853

Riteway Manufacturing Co., P.O. Box 6, Harrisonburg, VA 22801

Rohm and Haas Company, Independence Hall West, Philadelphia, PA 19105

Solar Energy Industries Association, 1001 Connecticut Avenue, NW, Suite 632, Washington, DC 20036

Solaron Corporation, Stapleton Field Industrial Park, 4850 Olive Street, Commerce City, CO 80022

Teledyne Mono-Thane, 1460 Industrial Parkway, Akron, OH 44310

Thermograte Enterprises Inc., 51 Iona Lane, St. Paul, MN 55117

Triangle Engineering Co., P.O. Box 38271, Houston, TX 77088

U.S. Department of Commerce, National Bureau of Standards, Washington, DC 20234

U.S. Department of Health, Education, and Welfare, 330 Independence Avenue, SW, Washington, DC 20201

U.S. Department of Housing and Urban Development, 451 Seventh Street SW, Washington, DC 20410

Velux-American Inc., 80 Cummings Park, Woburn, MA 01801

Vermont Woodstove Co., 307 Elm Street, Bennington, VT 05201

The West Bend Company, 400 Washington Street, West Bend, WI 53095

White-Rodgers Division, Emerson Electric Co., 9797 Reavis Road, St. Louis, MO 63123

Window Shade Manufacturers Association, Oak Brook Executive Plaza, 1211 W. 22nd Street, Oak Brook, IL 60521

Wind-Wonder, Inc., P.O. Box 36462, Houston, TX 77036

Winter Seal of Flint, Inc., 209 Elm Street, Holly, MI 48442

Woodmack Products, Inc., 850 Aldo Avenue, Santa Clara, CA 95050

York Division of Borg-Warner, P.O. Box 1592, York, PA 17405

WEATHERIZING
YOUR
HOME

Seven questions probably come immediately to your mind: What is home weatherization? What are its benefits? What's in it for me? When do I do it? How do I go about it? Where do I get the money to do it? Should I do the work myself or hire a contractor? These questions are answered in this chapter.

1-1. WHAT IS WEATHERIZATION?

Weatherization is the "buttoning up" of your home to prevent heat escapement in the winter and heat infiltration in the summer. Weatherization is the *efficient* use of heating and cooling equipment so that *less* fuel oil, gas, electricity, wood, coal, or solar energy is used to produce heated rooms in the winter and cooled rooms in the summer. Weatherization is the use of supplemental equipment to heat, cool, humidify, dehumidify, ventilate, and circulate air to decrease expenses for the primary fuel or to supplement the primary heater/air conditioner because of its inefficiency or inadequacy. Weatherization is the reclaiming of heated or cooled air normally lost; it is reusing the conditioned air for supplementary heating or cooling. And finally, weatherization is the effective use of nature to aid in heating and cooling homes. Weatherization *is not limited to the homeowner*; weatherization techniques are applicable for use by apartment renters, condominium owners, and new home builders as well.

1

1-2. WHAT ARE THE BENEFITS OF WEATHERIZATION— WHAT'S IN IT FOR ME?

You and your family derive three major benefits from weatherizing. First, you spend a little money to *save* a lot of money in the future. Your initial expenditures are recouped in a matter of one to several years. Once you've spent the dollars for weatherizing, you need not spend them again; nearly all weatherizing installations are permanent and require very little, if any, recurring maintenance. You prevent your home from losing heat in the winter and from gaining heat in the summer. You make effective use of the heating and cooling equipment already in the home. Therefore, you use *less* energy and save money. You prevent your money from "going up the chimney in smoke."

The second major benefit derived from weatherizing your home is family *comfort*. Gone are the cold drafts across the floor, the uneven heating of some rooms, the cold outer walls, the drafty windows, the frost on the inside of the windows, inefficient fireplaces, the dry air of winter, the humid air of summer, and the hot, breathless nights of summer. After weatherizing, your family will feel more comfortable and their health will also probably be somewhat better (Fig. 1-1).

Finally, by weatherizing your home, you'll use less oil, gas, coal, wood, or electricity. By using less, you will be contributing, along with countless others, to the saving of this country's fossil fuels. It has been said that even if oil and gas become more plentiful, the price will increase by an estimated 50 to 100 percent by the 1980s. Saving yourself money, increasing your living comfort, and helping prevent a fossil fuel depletion are three excellent reasons for weatherizing your home.

1-3. WHEN DO I WEATHERIZE?

You can start to weatherize during *any season* of the year. For procedures that are done on the outside, early fall is probably the nicest because of the temperatures, but as stated, any season is okay. Retuning furnaces and air conditioners is usually done just prior to the heating and cooling seasons, respectively. Each chapter in this book provides you with any special seasonal or periodic (such as monthly) procedures necessary.

1-4. HOW DO I WEATHERIZE?

You weatherize your home by following the procedures in this book and in the manufacturer's product instructions. First, you make an inspection to determine what must be done to weatherize your home, then you generate a plan of action (Section 2-4) and then you follow the procedures in this book which are written sequentially chapter by chapter to accomplish first things first (Fig. 1-2).

FIGURE 1-1. Perhaps you think a cozy little fireplace heats your home; actually, they are only about 10 percent efficient. If a heat-circulating fireplace such as this, with a heating chamber that recycles room air, is used along with other heat saving and reclaiming devices, the efficiency can be somewhat increased (*courtesy Heatilator Fireplace—Div. of Vega Industries, Inc.*).

To aid you in weatherizing your home in the most effective, efficient, and economical manner, this book contains quick reference charts (QRC), do-it-yourself (DIY) procedures, and more than 200 photographs and illustrations.

FIGURE 1-2. Insulating the attic is one of the major steps in weatherization. This homeowner is adding blanket insulation to loose fill insulation that previously existed in the attic (*courtesy Certain-Teed Corp.*).

The quick reference charts provide important information in summary form, tabulated data, selection guides, and so on. The do-it-yourself procedures are step-by-step guides on completing weatherizing installations, maintenance, additions, and similar jobs. Each do-it-yourself procedure lists the materials needed for the job, the tools needed to accomplish the job, and step-by-step procedures to accomplish the job; warnings, cautions, and notes are interspersed where applicable. The do-it-yourself procedure steps are detailed so that the inexperienced as well as the experienced do-it-yourself homeowner can satisfactorily complete the jobs. The numerous illustrations enable you to visualize how the job is to be accomplished.

In addition to the *permanent* weatherizing procedures presented in each chapter, there are also emergency quick-fix procedures, where applicable. For example, maybe you can't afford to buy storm windows this year; but you can afford some plastic to fashion into a storm window quite inexpensively which may last two or more seasons. Or, what are you going to do if a snowball is thrown through your window?

A comprehensive glossary at the end of the book will help you with unfamiliar weatherization and building terms and conversions.

1-5. SAVINGS

You can save money by weatherizing your home practically anywhere in the U.S. (Fig. 1-3). Save *every* year at no cost by setting the thermostat *down* (Chapter 4). Cut your heating bills in half by caulking and weather stripping your windows and doors (Chapters 5 and 6), insulating your attic (Chapter 7), and servicing your oil or gas burner (Chapter 12). Of course you might not need to do all of these things; the information in each chapter tells you what to look for before you do unnecessary work.

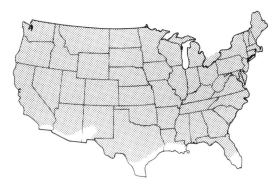

FIGURE 1-3. If you live in the shaded area, you can save money on your heating bill by following the energy-saving measures in this book (*courtesy U.S. Department of Housing and Urban Development*).

If you live in the southern part of the U.S. (Fig. 1-4), you can save on air-conditioning costs. Set your thermostat *up* in the summer from the usual setting, insulate your attic, and caulk and weather-strip your windows and doors.

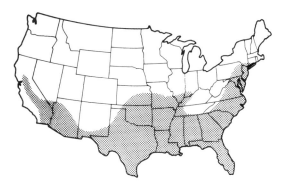

FIGURE 1-4. If you live in the shaded area of the U.S. shown on this map, you can save air-conditioning energy dollars by following the energy-saving measures in this book (*courtesy U.S. Department of Housing and Urban Development*).

Because of the energy crisis, a number of federal agencies including the Department of Housing and Urban Development and the Energy Research and Development Administration, The National Solar Heating and Cooling Information Center, power and utility companies, universities, private research companies, and magazine staffs have been making estimates on the savings in actual dollars or the savings in percent of dollars spent for energy that can be saved by the adoption of various weatherizing techniques. Of course these savings and percentages are averages, and they vary from place to place. The existing structure of your house, its location, and its heating system can all vary the actual dollars you can save on your home. Refer to QRC 1-1 and 1-2 and Fig. 1-5 for an idea of savings.

Quick Reference Chart 1-1
TYPICAL DOLLAR SAVINGS FROM WEATHERIZING

Weatherization Task	Savings Per Year or Season
Set thermostat down six degrees F during heating season	$20–65
Caulk and weather-strip windows and doors (for heating)	$30–70
(for cooling)	$20–50
Insulate the attic (for heating)	$35–120
(for air conditioning)	$25–50
Set thermostat up six degrees F for air conditioning	$ 5–15
Add clock thermostat	$24–48
Install storm doors and windows	$20–55
Yearly maintenance on heating and air-conditioning equipment	$25–65
Insulate ducts and hot water pipes	$20–160
Insulate hot water heater	$ 2–6
Set water heater temperature down to 120 to 140 degrees F	$ 5–45

The savings that you will realize after weatherizing your home are not taxable *and that is* a *saving*. You save much more, of course, if you do the work yourself, and most of the work is easy enough for anyone to do. Doing the work yourself eliminates the high cost of labor. The money that you spend now can be paid off in 1 to 10 years after which a permanent gain is realized every year. Over the long run, the money spent now will pay off in the greatest net savings. Putting money into energy savings now is a *hedge against inflation*.

Other savings that can be realized are federal and state tax credits for the installation of weatherizing materials and units and for solar heating applications. Most of these credits are applicable only on purchases of commercial equipment; this is particularly true of solar hot water and heating systems.

Quick Reference Chart 1-2
TYPICAL PERCENT SAVINGS FROM WEATHERIZING

Weatherization Task	Estimated Percent Saved on Energy Bill
Close up holes and cracks in window glass and foundations	20
Lower thermostat to 68 degrees F daytime and 60 degrees F nighttime (winter)	15
Raise thermostat to 78 degrees F (summer)	27
Caulk and weather-strip windows and doors	10
Insulate walls and ceilings	25
Add six inches of insulation to ceiling	20
Install storm windows and doors	15

Savings that make you feel warm all over

CONSTRUCTION	UNITS OF FUEL REQUIRED WITHOUT INSULATION OR STORM WINDOWS	WITH STORM WINDOWS AND DOORS, SUBTRACT:	WITH 6" ATTIC INSULATION, SUBTRACT:	WITH WALL INSULATION INDICATED, SUBTRACT:	TOTAL SAVING IN HEATING BILL IN UNITS AND PER CENT
TWO-STORY HOUSE					
Frame	100	18	16	3⅝"—19	53—53%
Frame with brick veneer	105	18	16	3⅝"—24	58—55%
Brick and concrete block	116	18	16	2"—32	66—57%
Stone	138	18	16	2"—53	87—63%
ONE-STORY HOUSE					
Frame	100	12	25	3⅝"—13	50—50%
Frame with brick veneer	104	12	25	3⅝"—16	53—51%
Brick and concrete block	111	12	25	2"—22	59—53%
Stone	126	12	25	2"—36	73—58%

Source: University of Illinois Small Homes Council-Building Research Council

Because wood is a better insulator than other exterior materials, wood houses are less costly to heat than houses made of brick or stone. These tables arbitrarily assign each wood frame house 100 units of heating cost; the additional fuel needed to heat other houses of the same size is figured from that base, as are the savings that result from insulation and from storm windows and doors. The brick or concrete block construction assumes eight inches of block with brick exterior; walls of the stone house are eight inches thick. Adding wall insulation to existing masonry houses is often impractical because it requires furring strips on the inside walls with insulation between the strips and then new interior wallboard or plaster. On most frame houses wall insulation can be blown in. Two-story houses cost less to heat than one-story houses of the same size because there is less roof area for heat to leak through.

FIGURE 1-5. The costs of heating can be reduced by adding storm windows and doors, attic and wall insulation, and other weatherizing materials.

Consequently, the do-it-yourselfer who enjoys designing, building, and installing his own system is not the beneficiary of some tax credits. Some states have passed legislation that prohibits local governments from adding the cost of solar equipment to property tax assessments.

Experts estimate that if all the changes and improvements recommended are made, a savings of 40 percent in fuel consumption can be realized. The cost of making the changes and improvements can be overcome in the years ahead by reduced heating and cooling costs. Savings continue permanently for the life of your home. Some utility companies have estimated that a $1,500 investment in energy-saving home improvements could save $15,000 in power bills by the time a mortgage is fully paid.

If you and your family live in an uninsulated house, relatively simple do-it-yourself energy-conserving home improvements can save up to 50 percent of your energy costs. Hopefully by the time we have a severe energy crunch in 1980–85, we'll all have better insulated homes and alternate means of heating, cooling, and powering our homes.

1-6. WHERE TO GET MONEY

The easiest place to get money is from your savings account. Remember, money in the bank is earning you 5 to 7 percent, but the amount of interest on the money may be relatively small if you can use the money to reduce your energy expenses by 10, 20, or up to 50 percent. Once you start completing some of the energy-saving procedures such as blocking infiltrating air, caulking, and weather stripping, you'll soon save enough money from energy you did *not* have to use to buy more weatherizing materials such as insulation. Conclusion: *Spending money now* to cut energy costs can be worth more than investing your money in the bank.

To get started without spending any money, wear warmer clothes, thereby keeping those 400 Btu of body heat generated per hour inside your clothes, and set the thermostat down in the winter; wear lighter clothes and set the thermostat up in the summer. The money you immediately save by *not* paying for excess energy can be used for caulking and weather stripping materials.

You can borrow money from a bank or other lending institution on a long-term, low-interest home improvement loan. Lending institutions are very willing to lend money for weatherizing projects because they know that you will be paying out less each month and you will therefore be able to pay back on your loan. Lending institutions are making money easier to obtain for fully insulated homes because the institutions realize that the monthly cost of heating and cooling will be about 40 percent less. Even some power and light companies are proposing to make loans toward the cost of insulating ceilings (attics).

The optimum level of your investment in energy conservation techniques increases considerably as the climatic conditions grow more severe and as more expensive heating and cooling energy forms are used. The rising real energy prices serve as an incentive for *energy consumers* to become *energy conservers*.

1-7. DO IT YOURSELF OR HIRE A CONTRACTOR?

Fortunately for all of us, most weatherizing procedures are relatively easy for anyone to do. No special skills are necessary and in most cases, common hand tools are the only tools necessary. The installation and maintenance procedures in this text and the accompanying illustrations should enable even the most inexperienced do-it-yourselfer to accomplish most weatherizing jobs. In addition, the manufacturers of weatherization products normally provide adequate instructions for installation by the do-it-yourselfer. Let's face it; with the rising costs of labor, most of us have to do it ourselves just to get the job done at a reasonable and affordable price.

One of the weatherization procedures, installing insulation in the outer walls of an existing home (Fig. 1-6), is not a do-it-yourself job because of the

FIGURE 1-6. One weatherization job to be left to a contractor is the insulation of walls of an existing home. Money may be saved by opening and closing the siding yourself (*courtesy Rapperswill Corp.*).

need for special training, materials, and special installation equipment. But you can save some money by doing part of the work. You may also decide to have someone install your storm windows. The point to be made is this: select a reputable contractor to do the work.

In selecting a reputable contractor, invite several contractors to your home to give you an estimate. Be sure all parties understand the scope of the work to be done, the brand of the material, the R value of the material, cost, and schedule. The R value is the resistance of the material to the flow of heat; estimates should be based on materials with like ratings. Be wary of any contractor's price that is significantly lower than the other contractors' prices and don't sign a contract immediately for a "special offer" or a "demonstrator in the neighborhood." Ask the contractors for names of people for whom they have done similar work; then check with those people to determine their satisfaction with the contractor and the job.

One last word about doing the work yourself. If you lack confidence in your own ability to do it yourself, be reminded that others may have felt the same way, but they are giving it a try. *Building Supply News* predicted in 1977 that because increasingly easier methods of installation have spurred industry sales, the sales of self-installed merchandise will equal the sales of materials for professional installation. If that prediction is true, it is the first in recent American history.

Keep one other thing in mind; it will cost about double the do-it-yourself price to have a contractor do the work.

chapter 2

HOME WEATHERIZATION— WHERE TO BEGIN

There is a definite course of action to be taken to weatherize your home. Basically, the *gross* problem areas are remedied first. These gross problem areas allow the maximum escapement of warm air from your home in winter and the maximum infiltration of warm air in the summer. For example, if you have broken window panes or unblocked holes through your masonry for the passage of electrical wires, you'd be foolish (and "fuelish" too) to insulate your attic. The major causes of the greatest escapement and infiltration of heat must be corrected first.

After the major causes of escapement and infiltration are resolved, a balanced combination of energy-conserving home improvements are installed for the most cost-effective means of weatherizing. A limited amount of money can be best spent by putting in some improvements of each type rather than by spending all of the money on a single or a few types of improvements. An unbalanced combination of improvements can become a waste of time and of dollars.

Fortunately, the most *effective* ways to keep your home comfortable and reduce costs because of heat losses are also the *least expensive*. For example, the first steps are to close up all holes and cracks in windows and walls, and add caulking and weather stripping around doors and windows. This is much more effective than replacing your furnace with a solar heater. You can't have an effective solar heater unless your home is already caulked, weather stripped, and well insulated. Some weatherization procedures cost nothing, take only seconds to do, and can save 10 percent of your heating and air-conditioning bill per year.

Once you've read through this book and have done some weatherizing, you may be concerned that you're "buttoning up" your home too well, that there will not be an adequate exchange of fresh air. Don't be concerned. In most homes, 70 to 100 percent of the inside air is exchanged with the outside air *every hour* because of natural drafts around windows and doors and through walls and from opening and closing entry doors. Only a 20 percent exchange is needed for normal ventilation purposes. The National Bureau of Standards says that it is unlikely that an existing house could be sealed up that tight by normal energy conservation improvements alone. The NBS further states that in *new* houses with tight construction, it may be necessary to provide a separate combustion air inlet to the furnace (Chapter 12).

Before weatherizing is started, it is necessary to present some preliminary information. Therefore, this chapter discusses weather and climate, the transfer of heat, definitions, construction of a weatherization plan, and safety.

2-1. WEATHER AND CLIMATE

Although this book is devoted to *weatherizing* your home, it is not a book on weather or weather forecasting; you can visit your local library to obtain books on this information. However, since the *climate* of your geographical area can help you make decisions on how best to weatherize a home, some pertinent information on climates is given in this section. Other sections in this book will refer to terms and data included in this section.

Weather is the condition of the atmosphere in a locality in terms of the elements of heat, atmospheric pressure, wind, and moisture. Heat is the element that *mixes* the atmosphere to cause changes in the weather—winds, air pressure changes, storms, rain, and snow. All weather changes are brought about by temperature changes in different parts of the atmosphere.

The sun is the source of the earth's heat. The sun is a ball of glowing gases, about 93 million miles from earth, generating solar energy into space. Most of this solar energy is lost in space. The sun's energy is transmitted as waves that are similar to radio waves; some are visible light waves while others are invisible. Some (although not heat waves) change to heat when absorbed by objects such as soil or our bodies. About 43 percent of the radiation reaching earth hits the earth's surface and is changed to heat; the rest stays in the atmosphere or is reflected into space.

For average weather, there is 52 percent cloudiness in the sky. A typical cloud reflects 75 percent of the sunlight striking it back into space; thus, on overcast days, only 25 percent of the sun's energy hits the ground and that energy is absorbed and reflected in varying degrees. For example, dense forests absorb about 95 percent of solar energy and change it to heat; snow absorbs about 25 percent; grassy fields, 80 to 90 percent; water, 60 to 96 percent (depending upon the angle of the sun); dry sand, 75 percent, and a plowed field absorbs about 75 to 95 percent of the solar energy striking it.

The earth's atmosphere admits most of the solar radiation. When this radiation is absorbed by the earth's surface, it is radiated again as heat waves, most of which are trapped by water vapor in the atmosphere keeping the earth warm. The atmosphere acts to moderate the daytime temperature and retard night heat losses. Clouds at night trap the earth's heat; on a clear night, more heat escapes.

Air is heated mainly by contact with the warm earth. When the air is warmed, it expands, becomes lighter, and rises. The rising air is replaced by colder, heavier air. This circulation of air is known as a *convection* current.

Convection currents cause local winds and breezes. As previously mentioned, different land and water surfaces absorb different amounts of heat. Mountains absorb heat faster during daylight than nearby valleys and lose heat faster at night. Land warms faster than water during the day and cools faster at night. The air above these surfaces is warmed and cooled accordingly, causing local winds.

Climate is the weather at a given locality over a period of time (usually several decades). It involves averages, totals, and extremes to set a picture of the weather pattern. Climate is affected by the same physical conditions that affect weather—latitude, prevailing winds, ocean currents, mountains, nearness to the sea, and the like. Climate might be called the *generalized weather* of an area. Climatic data is based on weather conditions reported by numerous weather stations over many years. From the data collected, a *mean* (a middle point between extremes) is determined for each of the weather factors such as inches of rain, snow, temperature, percent of sunshine, and so on.

Climatic data are important to you in weatherizing your home. The data can help you select the proper sized heating and air-conditioning equipment, solar energy heating equipment, and in determining the type and location of shrubbery and trees (more on this is included in the appropriate chapters). Complete climatic data for the United States are published in the *Climatic Atlas of the United States*, available for $6 from the National Climatic Center, Federal Building, Asheville, N.C. 28801, Attn: Publications. Checks are to be made payable to Department of Commerce, NOAA. The climatic maps in the atlas present, in uniform format, a series of analyses showing the national distribution of mean, normal, and/or extreme values of temperature, precipitation, wind, barometric pressure, relative humidity, dew point, sunshine, sky cover, heating degree days, solar radiation, and evaporation.

Of particular interest to homeowners, as well as the heating and air-conditioning industry, are the total number of *heating degree days* and *cooling hours* per year for local areas. Reference is made in chapters in this book about heating and cooling equipment sizing as related to heating degree days and cooling hours.

A *degree day* is the difference of the *average* temperature of the day from a standard reference of 65 degrees F. The *heating degree day* is the difference between the daily average temperature *below* the reference and 65 degrees F. Normally heating is not required in a home if the outdoor average temperature for the day is 65 degrees F. Heating degree days are determined by subtracting the average daily temperature from the reference temperature of 65 degrees F. For example, if the average temperature during 24 hours for a particular day is 50 degrees F, the day is a 15 heating degree day.

Because the heating degree days are accumulative, they represent the total heating load for that period. The relationship between degree days and fuel consumption is linear; doubling the degree days usually doubles the fuel consumption. Comparing normal seasonal degree days in different locations gives a rough estimate of seasonal fuel consumption. For example, it would require roughly 4½ times as much fuel to heat a home in Chicago, Ill., where the mean annual total heating degree days are about 6,200, than to heat a similar home in New Orleans, La., where the annual total heating degree days are around 1,400. Using degree days has the advantage that the consumption ratios are fairly constant, i.e., the fuel consumed per 100 degree days is about the same whether the 100 degree days occur in only three or four days or are spread over seven or eight days.

Cooling hours are the number of hours that the outdoor temperature is 80 degrees F or higher. The daily cooling hours are totaled for the year and data averages over a period of years are available. If the outdoor temperature is not 80 degrees F or higher, air conditioning is not normally needed.

The total number of heating degree days for a particular climatic area is obtained by summing the individual heating degree days (Fig. 2-1 and QRC 2-1). The total number of cooling hours is obtained by summing all of the cooling hours (QRC 2-2).

Sunshine

Sunshine certainly has a lot to do with weatherizing your home. During the winter in most parts of the country, you'd like to have as much sunshine as possible and you would like to use its energy to heat your home and domestic hot water. You want to let in as much sunshine as possible to warm the house, brighten the house, and make flowers and perhaps even vegetables grow in windows, enclosed porches, or attached greenhouses. In the summer, you'd like to prevent direct sunlight so that your home interior is cooler. Figures 2-2 through 2-5 and QRC 2-3 are included to help in planning for the use of sunshine in weatherizing your home.

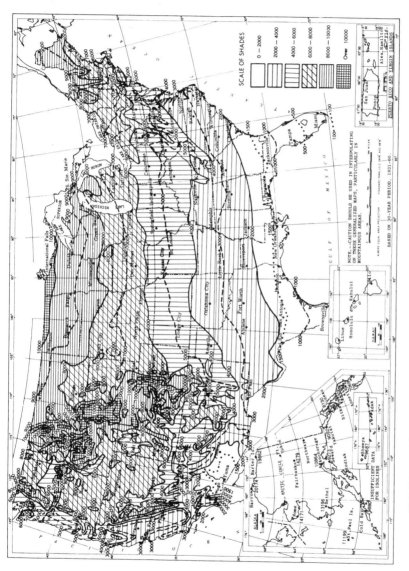

FIGURE 2-1. The normal total annual heating degree days are shown on this map (*from the Climatic Atlas of the U.S.*).

Quick Reference Chart 2-1
NORMAL TOTAL HEATING DEGREE DAYS

STATE AND STATION	JULY	AUG.	SEP.	OCT.	NOV.	DEC.	JAN.	FEB.	MAR.	APR.	MAY	JUNE	ANNUAL
ALA. BIRMINGHAM	0	0	6	93	363	555	592	462	363	108	9	0	2551
HUNTSVILLE	0	0	12	127	426	663	694	557	434	138	19	0	3070
MOBILE	0	0	0	22	213	357	415	300	211	42	0	0	1560
MONTGOMERY	0	0	0	68	330	527	543	417	316	90	0	0	2291
ALASKA ANCHORAGE	245	291	516	930	1284	1572	1631	1316	1293	879	592	315	10864
ANNETTE	242	208	327	567	738	899	949	837	843	648	490	321	7069
BARROW	803	840	1035	1500	1971	2362	2517	2332	2468	1944	1445	957	20174
BARTER IS.	735	775	987	1482	1944	2337	2536	2369	2477	1923	1373	924	19862
BETHEL	319	394	612	1042	1434	1866	1903	1590	1655	1173	806	402	13196
COLD BAY	474	425	525	772	918	1122	1153	1036	1122	951	791	591	9880
CORDOVA	366	391	522	781	1017	1221	1299	1086	1113	864	660	444	9764
FAIRBANKS	171	332	642	1203	1833	2254	2359	1901	1739	1068	555	222	14279
JUNEAU	301	338	483	725	921	1135	1237	1070	1073	810	601	381	9075
KING SALMON	313	322	513	908	1290	1606	1600	1333	1411	966	673	408	11343
KOTZEBUE	381	446	723	1249	1728	2127	2192	1932	2080	1554	1057	636	16105
MCGRATH	208	338	633	1184	1791	2232	2294	1817	1758	1122	648	258	14283
NOME	481	496	693	1094	1455	1820	1879	1666	1770	1314	930	573	14171
SAINT PAUL	605	539	612	862	963	1197	1228	1168	1265	1098	936	726	11199
SHEMYA	577	475	501	784	876	1042	1045	958	1011	885	837	696	9687
YAKUTAT	338	347	474	716	936	1144	1169	1019	1042	840	632	435	9092
ARIZ. FLAGSTAFF	46	68	201	558	867	1073	1169	991	911	651	437	180	7152
PHOENIX	0	0	0	22	234	415	474	328	217	75	0	0	1765
PRESCOTT	0	0	27	245	579	797	865	711	605	360	158	15	4362
TUCSON	0	0	0	25	231	406	471	344	242	75	6	0	1800
WINSLOW	0	0	6	245	711	1008	1054	770	601	291	96	0	4782
YUMA	0	0	0	0	148	319	363	228	130	29	0	0	1217
ARK. FORT SMITH	0	0	12	127	450	704	781	596	456	144	22	0	3292
LITTLE ROCK	0	0	9	127	465	716	756	577	434	126	9	0	3219
TEXARKANA	0	0	0	78	345	561	626	468	350	105	0	0	2533
CALIF. BAKERSFIELD	0	0	0	37	282	502	546	364	267	105	19	0	2122
BISHOP	0	0	42	248	576	797	874	666	539	306	143	36	4227
BLUE CANYON	34	50	120	347	579	766	865	781	791	582	397	195	5507
BURBANK	0	0	6	43	177	301	366	277	239	138	81	18	1646
EUREKA	270	257	258	329	414	499	546	470	505	438	372	285	4643
FRESNO	0	0	0	78	339	558	586	406	319	150	56	0	2492
LONG BEACH	0	0	12	40	156	288	375	297	267	168	90	18	1711
LOS ANGELES	28	22	42	78	180	291	372	302	288	219	158	81	2061
MT. SHASTA	25	34	123	406	696	902	983	784	738	525	347	159	5722
OAKLAND	53	50	45	127	309	481	527	400	353	255	180	90	2870
POINT ARGUELLO	202	186	162	205	291	400	474	392	403	339	298	243	3595
RED BLUFF	0	0	0	53	318	555	605	428	341	168	47	0	2515
SACRAMENTO	0	0	12	81	363	577	614	442	360	216	102	6	2773
SANDBERG	0	0	30	202	480	691	778	661	620	426	264	57	4209
SAN DIEGO	6	0	15	37	123	251	313	249	202	123	84	36	1439
SAN FRANCISCO	81	78	60	143	306	462	508	395	363	279	214	126	3015
SANTA CATALINA	16	0	9	50	165	279	353	308	326	249	192	105	2052
SANTA MARIA	99	93	96	146	270	391	459	370	363	282	233	165	2967
COLO. ALAMOSA	65	99	279	639	1065	1420	1476	1162	1020	696	440	168	8529
COLORADO SPRINGS	9	25	132	456	825	1032	1128	938	893	582	319	84	6423
DENVER	6	9	117	428	819	1035	1132	938	887	558	288	66	6283
GRAND JUNCTION	0	0	30	313	786	1113	1209	907	729	387	146	21	5641
PUEBLO	0	0	54	326	750	986	1085	871	772	429	174	15	5462
CONN. BRIDGEPORT	0	0	66	307	615	986	1079	966	853	510	208	27	5617
HARDFORT	0	0	99	372	711	1119	1209	1061	899	495	177	24	6172
NEW HAVEN	0	12	87	347	648	1011	1097	991	871	543	245	45	5897
DEL. WILMINGTON	0	0	51	270	588	927	980	874	735	387	112	6	4930
FLA. APALACHICOLA	0	0	0	16	153	319	347	260	180	33	0	0	1308
DAYTONA BEACH	0	0	0	0	75	211	248	190	140	15	0	0	879
FORT MYERS	0	0	0	0	24	109	146	101	62	0	0	0	442
JACKSONVILLE	0	0	0	12	144	310	332	246	174	21	0	0	1239
KEY WEST	0	0	0	0	0	28	40	31	9	0	0	0	108
LAKELAND	0	0	0	0	57	164	195	146	99	0	0	0	661
MIAMI BEACH	0	0	0	0	0	40	56	36	9	0	0	0	141
ORLANDO	0	0	0	0	72	198	220	165	105	6	0	0	766
PENSACOLA	0	0	0	19	195	353	400	277	183	36	0	0	1463
TALLAHASSEE	0	0	0	28	198	360	375	286	202	36	0	0	1485
TAMPA	0	0	0	0	60	171	202	148	102	0	0	0	683
WEST PALM BEACH	0	0	0	6	65	87	64	31	0	0	0	0	253
GA. ATHENS	0	0	12	115	405	632	642	529	431	141	22	0	2929
ATLANTA	0	0	18	127	414	626	639	529	437	168	25	0	2983
AUGUSTA	0	0	0	78	333	552	549	445	350	90	0	0	2397
COLUMBUS	0	0	0	87	333	543	552	434	338	96	0	0	2383
MACON	0	0	0	71	297	502	505	403	295	63	0	0	2136
ROME	0	0	24	161	474	701	710	577	468	177	34	0	3326
SAVANNAH	0	0	0	47	246	437	437	353	254	45	0	0	1819
THOMASVILLE	0	0	0	25	198	366	394	305	208	33	0	0	1529
IDAHO BOISE	0	0	132	415	792	1017	1113	854	722	438	245	81	5809
IDAHO FALLS 46W	16	34	270	623	1056	1370	1538	1249	1085	651	391	192	8475
IDAHO FALLS 42NW	16	40	282	648	1107	1432	1600	1291	1107	657	388	192	8760
LEWISTON	0	0	123	403	756	933	1063	815	694	426	239	90	5542
POCATELLO	0	0	172	493	900	1166	1324	1058	905	555	319	141	7033
ILL. CAIRO	0	0	36	164	513	791	856	680	539	195	47	0	3821
CHICAGO	0	0	81	326	753	1113	1209	1044	890	480	211	48	6155
MOLINE	0	0	99	335	774	1181	1314	1100	918	450	189	39	6408
PEORIA	0	0	87	326	759	1113	1218	1025	849	426	183	33	6025
ROCKFORD	0	9	114	400	837	1221	1333	1137	961	516	236	60	6830
SPRINGFIELD	0	0	72	291	696	1023	1135	935	769	354	136	18	5429
IND. EVANSVILLE	0	0	66	220	606	896	955	767	620	237	68	0	4435
FORT WAYNE	0	9	105	378	783	1135	1178	1028	890	471	189	39	6205
INDIANAPOLIS	0	0	90	316	723	1051	1113	949	809	432	177	39	5699
SOUTH BEND	0	6	111	372	777	1125	1221	1070	933	525	239	60	6439
IOWA Burlington	0	0	93	322	768	1135	1259	1042	859	426	177	33	6114
DES MOINES	0	0	99	363	837	1231	1398	1165	967	489	211	39	6808
DUBUQUE	12	31	156	450	906	1287	1420	1204	1026	546	260	78	7376
SIOUX CITY	0	9	108	369	867	1240	1435	1198	989	483	214	39	6951
WATERLOO	12	19	138	428	909	1296	1460	1221	1023	531	229	54	7320

STATE AND STATION	JULY	AUG.	SEP.	OCT.	NOV.	DEC.	JAN.	FEB.	MAR.	APR.	MAY	JUNE	ANNUAL
KANS. CONCORDIA	0	0	57	276	705	1023	1163	935	781	372	149	18	5479
DODGE CITY	0	0	33	251	666	939	1051	840	719	354	124	9	4986
GOODLAND	0	6	81	381	810	1073	1166	955	884	507	236	42	6141
TOPEKA	0	0	57	270	672	980	1122	893	722	330	124	12	5182
WICHITA	0	0	33	229	618	905	1023	804	645	270	87	6	4620
KY. COVINGTON	0	0	75	291	669	983	1035	893	756	390	149	24	5265
LEXINGTON	0	0	54	239	609	902	946	818	685	325	105	0	4683
LOUISVILLE	0	0	54	248	609	890	930	818	682	315	105	9	4660
LA. ALEXANDRIA	0	0	0	56	273	431	471	361	260	69	0	0	1921
BATON ROUGE	0	0	0	31	216	369	409	294	208	33	0	0	1560
BURRWOOD	0	0	0	0	96	214	298	218	171	27	0	0	1024
LAKE CHARLES	0	0	0	19	210	341	381	274	195	39	0	0	1459
NEW ORLEANS	0	0	0	19	192	322	363	258	192	39	0	0	1385
SHREVEPORT	0	0	0	47	297	477	552	426	304	81	0	0	2184
MAINE CARIBOU	78	115	336	682	1044	1535	1690	1470	1308	858	468	183	9767
PORTLAND	12	53	195	508	807	1215	1339	1182	1042	675	372	111	7511
MD. BALTIMORE	0	0	48	264	585	905	936	820	679	327	90	0	4654
FREDERICK	0	0	66	307	624	955	995	876	741	384	127	12	5087
MASS. BLUE HILL OBSY	0	22	108	381	690	1085	1178	1053	936	579	267	69	6368
BOSTON	0	9	60	316	603	983	1088	972	846	513	208	36	5634
NANTUCKET	12	22	93	332	573	896	992	941	896	621	384	129	5891
PITTSFIELD	25	59	219	524	831	1231	1339	1196	1063	660	326	105	7578
WORCESTER	6	34	147	450	774	1172	1271	1123	998	612	304	78	6969
MICH. ALPENA	68	105	273	580	912	1268	1404	1299	1218	777	446	156	8506
DETROIT (CITY)	0	0	87	360	738	1088	1181	1058	936	522	220	42	6232
ESCANABA	59	87	243	539	924	1293	1445	1296	1203	777	456	159	8481
FLINT	16	40	159	465	843	1212	1330	1198	1066	639	319	90	7377
GRAND RAPIDS	9	28	135	434	804	1147	1259	1134	1011	579	279	75	6894
LANSING	6	22	138	431	813	1163	1262	1142	1011	579	273	69	6909
MARQUETTE	59	81	240	527	936	1268	1411	1268	1187	771	468	177	8393
MUSKEGON	12	28	120	400	762	1088	1209	1100	995	594	310	78	6696
SAULT STE. MARIE	96	105	279	580	951	1367	1525	1380	1277	810	477	201	9048
MINN. DULUTH	71	109	330	632	1131	1581	1745	1518	1355	840	490	198	10000
INTERNATIONAL FALLS	71	112	363	701	1236	1724	1919	1621	1414	828	443	174	10606
MINNEAPOLIS	22	31	189	505	1014	1454	1631	1380	1166	621	288	81	8382
ROCHESTER	25	34	186	474	1005	1438	1593	1366	1150	630	301	93	8295
SAINT CLOUD	28	47	225	549	1065	1500	1702	1445	1221	666	326	105	8879
MISS. JACKSON	0	0	0	65	315	502	546	414	310	87	0	0	2239
MERIDIAN	0	0	0	81	339	518	543	417	310	81	0	0	2289
VICKSBURG	0	0	0	53	279	462	512	384	282	69	0	0	2041
MO. COLUMBIA	0	0	54	251	651	967	1076	874	716	324	121	12	5046
KANSAS	0	0	39	220	612	905	1032	818	682	294	109	0	4711
ST. JOSEPH	0	6	60	285	708	1039	1172	949	769	348	133	15	5484
ST. LOUIS	0	0	60	251	627	936	1026	848	704	312	121	15	4900
SPRINGFIELD	0	0	45	223	600	877	973	781	660	291	105	6	4561
MONT. BILLINGS	6	15	186	487	897	1135	1296	1100	970	570	285	102	7049
GLASGOW	31	47	270	608	1104	1466	1711	1439	1187	648	335	150	8996
GREAT FALLS	28	53	258	543	921	1169	1349	1154	1063	642	384	186	7750
HAVRE	28	53	306	595	1065	1367	1584	1364	1181	657	338	162	8700
HELENA	31	59	294	601	1002	1265	1438	1170	1042	651	381	195	8129
KALISPELL	50	99	321	654	1020	1240	1401	1134	1029	639	397	207	8191
MILES CITY	6	6	174	502	972	1296	1504	1252	1057	579	276	99	7723
MISSOULA	34	74	303	651	1035	1287	1420	1120	970	621	391	219	8125
NEBR. GRAND ISLAND	0	6	75	381	834	1172	1314	1089	908	462	211	45	6530
LINCOLN	0	6	75	301	726	1066	1237	1016	834	402	171	30	5864
NORFOLK	9	0	111	397	873	1234	1414	1179	983	498	233	48	6979
NORTH PLATTE	0	6	123	440	885	1166	1271	1039	930	519	248	57	6684
OMAHA	0	12	105	357	828	1175	1355	1126	939	465	208	42	6612
SCOTTSBLUFF	0	0	138	459	876	1128	1231	1008	921	552	285	75	6673
VALENTINE	9	12	165	493	942	1237	1395	1176	1045	579	288	84	7425
NEV. ELKO	9	34	225	561	924	1197	1314	1036	911	621	409	192	7433
ELY	28	43	234	592	939	1184	1308	1075	977	672	456	225	7733
LAS VEGAS	0	0	0	78	387	617	688	487	335	111	6	0	2709
RENO	43	87	204	490	801	1026	1073	823	729	510	357	189	6332
WINNEMUCCA	0	34	210	536	876	1091	1172	916	837	573	363	153	6761
N. H. CONCORD	6	50	177	505	822	1240	1358	1184	1032	636	298	75	7383
MT. WASH. OBSY.	493	536	720	1057	1341	1742	1820	1663	1652	1260	930	603	13817
N. J. ATLANTIC CITY	0	0	39	251	549	880	936	848	741	420	133	15	4812
NEWARK	0	0	30	248	573	921	983	876	729	381	118	0	4859
TRENTON	0	0	57	264	576	924	989	885	753	399	121	12	4980
N. MEX. ALBUQUERQUE	0	0	12	229	642	868	930	703	595	288	81	0	4348
CLAYTON	0	6	66	310	699	899	986	812	747	429	183	21	5158
RATON	9	28	126	431	825	1048	1116	904	834	543	301	63	6228
ROSWELL	0	0	18	202	573	806	840	641	481	201	31	0	3793
SILVER CITY	0	0	6	183	525	729	791	605	518	261	87	0	3705
N. Y. ALBANY	0	19	138	440	777	1194	1311	1156	992	564	239	45	6875
BINGHAMTON (AP)	22	65	201	471	810	1184	1277	1154	1045	645	313	99	7286
BINGHAMTON (PO)	0	28	141	406	732	1107	1190	1081	949	543	229	45	6451
BUFFALO	19	37	141	440	777	1156	1256	1145	1039	645	329	78	7062
CENTRAL PARK	0	0	30	233	540	902	986	885	760	408	118	9	4871
J. F. KENNEDY INTL	0	0	36	248	564	933	1029	935	815	480	167	12	5219
LAGUARDIA	0	0	27	223	528	887	973	879	750	414	124	6	4811
ROCHESTER	9	31	126	415	747	1125	1234	1123	1014	597	279	48	6748
SCHENECTADY	0	22	123	422	756	1159	1283	1131	970	543	211	30	6650
SYRACUSE	6	28	132	415	744	1153	1271	1140	1004	570	248	45	6756
N.C. ASHEVILLE	0	0	48	245	555	775	784	683	592	273	87	0	4042
CAPE HATTERAS	0	0	0	78	273	521	580	518	440	177	25	0	2612
CHARLOTTE	0	0	6	124	438	691	691	582	481	156	22	0	3191
GREENSBORO	0	0	33	192	513	778	784	672	552	234	47	0	3805
RALEIGH	0	0	21	164	450	716	725	616	487	180	34	0	3393
WILMINGTON	0	0	0	74	291	521	546	462	357	96	0	0	2347
WINSTON SALEM	0	0	21	171	483	747	753	652	524	207	37	0	3595
N. DAK. BISMARCK	34	28	222	577	1083	1463	1708	1442	1203	645	329	117	8851
DEVILS LAKE	40	53	273	642	1191	1634	1872	1579	1345	753	381	138	9901
FARGO	28	37	219	574	1107	1569	1789	1520	1262	690	332	99	9226
WILLISTON	31	43	261	601	1122	1513	1758	1473	1262	681	357	141	9243

STATE AND STATION	JULY	AUG.	SEP.	OCT.	NOV.	DEC.	JAN.	FEB.	MAR.	APR.	MAY	JUNE	ANNUAL
OHIO AKRON	0	9	96	381	726	1070	1138	1016	871	489	202	39	6037
CINCINNATI	0	0	54	248	612	921	970	837	701	336	118	9	4806
CLEVELAND	9	25	105	384	738	1088	1159	1047	918	552	260	66	6351
COLUMBUS	0	6	84	347	714	1039	1088	949	809	426	171	27	5660
DAYTON	0	6	78	310	696	1045	1097	955	809	429	167	30	5622
MANSFIELD	9	22	114	397	768	1110	1169	1042	924	543	245	60	6403
SANDUSKY	0	6	66	313	684	1032	1107	991	868	495	198	36	5796
TOLEDO	0	16	117	406	792	1138	1200	1056	924	543	242	60	6494
YOUNGSTOWN	6	19	120	412	771	1104	1169	1047	921	540	248	60	6417
OKLA. OKLAHOMA CITY	0	0	15	164	498	766	868	664	527	189	34	0	3725
TULSA	0	0	18	158	522	787	893	683	539	213	47	0	3860
OREG. ASTORIA	146	130	210	375	561	679	753	622	636	480	363	231	5186
BURNS	12	37	210	515	867	1113	1246	988	856	570	366	177	6957
EUGENE	34	34	129	366	585	719	803	627	589	426	279	135	4726
MEACHAM	84	124	288	580	918	1091	1209	1005	983	726	527	339	7874
MEDFORD	0	0	78	372	678	871	918	697	642	432	242	78	5008
PENDLETON	0	0	111	350	711	884	1017	773	617	396	205	63	5127
PORTLAND	25	28	114	335	597	735	825	644	586	396	245	105	4635
ROSEBURG	22	16	105	329	567	713	766	608	570	405	267	123	4491
SALEM	37	31	111	338	594	729	822	647	611	417	273	144	4754
SEXTON SUMMIT	81	81	171	443	666	874	958	809	818	609	465	279	6254
PA. ALLENTOWN	0	0	90	353	693	1045	1116	1002	849	471	167	24	5810
ERIE	0	25	102	391	714	1063	1169	1081	973	585	288	60	6451
HARRISBURG	0	0	63	298	648	992	1045	907	766	396	124	12	5251
PHILADELPHIA	0	0	60	291	621	964	1014	890	744	390	115	12	5101
PITTSBURGH	0	9	105	375	726	1063	1119	1002	874	480	195	39	5987
READING	0	0	54	257	597	939	1001	885	735	372	105	0	4945
SCRANTON	0	19	132	434	762	1104	1156	1028	893	498	195	33	6254
WILLIAMSPORT	0	9	111	375	717	1073	1122	1002	856	468	177	24	5934
R. I. BLOCK IS	0	16	78	307	594	902	1020	955	877	612	344	99	5804
PROVIDENCE	0	16	96	372	660	1023	1110	988	868	534	236	51	5954
S. C. CHARLESTON	0	0	0	59	282	471	487	389	291	54	0	0	2033
COLUMBIA	0	0	0	84	345	577	570	470	357	81	0	0	2484
FLORENCE	0	0	0	78	315	552	552	459	347	84	0	0	2387
GREENVILLE	0	0	0	112	387	636	648	535	434	120	12	0	2884
SPARTANBURG	0	0	15	130	417	667	663	560	453	144	25	0	3074
S. DAK. HURON	9	12	165	508	1014	1432	1628	1355	1125	600	288	87	8223
RAPID CITY	22	12	165	481	897	1172	1333	1145	1051	615	326	126	7345
SIOUX FALLS	19	25	168	462	972	1361	1544	1285	1082	573	270	78	7839
TENN. BRISTOL	0	0	51	236	573	828	828	700	598	261	68	0	4143
CHATTANOOGA	0	0	18	143	468	698	722	577	453	150	25	0	3254
KNOXVILLE	0	0	30	171	489	725	732	613	493	198	43	0	3494
MEMPHIS	0	0	18	130	447	698	729	585	456	147	22	0	3232
NASHVILLE	0	0	30	158	495	732	778	644	512	189	40	0	3578
OAK RIDGE (CO)	0	0	39	192	531	772	778	669	552	228	56	0	3817
TEX. ABILENE	0	0	0	99	366	586	642	470	347	114	0	0	2624
AMARILLO	0	0	18	205	570	797	877	664	546	252	56	0	3985
AUSTIN	0	0	0	31	225	388	468	325	223	51	0	0	1711
BROWNSVILLE	0	0	0	0	66	149	205	106	74	0	0	0	600
CORPUS CHRISTI	0	0	0	0	120	220	291	174	109	0	0	0	914
DALLAS	0	0	0	62	321	524	601	440	319	90	6	0	2363
EL PASO	0	0	0	84	414	648	685	445	319	105	0	0	2700
FORT WORTH	0	0	0	65	324	536	614	448	319	99	0	0	2405
GALVESTON	0	0	0	0	138	270	350	258	189	30	0	0	1235
HOUSTON	0	0	0	6	183	307	384	288	192	36	0	0	1396
LAREDO	0	0	0	0	105	217	267	134	74	0	0	0	797
LUBBOCK	0	0	18	174	513	744	800	613	484	201	31	0	3578
MIDLAND	0	0	0	87	381	592	651	468	322	90	0	0	2591
PORT ARTHUR	0	0	0	22	207	329	384	274	192	39	0	0	1447
SAN ANGELO	0	0	0	68	318	536	567	412	288	66	0	0	2255
SAN ANTONIO	0	0	0	31	207	363	428	286	195	39	0	0	1549
VICTORIA	0	0	0	6	150	270	344	230	152	21	0	0	1173
WACO	0	0	0	43	270	456	536	389	270	66	0	0	2030
WICHITA FALLS	0	0	0	99	381	632	698	518	378	120	6	0	2832
UTAH MILFORD	0	0	99	443	867	1141	1252	988	822	519	279	87	6497
SALT LAKE CITY	0	0	81	419	849	1082	1172	910	763	459	233	84	6052
WENDOVER	0	0	48	372	822	1091	1178	902	729	408	177	51	5778
VT. BURLINGTON	28	65	207	539	891	1349	1513	1333	1187	714	353	90	8269
VA. CAPE HENRY	0	0	0	112	360	645	694	633	536	246	53	0	3279
LYNCHBURG	0	0	51	223	540	822	849	731	605	267	78	0	4166
NORFOLK	0	0	0	136	408	698	738	655	533	216	37	0	3421
RICHMOND	0	0	36	214	495	784	815	703	546	219	53	0	3865
ROANOKE	0	0	51	229	549	825	834	722	614	261	65	0	4150
WASH. NAT'L. AP.	0	0	33	217	519	834	871	762	626	288	74	0	4224
WASH. OLYMPIA	68	71	198	422	636	753	834	675	645	450	307	177	5236
SEATTLE	50	47	129	329	543	657	738	599	577	396	242	117	4424
SEATTLE BOEING	34	40	147	384	624	763	831	655	608	411	242	99	4838
SEATTLE TACOMA	56	62	162	391	633	750	828	678	657	474	295	159	5145
SPOKANE	9	25	168	493	879	1082	1231	980	834	531	288	135	6655
STAMPEDE PASS	273	291	393	701	1008	1178	1287	1075	1085	855	654	483	9283
TATOOSH IS.	295	279	306	406	534	639	713	613	645	525	431	333	5719
WALLA WALLA	0	0	87	310	681	843	986	745	589	342	177	45	4805
YAKIMA	0	12	144	450	828	1039	1163	868	713	435	220	69	5941
W. VA. CHARLESTON	0	0	63	254	591	865	880	770	648	300	96	9	4476
ELKINS	9	25	135	400	729	992	1008	896	791	444	198	48	5675
HUNTINGTON	0	0	63	257	585	856	880	764	636	294	99	12	4446
PARKERSBURG	0	0	60	264	606	905	942	826	691	339	115	6	4754
WIS. GREEN BAY	28	50	174	484	924	1333	1494	1313	1141	654	335	99	8029
LA CROSSE	12	19	153	437	924	1339	1504	1277	1070	540	245	69	7589
MADISON	25	40	174	474	930	1330	1473	1274	1113	618	310	102	7863
MILWAUKEE	43	47	174	471	876	1252	1376	1193	1054	642	372	135	7635
WYO. CASPER	6	16	192	524	942	1169	1290	1084	1020	657	381	129	7410
CHEYENNE	19	31	210	543	924	1101	1228	1056	1011	672	381	102	7278
LANDER	6	19	204	555	1020	1299	1417	1145	1017	654	381	153	7870
SHERIAN	25	31	219	539	948	1200	1355	1154	1054	642	366	150	7683

Quick Reference Chart 2-2
COOLING HOURS (ROUNDED TO NEAREST 50)

State and City	Cooling Hours
Alabama, Birmingham	1500
Huntsville	1250
Mobile	1850
Arizona, Flagstaff	200
Phoenix	2750
Tucson	2450
Arkansas, Little Rock	2000
Pine Bluff	1605
Walnut Ridge	1400
California, Los Angeles	550
Sacramento	1000
San Francisco	<50
Colorado, Colorado Springs	500
Denver	650
Pueblo	900
Connecticut, Bridgeport	300
Danbury	350
Hartford	500
Delaware, Dover	700
Newark	600
Wilmington	600
District of Columbia, Washington	1000
Florida, Daytona Beach	1600
Miami	2400
West Palm Beach	2400
Georgia, Atlanta	1000
Brunswick	1600
Savannah	1450
hawaii, Honolulu	1350
Idaho, Boise	700
Idaho Falls	350
Twin Falls	850
Illinois, Chicago	750
Peoria	600
Springfield	950
Indiana, Fort Wayne	600
Gary	550
Indianapolis	750
Iowa, Cedar Rapids	550
Des Moines	600
Sioux City	800
Kansas, Kansas City	1100
Salina	1400
Wichita	1300

State and City	Cooling Hours
Kentucky, Covington	750
Lexington	950
Louisville	1050
Louisiana, Baton Rouge	1650
Monroe	1850
New Orleans	1750
Maine, Bangor	200
Lewiston	300
Portland	250
Maryland, Baltimore	850
Cumberland	650
Hagerstown	550
Massachusetts, Boston	400
Lowell	500
Springfield	400
Michigan, Detroit	500
Grand Rapids	400
Kalamazoo	500
Minnesota, Duluth	<50
Minneapolis	500
St. Paul	500
Mississippi, Biloxi	2050
Gulfport	2050
Jackson	1600
Missouri, Jefferson City	1100
Kansas City	1100
Springfield	950
Montana, Billings	500
Great Falls	300
Missoula	300
Nebraska, Lincoln	1000
Omaha	900
Scottsbluff	850
Nevada, Elko	650
Las Vegas	2350
Reno	650
New Hampshire, Concord	400
Nashua	500
Portsmouth	300
New Jersey, Atlantic City	450
Newark	600
Trenton	450
New Mexico, Alamogordo	1500
Albuquerque	1150
Santa Fe	700

State and City	Cooling Hours
New York, Albany	400
Buffalo	350
New York	650
North Carolina, Fayetteville	1200
Raleigh	1050
Winston-Salem	850
North Dakota, Bismarck	450
Fargo	400
Grand Forks	350
Ohio, Akron	400
Dayton	700
Toledo	600
Oklahoma, Bartlesville	1350
Oklahoma City	1450
Tulsa	1600
Oregon, Baker	200
Grants Pass	850
Portland	200
Pennsylvania, Harrisburg	750
Philadelphia	700
Pittsburgh	450
Rhode Island, Newport	250
Providence	300
Westerly	250
South Carolina, Charleston	1250
Greenville	1100
Myrtle Beach	1200
South Dakota, Huron	650
Rapid City	550
Sioux Falls	500
Tennessee, Chattanooga	1250
Memphis	1500
Nashville	1300
Texas, Dallas	2300
Houston	1900
San Antonio	2000
Utah, Ogden	800
Salt Lake City	900
Vermont, Burlington	300
Montpelier	300
Rutland	300
Virginia, Norfolk	1000
Richmond	1000
Roanoke	800
Washington, Seattle	100

Quick Reference Chart 2-2 (continued)
COOLING HOURS (ROUNDED TO NEAREST 50)

State and City	Cooling Hours
Spokane	350
Tacoma	100
West Virginia, Charleston	800
Huntington	1000
Wheeling	450
Wisconsin, Appleton	350
Green Bay	250
Milwaukee	350
Wyoming, Casper	550
Cheyenne	350
Lander	200

NOTE: The symbol < means *less than*.

(Source: Air Force Publication AFM88–8)

Quick Reference Chart 2-3
MEAN NUMBER OF HOURS OF SUNSHINE

STATE AND STATION	YEARS	JAN.	FEB.	MAR.	APR.	MAY	JUNE	JULY	AUG.	SEPT.	OCT.	NOV.	DEC.	ANNUAL
ALA. BIRMINGHAM	30	138	152	207	248	293	294	269	265	244	234	182	136	2662
MOBILE	22	157	158	212	253	301	289	249	259	235	254	195	146	2708
MONTGOMERY	30	160	168	227	267	317	311	288	290	260	250	200	158	2894
ALASKA ANCHORAGE	19	78	114	210	254	268	288	255	184	128	96	68	49	1992
FAIRBANKS	20	54	120	224	302	319	334	274	164	122	85	71	36	2105
JUNEAU	29	71	102	171	200	230	251	193	161	123	67	60	51	1680
NOME	27	72	109	193	226	285	297	204	146	142	101	67	42	1884
ARIZ. PHOENIX	30	248	244	314	346	404	404	377	351	334	307	267	236	3832
PRESCOTT	14	222	230	293	323	378	392	323	305	315	286	254	228	3549
TUCSON	13	255	266	317	350	399	394	329	329	335	317	280	258	3829
YUMA	30	258	266	337	365	418	420	404	380	351	330	285	262	4077
ARK. FT. SMITH	30	146	156	202	234	268	303	321	305	261	230	174	147	2747
LITTLE ROCK	30	143	158	213	243	291	316	321	316	265	251	181	142	2840
CALIF. EUREKA	30	120	138	180	209	247	261	244	205	195	164	127	108	2198
FRESNO	29	153	192	283	330	389	418	435	406	355	306	221	144	3632
LOS ANGELES	30	224	217	273	264	292	299	352	336	295	263	249	220	3284
RED BLUFF	15	156	186	246	302	366	396	438	407	341	277	199	154	3468
SACRAMENTO	30	134	169	255	300	367	405	437	406	347	283	197	122	3422
SAN DIEGO	30	216	212	262	242	261	253	293	277	255	234	236	217	2958
SAN FRANCISCO	30	165	182	251	281	314	330	300	272	267	243	198	156	2959
COLO. DENVER	30	207	205	247	252	281	311	321	297	274	246	200	192	3033
GRAND JUNCTION	30	169	182	243	265	314	350	349	311	291	255	198	168	3095
PUEBLO	30	224	217	261	271	299	340	349	318	290	265	225	211	3270
CONN. HARTFORD	30	141	164	206	223	267	285	299	268	220	193	137	136	2541
NEW HAVEN	30	155	178	215	234	274	291	309	284	238	215	157	154	2704
D. C. WASHINGTON	30	138	160	205	226	267	288	291	264	233	207	162	153	2576
FLA. APALACHICOLA	26	193	195	233	274	328	296	273	259	236	236	216	175	2941
JACKSONVILLE	30	192	189	241	267	296	260	255	248	199	205	191	170	2713
KEY WEST	30	229	238	285	296	307	273	277	269	236	237	226	225	3098
LAKELAND	7	204	186	222	251	285	268	252	242	203	209	212	198	2732
MIAMI	30	222	227	266	275	280	251	267	263	216	215	212	209	2903
PENSACOLA	30	175	180	232	270	311	302	278	284	249	265	206	166	2918
TAMPA	30	223	220	260	283	320	275	257	252	232	243	227	209	3001
GA. ATLANTA	25	154	165	218	266	309	304	284	285	247	241	188	160	2821
MACON	30	177	178	235	279	321	314	292	295	253	236	202	168	2950
SAVANNAH	30	175	173	229	274	307	279	267	256	212	216	197	167	2752
HAWAII HILO	7	153	135	161	112	106	158	184	134	137	153	106	131	1670
HONOLULU	30	227	202	250	255	276	280	293	290	279	257	221	211	3041
LIHUE	10	171	162	176	176	211	246	246	236	246	210	170	161	2411
IDAHO BOISE	30	116	144	218	274	322	352	412	378	311	232	143	104	3006
POCATELLO	30	111	143	211	255	300	338	380	347	298	250	145	108	2884
ILL. CAIRO	15	124	160	218	254	298	324	345	336	279	254	181	145	2918
CHICAGO	30	126	142	199	221	274	300	333	299	247	216	136	118	2611
MOLINE	18	132	139	189	214	255	279	337	300	251	214	130	123	2563
PEORIA	30	134	149	198	229	273	303	336	299	259	222	149	122	2673
SPRINGFIELD	30	127	149	193	224	282	304	344	312	266	225	152	122	2702
IND. EVANSVILLE	30	123	145	199	237	294	322	342	318	274	236	156	120	2766
FT. WAYNE	30	113	136	191	217	281	310	342	306	242	210	120	102	2370
INDIANAPOLIS	30	118	140	193	227	278	313	342	313	265	222	139	118	2668
TERRE HAUTE	24	125	148	189	231	274	302	341	305	253	235	150	122	2675
IOWA BURLINGTON	19	148	165	217	241	284	315	353	327	270	243	175	147	2885
CHARLES CITY	22	137	157	190	226	258	285	336	290	241	207	130	115	2572
DES MOINES	30	155	170	203	236	276	303	346	299	263	227	156	136	2770
SIOUX CITY	30	164	177	216	254	300	320	363	320	270	236	160	146	2926
KAN. CONCORDIA	30	180	172	214	243	281	315	348	308	249	245	189	172	2916

Location	Yrs	Jan	Feb	Mar	Apr	May	Jun	Jul	Aug	Sep	Oct	Nov	Dec	Ann
SHREVEPORT	19	151	172	214	240	298	332	339	322	289	273	206	177	3015
MAINE EASTPORT	22	133	151	196	201	245	248	275	260	205	175	105	115	2309
PORTLAND	30	155	174	213	226	268	286	312	294	229	202	146	148	2653
MD. BALTIMORE OBS.	30	148	170	211	229	270	295	299	272	238	212	164	145	2653
MASS. BLUE HILL OBS.	10	125	136	165	182	233	248	266	241	211	181	134	135	2257
BOSTON	30	148	168	212	222	263	283	300	280	232	207	152	148	2615
NANTUCKET	22	128	156	-.214	227	278	284	291	279	242	208	149	129	2585
MICH. ALPENA	24	86	124	198	228	261	303	339	285	204	159	70	67	2324
DETROIT	30	90	128	180	212	263	295	321	284	226	189	98	89	2375
LANSING	30	84	119	175	215	272	305	344	294	228	182	87	73	2378
ESCANABA	30	112	148	204	226	266	283	316	267	198	162	90	90	2366
GRAND RAPIDS	30	74	117	178	218	277	308	349	304	231	188	92	70	2406
MARQUETTE	30	78	113	172	207	248	268	305	251	186	142	68	66	2104
SAULT STE. MARIE	30	83	123	187	217	252	269	309	256	165	133	61	62	2117
MINN. DULUTH	30	125	163	221	235	268	282	328	277	203	166	100	107	2475
MINNEAPOLIS	30	140	166	200	231	272	302	343	296	237	192	115	112	2607
MISS. JACKSON	12	130	147	199	244	280	287	279	287	235	223	185	150	2646
VICKSBURG	30	136	141	199	232	284	304	291	297	254	244	183	140	2705
MO. COLUMBIA	30	147	164	207	232	281	296	341	298	262	225	168	138	2757
KANSAS CITY	30	154	170	211	235	278	313	347	308	266	235	178	151	2846
ST. JOSEPH	23	154	165	211	231	274	301	347	287	260	224	168	144	2766
ST. LOUIS	30	137	152	202	235	283	301	325	289	256	223	166	125	2694
SPRINGFIELD	30	145	164	213	238	278	305	342	310	269	233	183	140	2820
MONT. BILLINGS	21	140	154	208	236	283	301	372	332	258	213	136	129	2762
GREAT FALLS	19	154	176	245	261	299	299	381	342	256	206	132	133	2884
HAVRE	30	136	174	234	268	311	312	384	339	260	202	132	122	2874
HELENA	30	138	168	215	241	292	292	342	336	258	202	137	121	2742
MISSOULA	25	85	109	167	209	261	260	378	328	246	178	90	66	2377
NEBR. LINCOLN	30	173	172	213	244	287	316	356	309	266	237	174	160	2907
NORTH PLATTE	30	181	179	221	246	282	310	343	304	264	242	184	169	2925
OMAHA	30	172	188	222	259	305	332	379	311	270	248	166	145	2997
VALENTINE	30	185	194	229	252	296	323	369	326	275	242	174	172	3037
NEV. ELY	22	186	197	262	260	300	354	359	344	303	255	204	187	3211
LAS VEGAS	8	239	251	314	336	386	411	383	364	345	301	258	250	3838
RENO	30	185	199	267	306	354	376	414	391	336	273	212	170	3483
WINNEMUCCA	30	142	155	207	255	312	346	395	375	316	242	177	139	3061
N. H. CONCORD	23	136	153	192	196	229	261	286	260	214	179	122	126	2354
MT. WASHINGTON OBS.	18	94	98	133	141	162	145	150	143	139	159	89	97	1540
N. J. ATLANTIC CITY	30	151	173	210	233	273	287	298	271	239	218	177	153	2683
TRENTON	30	145	168	203	235	277	294	309	273	239	208	160	142	2653
N. MEX. ALBUQUERQUE	30	221	218	273	299	343	365	340	317	299	279	245	219	3418
ROSWELL	21	218	223	286	306	330	333	341	317	286	266	242	216	3340
N. Y. ALBANY	30	125	151	194	213	266	301	317	286	224	192	115	112	2496
BINGHAMTON	30	94	119	151	170	226	256	266	230	184	158	92	79	2025
BUFFALO	30	110	125	180	212	274	319	338	297	239	183	97	84	2458
NEW YORK	30	154	171	213	237	268	289	302	271	253	213	169	155	2677
ROCHESTER	30	93	123	172	209	274	314	333	294	224	173	97	86	2392
SYRACUSE	30	87	115	165	197	261	295	316	276	211	163	81	74	2241
N. C. ASHEVILLE	30	146	161	211	247	289	292	268	250	235	222	179	146	2646
CAPE HATTERAS	9	152	168	206	259	293	301	286	265	214	202	169	154	2669
CHARLOTTE	30	165	177	230	267	313	316	291	277	247	243	198	167	2891
GREENSBORO	30	157	171	217	231	298	302	287	272	243	236	190	163	2767
RALEIGH	29	154	168	220	255	290	284	277	253	224	215	184	156	2680
WILMINGTON	30	179	180	237	279	314	312	286	273	237	238	206	178	2919
N. DAK. BISMARCK	30	141	170	205	236	279	294	358	307	243	198	130	125	2686
DEVILS LAKE	30	150	177	220	250	291	297	352	302	230	198	123	124	2714
FARGO	30	132	170	210	232	283	288	343	293	222	187	112	114	2586
WILLISTON	29	141	168	215	260	305	312	377	328	247	206	131	129	2819
OHIO CINCINNATI (ABBE)	30	115	137	186	222	273	309	323	295	253	205	138	118	2574
CLEVELAND	30	79	111	167	209	274	301	325	298	233	187	99	77	2352
COLUMBUS	30	112	132	177	215	270	296	323	291	250	210	131	101	2508
DAYTON	10	114	136	195	222	281	313	323	307	268	229	152	124	2664
SANDUSKY	30	100	128	183	229	285	312	343	302	248	201	111	91	2533
TOLEDO	30	93	120	170	203	263	296	331	298	241	196	106	92	2409
OKLA. OKLAHOMA CITY	29	175	182	235	253	290	329	352	331	282	243	201	171	3048
TULSA	18	152	164	200	213	244	287	314	308	281	241	207	172	2783
OREG. BAKER	22	118	143	198	251	302	313	406	368	289	215	132	100	2835
PORTLAND	30	77	97	142	203	246	249	329	275	218	134	87	65	2122
ROSEBURG	30	69	96	148	205	257	278	369	329	255	146	81	50	2283
PA. HARRISBURG	30	132	160	203	230	277	297	319	282	233	200	140	131	2604
PHILADELPHIA	30	142	168	203	231	270	281	288	253	225	205	158	142	2564
PITTSBURGH	25	89	114	163	200	239	260	283	250	234	180	114	76	2202
READING	30	133	151	195	220	259	277	293	259	219	198	144	127	2473
SCRANTON	30	108	138	178	199	251	269	290	249	213	183	120	105	2303
R. I. PROVIDENCE	30	145	168	211	221	271	285	292	267	226	207	153	143	2589
S. C. CHARLESTON	30	188	189	243	284	323	308	297	281	244	239	210	187	2993
COLUMBIA	30	173	183	233	274	312	312	291	283	243	242	202	166	2914
GREENVILLE	26	166	176	227	274	307	300	278	274	239	232	192	157	2822
S. DAK. HURON	30	153	177	213	250	295	321	367	320	260	212	142	134	2844
RAPID CITY	30	164	182	222	245	278	300	348	317	266	228	164	144	2858
TENN. CHATTANOOGA	30	126	146	187	239	290	295	278	266	247	220	169	128	2591
KNOXVILLE	30	124	144	189	237	281	288	277	248	237	213	157	120	2515
MEMPHIS	30	135	152	204	244	296	321	319	314	261	243	180	139	2808
NASHVILLE	30	123	142	196	241	285	308	292	279	250	224	168	126	2634
TEX. ABILENE	13	190	199	250	259	290	347	335	322	276	245	223	201	3137
AMARILLO	30	207	199	258	276	305	338	350	328	288	260	229	205	3243
AUSTIN	30	148	152	207	221	266	302	331	320	260	261	242	180	2790
BROWNSVILLE	30	147	152	187	210	272	297	326	311	246	252	165	151	2716
CORPUS CHRISTI	24	160	165	212	237	295	329	366	341	276	264	194	164	3003
DALLAS	30	155	159	220	238	279	326	341	325	274	240	191	163	2911
DEL RIO	27	173	173	230	237	259	279	331	319	252	240	195	178	2866
EL PASO	30	234	236	299	329	373	363	336	327	300	287	253	203	3583
GALVESTON	30	151	149	203	230	288	322	305	292	257	264	199	151	2811
HOUSTON	30	144	141	193	212	266	298	294	281	238	239	181	146	2633
PORT ARTHUR	30	153	149	209	235	292	317	285	281	252	256	191	148	2768
SAN ANTONIO	30	148	153	213	224	258	292	325	307	261	241	183	160	2765
UTAH SALT LAKE CITY	30	137	155	227	269	329	358	377	346	306	249	171	135	3059
VT. BURLINGTON	30	103	127	184	185	244	270	291	266	199	152	77	80	2178
VA. LYNCHBURG	26	153	169	216	243	288	297	288	264	235	217	177	158	2705
NORFOLK	30	156	174	223	257	304	311	296	282	237	220	182	161	2803
RICHMOND	30	144	166	211	248	280	296	286	263	231	211	176	152	2663
WASH. NORTH HEAD	22	76	97	135	182	221	214	226	186	170	123	87	66	1783
SEATTLE	30	74	99	154	201	247	234	304	248	197	122	77	62	2019
SPOKANE	30	78	120	197	262	308	309	397	350	264	177	86	57	2605
TATOOSH ISLAND	30	70	100	135	182	229	217	235	199	175	129	71	60	1793
WALLA WALLA	30	72	106	194	262	317	335	411	367	280	198	93	51	2685
W. VA. ELKINS	24	110	119	158	198	227	256	225	236	211	186	131	103	2160
PARKERSBURG	30	91	111	155	200	252	277	286	264	230	189	117	93	2265
WIS. GREEN BAY	30	121	148	194	210	251	279	314	266	203	176	110	106	2388
MADISON	30	126	147	196	214	258	285	336	288	225	203	130	108	2502
MILWAUKEE	30	116	134	191	218	267	293	340	292	215	193	126	106	2510
WYO. CHEYENNE	30	191	197	243	237	259	304	318	286	265	242	188	170	2900
LANDER	30	200	208	260	264	301	340	361	326	280	233	186	185	3144
SHERIDAN	30	160	179	226	245	286	303	367	333	266	221	153	145	2884
P. R. SAN JUAN	30	231	229	273	252	240	245	264	257	219	229	217	222	2878

(Source: *Climatic Atlas of the U.S.*)

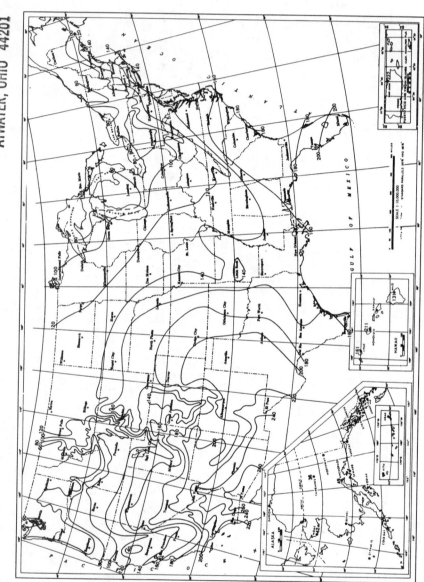

FIGURE 2-2. Mean monthly total hours of sunshine, December *(from the Climatic Atlas of the U.S.).*

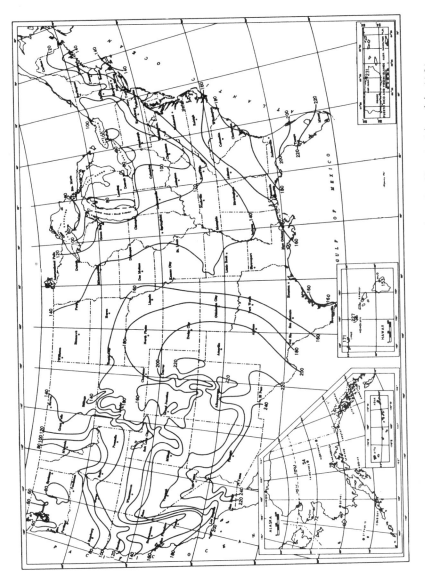

FIGURE 2-3. Mean monthly total hours of sunshine, January *(from the Climatic Atlas of the U.S.).*

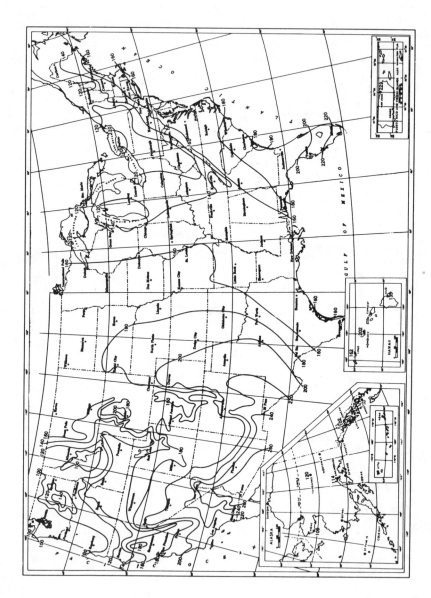

FIGURE 2-4. Mean monthly total hours of sunshine, February *(from the Climatic Atlas of the U.S.).*

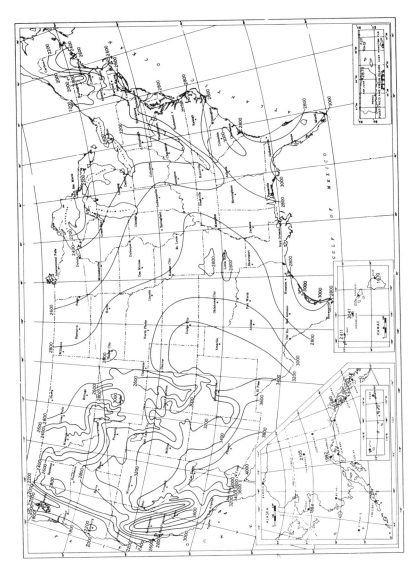

FIGURE 2-5. Annual mean total hours of sunshine *(from the Climatic Atlas of the U.S.).*

Temperatures

Knowing the normal daily maximum, average, minimum, and extreme temperatures in January (Fig. 2-6) and July (Fig. 2-7), the mean annual number of days that the maximum temperature is 90 degrees F and above (Fig. 2-8 and QRC 2-4) and the mean annual number of days that the minimum temperature is 32 degrees F and below (Fig. 2-9 and QRC 2-5) will also aid in your weatherization planning. Observation of these maps and charts may make you realize that your area is not as cold or as hot as you thought. Remember these maps are generalized. Sharp changes in the means may occur in short distances due to differences in altitude, slope of land, type of soil, vegetative cover, bodies of water, air currents, urban heat effects, and so on.

Surface Wind

Knowing the direction of surface winds is important in weatherizing your home. You'll want to grow natural windbreaks or build windbreaks against the prevailing surface winds in the winter, and you'll want to funnel the summer prevailing surface winds to your home for natural summer cooling. Use of winds for cooling is discussed in Section 10-1 and Chapter 11.

Figures 2-10 and 2-11 illustrate the mean resultant surface wind direction and speed for midwinter (January) and midsummer (July). Use them to help determine the prevailing surface winds in your area. Make your own observations, however, because nearby houses, buildings, hills, and valleys can funnel the winds in other directions. You might also contact your local weather bureau for information.

Relative Humidity

Relative humidity is the ratio of the amount of water vapor actually present in the air to the greatest amount possible at the same temperature. When the relative humidity is low, our bodies feel cooler because the perspiration on our skin is able to quickly and easily evaporate into the surrounding air. We generally feel quite comfortable with a temperature of 65 to 70 degrees F and a relative humidity of 30 to 35 percent.

Figure 2-12 illustrates the annual mean relative humidity (percentage) across the country. Mean relative humidities during midseasons and annually for selected cities are listed in QRC 2-6. If you're planning on moving, you might consider relocating to a locality having a more ideal relative humidity than your present location. Incidentally, the map represents *outdoor* relative humidities. When you weatherize your home, your indoor relative humidity will probably be less than outdoors during the heating season and more during the cooling season. Humidity can be added to your indoor air during the

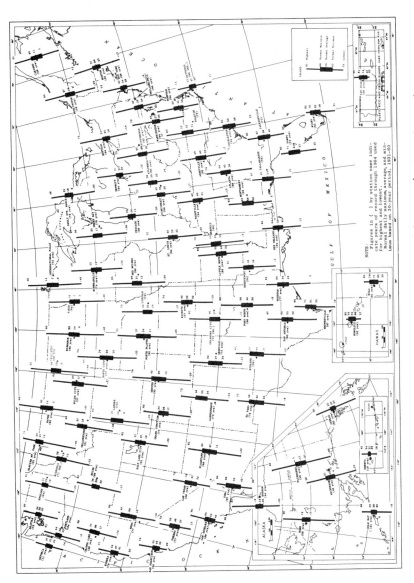

FIGURE 2-6. January normal daily maximum, average, minimum, and extreme temperatures *(from the Climatic Atlas of the U.S.).*

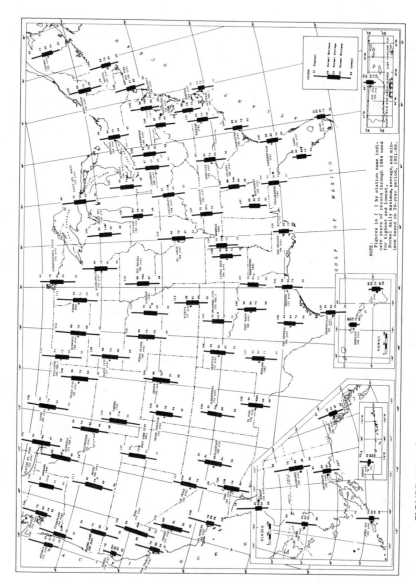

FIGURE 2-7. July normal daily maximum, average, minimum, and extreme temperatures *(from the Climatic Atlas of the U.S.).*

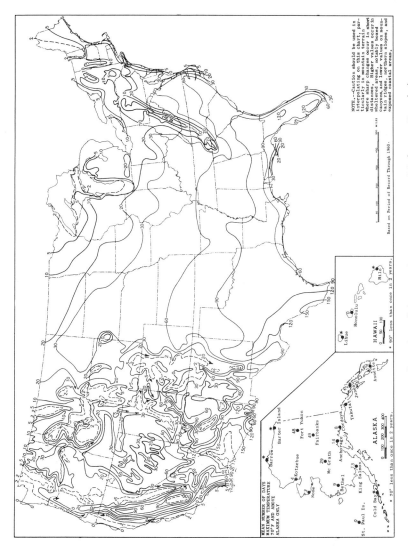

FIGURE 2-8. Mean annual number of days maximum temperature 90 degrees F and above (*from the Climatic Atlas of the U.S.*).

Quick Reference Chart 2-4

MEAN NUMBER OF DAYS MAXIMUM TEMPERATURE 90 DEGREES F AND ABOVE EXCEPT 70 DEGREES F AND ABOVE IN ALASKA

MEAN NUMBER OF DAYS MAXIMUM TEMPERATURE 90°F. AND ABOVE EXCEPT 70°F. AND ABOVE IN ALASKA

States and Stations	Yrs.	Jan.	Feb.	Mar.	Apr.	May	June	July	Aug.	Sept.	Oct.	Nov.	Dec.	Annual
ALA. BIRMINGHAM	17	0	0	0	*	4	15	19	20	8	1	0	0	67
MOBILE	19	0	0	*	*	5	17	18	20	9	1	0	0	70
MONTGOMERY	16	0	0	0	*	8	19	22	24	13	1	0	0	87
ARIZ. FLAGSTAFF U	43	0	0	0	0	0	1	1	*	*	0	0	0	2
PHOENIX	21	0	0	*	8	22	29	30	31	28	14	*	0	163
PRESCOTT	18	0	0	0	0	1	10	17	11	8	*	0	0	48
TUCSON	20	0	*	*	5	18	28	29	29	26	10	*	0	146
WINSLOW	29	0	0	0	*	2	18	25	18	9	*	0	0	72
YUMA	10	0	1	3	14	24	30	31	31	29	22	2	0	187
ARK. FORT SMITH	15	0	0	*	*	4	17	24	24	14	1	0	0	84
LITTLE ROCK	19	0	0	0	*	4	16	22	21	10	1	0	0	74
TEXARKANA	18	0	0	*	*	3	19	25	24	13	1	0	0	85
CALIF. BAKERSFIELD	23	0	0	0	2	9	19	30	27	18	5	*	0	110
BISHOP	13	0	0	0	1	5	19	30	28	17	1	0	0	101
BLUE CANYON	17	0	0	0	0	0	*	*	*	*	0	0	0	1
BURBANK	18	0	*	0	1	2	3	12	12	13	4	1	*	47
EUREKA	50	0	0	0	0	0	0	0	0	0	0	0	0	0
FRESNO	21	0	0	0	1	7	16	29	26	17	4	0	0	100
LONG BEACH	17	0	0	0	*	1	*	2	2	3	2	1	*	11
LOS ANGELES U	20	0	*	0	1	1	1	4	3	6	2	1	0	19
MT. SHASTA	18	0	0	0	0	*	1	8	6	4	*	0	0	20
OAKLAND	30	0	0	0	0	*	1	*	*	2	*	0	0	4
RED BLUFF	16	0	0	*	2	7	15	28	25	17	4	*	0	98
SACRAMENTO	27	0	0	0	*	4	11	20	18	12	2	0	0	67
SANDBERG	28	0	0	0	0	*	2	8	7	4	0	0	0	21
SAN DIEGO	20	0	0	0	*	*	*	*	*	1	1	*	0	3
SAN FRANCISCO	24	0	0	0	0	*	*	*	*	1	*	0	0	1
SANTA MARIA	18	0	0	0	*	*	*	*	*	1	1	1	*	4
COLO. ALAMOSA	15	0	0	0	0	0	*	1	0	0	0	0	0	1
COLORADO SPRINGS	12	0	0	0	0	0	4	8	5	2	0	0	0	19
DENVER	26	0	0	0	0	*	7	14	11	3	0	0	0	35
GRAND JUNCTION U	64	0	0	0	0	1	11	22	15	2	0	0	0	51
PUEBLO	20	0	0	0	0	1	12	19	17	7	*	0	0	57
CONN. BRIDGEPORT	12	0	0	0	0	0	2	4	2	*	0	0	0	8
HARTFORD	51	0	0	0	*	*	2	4	3	1	*	0	0	10
NEW HAVEN	17	0	0	0	0	0	1	1	1	*	0	0	0	3
DEL. WILMINGTON	13	0	0	0	0	1	5	9	5	1	*	0	0	21
D.C. WASHINGTON U	88	0	0	*	*	2	6	10	7	3	*	0	0	28
FLA. APALACHICOLA	31	0	0	0	*	1	4	5	5	2	*	0	0	17
DAYTONA BEACH	17	0	0	*	1	6	13	17	16	8	1	0	0	62
EVERGLADES	26	0	*	1	5	15	25	27	28	22	9	1	0	133
FORT MYERS	20	0	*	1	4	16	22	25	27	21	5	*	0	121
JACKSONVILLE	19	0	0	*	1	10	19	24	22	11	1	0	0	88
KEY WEST	12	0	0	0	0	1	10	19	20	10	1	0	0	61
KEY WEST U	87	0	0	0	*	1	5	10	12	5	*	*	0	33
LAKELAND U	20	0	0	*	1	9	18	21	22	12	2	0	0	85
MIAMI	18	0	0	*	1	3	12	17	22	11	1	0	0	67
MIAMI BEACH	17	0	0	*	*	1	2	2	6	2	*	0	0	13
ORLANDO	18	0	0	1	3	13	21	24	25	17	3	0	0	107
PENSACOLA	21	0	0	0	*	2	7	12	14	4	*	0	0	39
TALLAHASSEE	21	0	0	0	1	9	18	20	21	11	1	0	0	81
TAMPA	14	0	0	*	1	9	17	20	20	14	3	0	0	84
WEST PALM BEACH	20	0	*	1	2	5	16	23	24	15	3	*	*	89
GA. ATHENS	17	0	0	0	*	4	14	16	17	6	*	0	0	57
ATLANTA	12	0	0	0	0	3	12	14	15	4	*	0	0	48
AUGUSTA	10	0	0	0	1	9	19	24	25	9	1	0	0	88
COLUMBUS	15	0	0	0	*	8	18	21	22	10	1	0	0	80
MACON	12	0	0	0	1	12	21	24	25	12	2	0	0	97
ROME	15	0	0	0	*	4	15	20	21	8	1	0	0	69
SAVANNAH	10	0	0	*	1	7	15	20	21	6	1	0	0	71
THOMASVILLE	36	0	0	*	*	7	17	19	20	12	1	0	0	76
HAWAII HILO	15	0	0	0	0	0	0	0	*	*	*	0	0	*
HONOLULU	38	0	0	0	0	0	0	0	0	0	0	0	0	0
LIHUE	11	0	0	0	0	0	0	0	0	0	*	0	0	*
IDAHO BOISE	21	0	0	0	*	1	4	19	14	3	0	0	0	41
IDAHO FALLS	11	0	0	0	0	*	2	13	10	2	0	0	0	27
LEWISTON	14	0	0	0	0	1	3	16	12	4	0	0	0	36
POCATELLO	22	0	0	0	0	*	2	16	12	2	0	0	0	32
ILL. CAIRO U	18	0	0	0	0	2	13	18	17	5	*	0	0	55
CHICAGO	18	0	0	0	0	1	6	8	8	3	*	0	0	26
CHICAGO U	85	0	0	0	*	1	3	5	3	1	*	0	0	13
MOLINE	26	0	0	0	*	1	6	10	8	3	*	0	0	28
PEORIA	21	0	0	0	0	1	6	8	8	3	*	0	0	26
ROCKFORD	10	0	0	0	0	1	5	7	6	3	0	0	0	22
SPRINGFIELD	13	0	0	0	0	1	7	10	8	4	*	0	0	30
IND. EVANSVILLE	20	0	0	0	0	1	10	14	13	5	*	0	0	43
FORT WAYNE	14	0	0	0	0	*	4	6	6	2	*	0	0	18
INDIANAPOLIS	20	0	0	0	0	*	5	7	7	2	*	0	0	21
SOUTH BEND	20	0	0	0	*	*	4	6	6	2	0	0	0	18
IOWA BURLINGTON U	62	0	0	0	*	1	7	12	10	4	*	0	0	34
DES MOINES	27	0	0	0	0	1	5	10	9	3	*	0	0	28
DUBUQUE	10	0	0	0	0	*	2	4	3	2	0	0	0	11
SIOUX CITY	20	0	0	0	*	1	7	12	10	4	*	0	0	34
WATERLOO	12	0	0	0	*	*	3	6	6	2	*	0	0	17
KANS. CONCORDIA U	76	0	0	*	*	1	9	16	15	7	*	0	0	48
DODGE CITY	18	0	0	*	1	3	13	18	19	8	1	0	0	63
GOODLAND U	40	0	0	0	*	2	11	20	18	8	1	0	0	60
TOPEKA	14	0	0	0	1	1	10	15	17	8	1	0	0	53
WICHITA	7	0	0	0	1	2	12	21	23	10	1	0	0	70
KY. LEXINGTON	16	0	0	0	0	1	6	10	9	4	*	0	0	30
LOUISVILLE	13	0	0	0	*	2	10	16	14	6	1	0	0	49
LA. BATON ROUGE	9	0	0	0	1	7	22	26	24	15	2	0	0	97
LAKE CHARLES	22	0	0	*	*	3	19	23	24	15	2	0	0	86
NEW ORLEANS	45	0	0	*	*	3	16	20	19	10	1	0	0	69
SHREVEPORT U	80	0	0	*	*	4	18	24	23	14	2	0	0	85
MAINE CARIBOU	21	0	0	0	0	*	1	1	1	*	0	0	0	3
PORTLAND	20	0	0	0	0	*	1	2	2	*	0	0	0	5
MD. BALTIMORE U	88	0	0	*	*	1	5	10	6	2	*	0	0	24
FREDERICK	11	0	0	0	*	1	7	10	8	2	*	0	0	28
MASS. BLUE HILL OBS.	75	0	0	0	0	*	1	2	1	*	0	0	0	4
BOSTON	9	0	0	0	0	*	3	5	4	1	*	0	0	13
NANTUCKET	14	0	0	0	0	0	0	*	*	0	0	0	0	*
PITTSFIELD	22	0	0	0	0	0	*	1	1	*	0	0	0	2
WORCESTER U	10	0	0	0	0	0	2	2	2	*	0	0	0	6
MICH. ALPENA U	45	0	0	0	0	*	1	1	1	*	0	0	0	3
DETROIT	27	0	0	0	0	*	3	6	4	1	0	0	0	14
ESCANABA U	51	0	0	0	0	*	*	*	*	*	0	0	0	*
FLINT	19	0	0	0	0	*	3	4	4	1	0	0	0	12
GRAND RAPIDS	21	0	0	0	0	*	2	5	5	1	0	0	0	13
LANSING	43	0	0	0	0	*	2	5	2	1	0	0	0	10
MARQUETTE	23	0	0	0	0	*	1	2	2	1	0	0	0	6
MUSKEGON	21	0	0	0	0	0	1	2	2	1	0	0	0	6
SAULT STE. MARIE	19	0	0	0	0	0	*	*	1	*	0	0	0	1
MINN. DULUTH	19	0	0	0	0	0	*	1	1	0	0	0	0	2
INTERNATIONAL FALLS	21	0	0	0	*	*	1	2	1	*	0	0	0	4
MINNEAPOLIS	22	0	0	0	*	*	3	7	6	2	0	0	0	18
ROCHESTER	10	0	0	0	0	*	3	5	5	1	0	0	0	14
ST. CLOUD	21	0	0	0	0	*	2	4	4	1	*	0	0	11
MISS. JACKSON U	58	0	0	*	1	7	21	24	25	17	3	0	0	98
MERIDIAN	15	0	0	0	1	8	21	24	24	12	2	0	0	92
VICKSBURG	23	0	0	0	*	2	15	20	21	9	1	0	0	68
MO. COLUMBIA	21	0	0	0	*	1	8	14	14	6	*	0	0	43
KANSAS CITY	27	0	0	0	*	2	10	18	17	7	1	0	0	55
ST. JOSEPH U	43	0	0	0	*	1	8	16	13	5	*	0	0	43
ST. LOUIS U	23	0	0	0	*	2	9	16	13	6	*	0	0	46
SPRINGFIELD	15	0	0	0	*	*	8	13	14	6	*	0	0	41
MONT. BILLINGS	26	0	0	0	*	1	3	13	11	3	0	0	0	31
BUTTE	29	0	0	0	0	0	*	3	1	*	0	0	0	4
GLASGOW	17	0	0	0	*	1	2	13	9	2	*	0	0	28
GREAT FALLS	23	0	0	0	0	*	1	8	5	1	*	0	0	15
HAVRE U	57	0	0	0	*	1	2	10	7	1	*	0	0	21
HELENA	20	0	0	0	0	*	1	8	5	1	0	0	0	15
KALISPELL	11	0	0	0	0	0	*	5	3	1	0	0	0	9
MILES CITY	16	0	0	0	*	1	3	16	14	4	*	0	0	38
MISSOULA	16	0	0	0	0	*	1	11	6	1	0	0	0	19
NEBR. GRAND ISLAND	22	0	0	0	*	1	8	16	14	6	0	0	0	45
LINCOLN U	66	0	0	*	*	2	7	14	12	5	*	0	0	40
NORFOLK	15	0	0	0	*	1	7	11	11	5	*	0	0	35
NORTH PLATTE	9	0	0	0	0	*	8	15	14	6	*	0	0	43
OMAHA	25	0	0	0	*	1	8	14	12	5	*	0	0	40
SCOTTSBLUFF	17	0	0	0	0	1	6	15	13	5	*	0	0	40
VALENTINE U	64	0	0	0	*	1	5	12	10	4	*	0	0	32
NEV. ELKO	30	0	0	0	0	*	4	20	16	3	0	0	0	44
ELY	22	0	0	0	0	0	1	9	5	1	0	0	0	16
LAS VEGAS	12	0	0	0	4	15	27	31	30	25	8	0	0	140
RENO	18	0	0	0	0	1	5	20	15	6	0	0	0	47
WINNEMUCCA	11	0	0	0	0	1	7	22	17	6	0	0	0	53
N. H. CONCORD	19	0	0	0	0	1	4	5	3	1	0	0	0	14
MT. WASHINGTON	28	0	0	0	0	0	0	0	0	0	0	0	0	0
N. J. ATLANTIC CITY U	79	0	0	0	0	*	1	1	1	*	*	0	0	3
NEWARK	19	0	0	0	*	1	6	9	7	2	*	0	0	25
TRENTON	28	0	0	0	*	1	4	7	5	1	*	0	0	18
N. MEX. ALBUQUERQUE	21	0	0	0	0	2	18	23	18	6	0	0	0	67
CLAYTON	15	0	0	0	*	1	11	14	11	4	*	0	0	41
RATON	15	0	0	0	0	0	5	7	4	1	0	0	0	17
ROSWELL	13	0	0	0	3	11	25	26	26	15	2	0	0	108
N. Y. ALBANY	14	0	0	0	0	*	3	6	4	1	0	0	0	14
BINGHAMTON U	64	0	0	0	*	*	2	4	3	1	*	0	0	10
BUFFALO	17	0	0	0	0	*	1	3	3	1	0	0	0	8
NEW YORK U	44	0	0	0	*	1	3	6	4	1	*	0	0	15
NEW YORK	20	0	0	0	0	1	4	7	5	1	*	0	0	18
ROCHESTER	20	0	0	0	0	*	3	5	4	2	*	0	0	14
SYRACUSE	11	0	0	0	0	0	2	4	3	1	0	0	0	10
N. C. ASHEVILLE	30	0	0	0	0	*	4	6	3	1	*	0	0	14
CAPE HATERAS U	78	0	0	0	0	0	*	1	*	*	*	0	0	1
CHARLOTTE	21	0	0	*	*	4	14	16	15	6	1	0	0	56
GREENSBORO	32	0	0	*	*	2	10	12	9	4	*	0	0	37
RALEIGH	16	0	0	*	1	2	12	15	12	5	1	0	0	48
WILMINGTON	9	0	0	0	*	3	10	15	14	4	*	0	0	46
WINSTON-SALEM U	57	0	0	*	*	3	10	13	10	5	*	0	0	41
N. DAK. BISMARCK	21	0	0	0	*	1	2	9	8	3	*	0	0	23
DEVILS LAKE U	56	0	0	0	*	*	1	4	4	1	0	0	0	10
FARGO	19	0	0	0	*	1	2	5	5	1	*	0	0	14
WILLISTON U	44	0	0	0	*	1	2	8	6	1	*	0	0	18
OHIO AKRON-CANTON	12	0	0	0	0	0	2	3	3	2	0	0	0	10
CINCINNATI (ABBE.)	45	0	0	0	*	1	5	11	8	4	*	0	0	29
CLEVELAND	19	0	0	0	0	*	5	6	5	2	*	0	0	18
COLUMBUS	21	0	0	0	0	1	6	9	7	3	*	0	0	26
DAYTON	17	0	0	0	0	*	4	6	6	2	0	0	0	18
SANDUSKY	33	0	0	0	*	1	4	7	5	2	*	0	0	19
TOLEDO U	81	0	0	0	0	*	3	5	3	1	0	0	0	12
YOUNGSTOWN	17	0	0	0	0	*	2	4	4	1	0	0	0	11
OKLA. OKLAHOMA CITY U	62	0	*	*	*	2	11	20	20	10	1	0	0	64
TULSA	22	0	0	*	*	2	13	21	22	11	2	0	0	71
OREG. ASTORIA	7	0	0	0	0	0	*	*	0	*	0	0	0	1
BURNS	10	0	0	0	0	*	1	11	7	2	0	0	0	21
EUGENE	18	0	0	0	0	*	1	6	4	3	0	0	0	14
MEACHAM	16	0	0	0	0	0	0	2	1	*	0	0	0	3
MEDFORD	31	0	0	0	*	2	5	16	15	8	1	0	0	47
PENDLETON	25	0	0	0	0	1	4	14	9	3	0	0	0	31
PORTLAND U	58	0	0	0	*	*	1	3	2	1	*	0	0	8
ROSEBURG	8	0	0	0	*	1	2	7	6	3	*	0	0	20
SALEM	23	0	0	0	0	*	2	6	5	3	0	0	0	16
SEXTON SUMMIT	16	0	0	0	0	0	*	1	1	*	0	0	0	2
PA. ALLENTOWN	17	0	0	0	0	*	5	8	4	1	*	0	0	18
ERIE U	77	0	0	0	0	*	1	2	1	*	0	0	0	4
HARRISBURG	22	0	0	0	*	1	5	9	7	2	*	0	0	24
PHILADELPHIA	20	0	0	0	*	1	6	10	6	2	*	0	0	25
PITTSBURGH	8	0	0	0	0	0	1	3	3	2	0	0	0	9
READING	20	0	0	0	*	1	6	9	6	2	*	0	0	24
WILLIAMSPORT	16	0	0	0	*	*	5	8	4	1	*	0	0	18
R. I. BLOCK ISLAND	10	0	0	0	0	0	*	*	0	0	0	0	0	*
PROVIDENCE	7	0	0	0	0	*	2	2	2	*	0	0	0	6
S. C. CHARLESTON	18	0	0	0	*	4	13	13	14	5	*	0	0	49
COLUMBIA	13	0	0	0	1	9	19	24	23	9	1	0	0	86
FLORENCE	12	0	0	0	1	7	16	21	21	7	1	0	0	74
GREENVILLE	7	0	0	0	*	3	11	17	18	6	0	0	0	55
SPARTANBURG	17	0	0	0	*	3	13	15	13	4	1	0	0	49
S. DAK. HURON	21	0	0	0	*	1	5	11	10	3	*	0	0	30
RAPID CITY	18	0	0	0	0	1	3	12	12	4	*	0	0	32
Sioux Falls	15	0	0	0	*	1	4	10	9	3	*	0	0	27
TENN. BRISTOL	15	0	0	0	0	*	5	5	7	2	*	0	0	19
CHATTANOOGA	20	0	0	0	*	4	14	18	17	6	1	0	0	60
KNOXVILLE	18	0	0	0	*	2	11	15	14	5	*	0	0	47
MEMPHIS	19	0	0	0	*	3	16	22	21	9	1	0	0	72
NASHVILLE	19	0	0	0	*	2	14	19	17	8	*	0	0	60
OAK RIDGE	15	0	0	0	0	2	9	12	11	5	*	0	0	39
TEX. ABILENE	21	0	*	1	3	10	22	27	27	16	3	*	0	109
AMARILLO	20	0	0	*	1	4	16	20	20	9	1	0	0	71
AUSTIN	19	0	*	1	1	9	23	29	28	18	5	*	*	114
BROWNSVILLE	18	0	*	1	2	9	23	29	28	18	5	*	0	115
CORPUS CHRISTI	22	0	*	*	1	4	19	29	28	16	3	*	0	100
DALLAS	20	0	0	*	1	7	21	27	27	17	2	0	0	102
EL PASO	21	0	0	0	1	10	25	26	24	14	1	0	0	101
FT. WORTH	13	0	0	*	1	6	20	26	27	15	2	0	*	98
GALVESTON	10	0	0	0	*	1	4	13	14	3	*	0	0	35
HOUSTON	22	0	*	*	*	4	19	25	25	15	2	0	0	90
LAREDO	17	*	2	6	13	23	28	31	29	25	12	1	*	171
LUBBOCK	14	0	0	0	2	8	21	23	22	11	1	0	0	88
MIDLAND	13	0	0	*	4	12	25	27	27	15	3	0	0	113
PORT ARTHUR	7	0	0	0	*	3	19	24	22	13	2	0	0	83
SAN ANGELO	13	0	*	1	4	12	23	27	27	16	3	*	*	113
SAN ANTONIO	18	0	*	1	3	10	24	30	28	18	5	*	*	119
VICTORIA	14	0	*	*	1	7	25	30	28	19	5	*	0	115
WACO	18	0	*	*	1	7	23	28	28	18	4	*	*	109
WICHITA FALLS	17	0	0	1	3	8	21	26	27	17	3	0	0	106
UTAH MILFORD	12	0	0	0	0	1	11	24	18	7	0	0	0	61
SALT LAKE CITY	32	0	0	0	0	1	8	22	18	4	0	0	0	53
WENDOVER	11	0	0	0	0	1	8	22	19	4	0	0	0	54
VT. BURLINGTON	17	0	0	0	0	*	2	3	3	1	0	0	0	9
VA. LYNCHBURG	16	0	0	0	*	1	7	10	8	3	*	0	0	29
NORFOLK	12	0	0	0	1	2	9	13	9	3	*	0	0	37
RICHMOND	31	0	0	*	1	3	10	14	11	5	1	0	0	45
ROANOKE	13	0	0	0	2	2	9	14	11	4	*	0	0	42
WASH. OLYMPIA	19	0	0	0	0	*	1	3	2	*	0	0	0	6
SEATTLE U	27	0	0	0	0	*	*	1	*	*	0	0	0	2
SPOKANE	13	0	0	0	0	*	1	9	5	1	0	0	0	16
STAMPEDE PASS	17	0	0	0	0	0	0	0	*	0	0	0	0	*
TATOOSH ISLAND U	58	0	0	0	0	0	0	0	0	0	0	0	0	0
WALLA WALLA U	46	0	0	0	*	1	4	16	13	3	0	0	0	37
YAKIMA	14	0	0	0	0	2	4	15	10	4	0	0	0	35
W.I. SAN JUAN P. R.	62	*	*	*	1	2	1	*	1	2	2	*	*	9
W. VA. CHARLESTON	13	0	0	0	*	1	5	9	7	3	*	0	0	25
HUNTINGTON	11	0	0	0	1	3	10	14	12	6	1	0	0	47
PARKERSBURG	72	0	0	0	*	1	5	8	6	3	*	0	0	23
WIS. GREEN BAY	11	0	0	0	0	*	1	1	2	2	0	0	0	6
LA CROSSE	10	0	0	0	*	*	3	5	4	2	0	0	0	14
MADISON	21	0	0	0	*	*	4	7	6	2	0	0	0	19
MILWAUKEE	20	0	0	0	0	*	2	3	4	1	0	0	0	10
WYO. CASPER	21	0	0	0	0	0	3	12	8	2	0	0	0	25
CHEYENNE	25	0	0	0	0	0	1	5	4	1	0	0	0	11
LANDER	14	0	0	0	0	*	2	10	7	2	0	0	0	21
SHERIDAN	20	0	0	0	0	*	2	12	11	3	0	0	0	28
YELLOWSTONE	30	0	0	0	0	0	0	1	*	*	0	0	0	1

MEAN NUMBER OF DAYS MAXIMUM TEMPERATURE 70°F. AND ABOVE

States and Stations	Yrs.	Jan.	Feb.	Mar.	Apr.	May	June	July	Aug.	Sept.	Oct.	Nov.	Dec.	Annual
ALASKA ANCHORAGE	35	0	0	0	0	*	4	6	4	*	0	0	0	14
ANNETTE	13	0	0	0	*	1	3	4	5	2	0	0	0	15
BARROW	40	0	0	0	0	0	*	*	*	0	0	0	0	*
BARTER ISLAND	13	0	0	0	0	0	0	*	*	0	0	0	0	*
Bethel	18	0	0	0	0	*	3	4	1	*	0	0	0	9
COLD BAY	16	0	0	0	0	0	0	*	*	*	0	0	0	*
CORDOVA	15	0	0	0	0	*	2	3	3	*	*	0	0	8
FAIRBANKS	31	0	0	0	*	3	17	19	9	1	0	0	0	49
JUNEAU	17	0	0	0	*	2	7	7	5	*	0	0	0	21
KING SALMON	15	0	0	0	0	*	3	6	3	*	0	0	0	13
KOTZEBUE	17	0	0	0	0	*	1	2	1	0	0	0	0	4
MCGRATH	18	0	0	0	0	1	11	13	4	*	0	0	0	29
NOME	14	0	0	0	0	0	1	1	1	0	0	0	0	3
ST. PAUL ISLAND	43	0	0	0	0	0	0	0	0	0	0	0	0	0
SHEMYA	15	0	0	0	0	0	0	0	0	0	0	0	0	0
YAKUTAT	14	0	0	0	0	1	1	2	1	*	0	0	0	5

*LESS THAN ONCE IN 2 YEARS

DATA FROM AIRPORT, EXCEPT THOSE MARKED WITH U FOR URBAN.

THESE CHARTS AND TABULATIONS WERE DERIVED FROM "NORMALS, MEANS, AND EXTREMES" TABLE IN U. S. WEATHER BUREAU PUBLICATION LOCAL CLIMATOLOGICAL DATA 1960.

(Source: Climatic Atlas of the U.S.)

Quick Reference Chart 2-5

MEAN NUMBER OF DAYS MINIMUM TEMPERATURE 32 DEGREES F AND BELOW

MEAN NUMBER OF DAYS MINIMUM TEMPERATURE 32°F AND BELOW

State and Stations	Yrs.	Jan.	Feb.	Mar.	Apr.	May	June	July	Aug.	Sept.	Oct.	Nov.	Dec.	Annual
ALA. BIRMINGHAM	18	14	9	7	1	0	0	0	0	0	1	8	14	53
MOBILE	20	7	4	1	0	0	0	0	0	0	*	2	5	18
MONTGOMERY U	83	7	4	1	*	0	0	0	0	0	*	2	6	20
ALASKA ANCHORAGE	35	31	28	30	24	5	*	0	*	4	20	29	31	201
ANNETTE	13	20	16	15	5	*	0	0	0	0	1	7	15	79
BARROW	41	31	28	31	30	31	24	15	15	26	31	30	31	323
BARTER ISLAND	14	31	28	31	30	31	23	8	9	24	31	30	31	307
BETHEL	18	31	28	31	28	14	*	*	*	6	26	29	31	225
COLD BAY	18	26	25	28	24	7	*	0	0	*	7	20	27	163
CORDOVA	16	29	26	29	25	10	1	0	*	6	17	22	27	191
FAIRBANKS	32	31	28	31	28	9	*	0	1	13	29	30	31	232
FT. YUKON	30	31	28	31	29	17	1	0	2	15	30	30	31	245
JUNEAU	18	26	23	25	16	4	*	0	*	2	9	18	23	146
KING SALMON	16	29	27	29	26	10	1	0	*	6	21	26	30	205
KOTZEBUE	18	31	28	31	30	26	7	0	*	9	29	30	31	252
MCGRATH	18	31	28	31	29	11	*	0	1	10	28	30	31	231
NOME	39	31	28	31	29	22	4	0	1	10	25	29	31	241
ST. PAUL ISLAND	44	27	27	30	28	20	3	*	*	1	9	19	26	189
SHEMYA	16	23	24	25	14	2	0	0	0	0	1	12	22	124
YAKUTAT	15	28	26	28	23	8	*	0	*	3	12	19	25	171
ARIZ. FLAGSTAFF	12	31	27	30	24	15	3	0	0	3	18	28	30	211
PHOENIX	21	6	4	0	0	0	0	0	0	0	0	1	6	17
PRESCOTT	19	27	24	21	7	1	0	0	0	0	3	20	27	131
TUCSON	21	7	5	1	*	0	0	0	0	0	0	2	6	21
WINSLOW	29	28	23	20	7	1	0	0	0	0	3	22	28	132
YUMA	11	1	1	*	0	0	0	0	0	0	0	*	*	2
ARK. FT. SMITH	15	20	14	7	1	0	0	0	0	0	1	10	18	71
LITTLE ROCK	17	16	9	4	*	0	0	0	0	0	*	6	13	48
TEXARKANA	19	13	7	3	0	0	0	0	0	0	*	5	11	39
CALIF. BAKERSFIELD	24	7	2	*	0	0	0	0	0	0	0	1	5	15
BISHOP	14	30	25	22	7	1	*	0	0	*	6	24	30	146
BLUE CANYON	18	20	18	20	10	5	1	0	0	*	3	11	15	102
BURBANK	19	2	1	*	0	0	0	0	0	0	0	*	*	4
EUREKA	51	2	1	*	*	0	0	0	0	0	0	*	1	4
FRESNO	21	9	5	2	*	0	0	0	0	0	0	2	8	26
LONG BEACH	17	1	*	0	0	0	0	0	0	0	0	0	0	1
LOS ANGELES U	21	*	0	0	0	0	0	0	0	0	0	0	*	*
MT. SHASTA	19	26	22	21	11	4	*	*	0	1	6	19	26	135
OAKLAND	31	4	1	*	0	0	0	0	0	0	0	*	2	7
RED BLUFF	17	10	3	1	*	0	0	0	0	0	*	1	6	23
SACRAMENTO	28	5	1	*	0	0	0	0	0	0	0	1	3	9
SANDBERG	29	13	11	10	4	1	*	0	0	0	1	5	9	54
SAN DIEGO	21	*	0	0	0	0	0	0	0	0	0	0	0	*
SAN FRANCISCO	34	3	1	*	*	0	0	0	0	0	0	1	1	6
SANTA MARIA	19	7	3	1	*	0	0	0	0	0	*	1	3	15
COLO. ALAMOSA	16	31	28	31	26	13	2	0	*	10	26	30	31	226
COLORADO SPRINGS	12	30	27	27	16	2	*	0	0	*	9	25	30	166
DENVER U	84	27	24	22	10	2	*	0	0	*	6	20	27	139
GRAND JUNCTION	15	31	25	20	6	*	0	0	0	0	3	24	30	138
PUEBLO	21	30	27	26	10	1	0	0	0	*	8	26	30	158
CONN. BRIDGEPORT	13	25	22	18	3	0	0	0	0	0	*	9	23	101
HARTFORD U	51	27	25	20	6	*	0	0	0	*	3	13	25	119
NEW HAVEN	18	26	24	20	4	*	0	0	0	0	2	11	25	112
DEL. WILMINGTON	14	25	22	17	3	0	0	0	0	0	1	12	23	102
D.C. WASHINGTON	20	20	17	10	1	0	0	0	0	0	*	6	19	73
FLA. APALACHICOLA	31	2	1	*	0	0	0	0	0	0	0	*	1	4
DAYTONA BEACH	18	2	1	*	0	0	0	0	0	0	0	*	2	5
EVERGLADES	26	*	0	0	0	0	0	0	0	0	0	0	*	*
FT. MYERS	20	*	*	0	0	0	0	0	0	0	0	0	*	*
JACKSONVILLE	20	4	3	1	0	0	0	0	0	0	0	1	4	11
KEY WEST	13	0	0	0	0	0	0	0	0	0	0	0	0	0
KEY WEST U	87	0	0	0	0	0	0	0	0	0	0	0	0	0
LAKELAND U	21	*	*	*	0	0	0	0	0	0	0	*	*	1
MIAMI	19	0	*	0	0	0	0	0	0	0	0	0	0	*
MIAMI BEACH	20	0	0	0	0	0	0	0	0	0	0	0	0	0
ORLANDO	18	1	1	*	0	0	0	0	0	0	0	*	1	3
PENSACOLA	22	4	2	*	0	0	0	0	0	0	0	1	2	9
TALLAHASSEE	22	7	4	1	*	0	0	0	0	0	0	2	6	20
TAMPA	15	1	*	0	0	0	0	0	0	0	0	*	*	2
WEST PALM BEACH	21	*	0	0	0	0	0	0	0	0	0	0	*	*
GA. ATHENS	18	13	9	6	1	0	0	0	0	0	*	6	14	49
ATLANTA U	77	11	9	4	*	0	0	0	0	0	*	4	10	38
AUGUSTA	11	15	9	5	*	0	0	0	0	0	1	9	14	54
COLUMBUS	16	12	8	5	*	0	0	0	0	0	*	6	12	44
MACON	13	10	6	3	*	0	0	0	0	0	*	5	10	34
ROME	16	17	14	11	2	0	0	0	0	0	2	13	19	79
SAVANNAH	11	10	5	3	0	0	0	0	0	0	*	3	9	30
THOMASVILLE	39	5	3	1	*	0	0	0	0	0	0	1	4	14
HAWAII Hilo	16	0	0	0	0	0	0	0	0	0	0	0	0	0
HONOLULU	39	0	0	0	0	0	0	0	0	0	0	0	0	0
LIHUE	12	0	0	0	0	0	0	0	0	0	0	0	0	0
IDAHO BOISE	21	27	23	19	7	1	0	0	0	*	5	20	26	128
IDAHO FALLS	12	31	28	31	22	10	2	0	*	7	23	29	31	213
LEWSITON	15	24	18	14	4	*	0	0	0	*	4	15	21	101
POCATELLO	23	29	25	25	14	3	*	0	0	2	12	23	29	161
ILL. CAIRO U	19	19	13	8	*	0	0	0	0	0	*	7	17	64
CHICAGO	19	29	24	20	6	*	0	0	0	0	2	16	26	123
CHICAGO U	85	27	24	19	4	*	0	0	0	*	1	12	23	110
MOLINE	21	31	26	22	8	*	0	0	0	*	5	19	28	139
PEORIA	19	29	25	21	5	*	0	0	0	*	2	18	27	127
ROCKFORD	11	30	27	27	9	1	0	0	0	*	6	21	29	150
SPRINGFIELD	12	28	23	20	6	*	0	0	0	0	3	17	24	121
IND. EVANSVILLE	20	24	20	15	2	0	0	0	0	*	2	15	23	101
FT. WAYNE	10	28	25	23	8	*	0	0	0	*	5	18	27	134
INDIANAPOLIS	18	27	23	20	6	*	0	0	0	*	3	16	25	120
SOUTH BEND	21	29	26	23	10	1	0	0	0	*	4	17	27	137
IOWA BURLINGTON U	61	29	25	20	6	*	0	0	0	*	4	17	27	128
DES MOINES	21	30	27	23	7	1	0	0	0	*	4	20	28	140
DUBUQUE	11	30	27	28	12	1	0	0	0	*	6	22	29	155
SIOUX CITY	20	31	27	24	10	1	0	0	0	*	6	23	30	152
WATERLOO	12	31	27	26	12	1	0	0	0	1	7	23	30	158
KANS. CONCORDIA U	76	28	23	18	4	*	0	0	0	*	3	16	26	118
DODGE CITY	19	29	24	20	6	*	0	0	0	0	2	18	28	127
GOODLAND	41	30	26	25	11	1	*	0	0	1	7	25	30	156
TOPEKA	15	29	25	20	6	*	0	0	0	0	3	19	27	129
WICHITA	8	29	23	16	3	0	0	0	0	0	1	14	26	112
KY. LEXINGTON	17	23	19	15	3	*	0	0	0	0	2	13	22	97
LOUISVILLE	13	23	17	15	2	0	0	0	0	0	2	13	20	92
LA. BATON ROUGE	9	6	3	1	0	0	0	0	0	0	*	2	6	18
LAKE CHARLES	22	5	2	1	0	0	0	0	0	0	0	1	3	12
NEW ORLEANS	46	2	1	*	0	0	0	0	0	0	0	*	1	4
SHREVEPORT	9	10	6	3	0	0	0	0	0	0	*	4	8	31
MAINE CARIBOU	22	31	28	30	23	6	*	0	0	3	15	25	30	192
PORTLAND	21	30	27	27	14	3	0	0	0	1	10	20	29	160
MD. BALTIMORE U	89	20	18	10	1	0	0	0	0	0	*	5	17	71
FREDERICK	12	27	22	20	6	*	0	0	0	*	3	17	25	119
MASS. BLUE HILL OBS.	76	29	26	24	10	*	0	0	0	*	3	14	26	132
BOSTON	10	26	22	17	2	0	0	0	0	0	*	6	21	94
NANTUCKER	15	24	22	18	4	*	0	0	0	0	*	5	19	93
PITTSFIELD	23	30	25	28	15	2	0	0	0	2	10	20	28	160
WORCESTER	5	30	27	28	12	1	0	0	0	*	6	15	29	148
MICH. ALPENA U	45	30	28	28	16	2	0	0	0	*	5	18	28	155
DETROIT	28	28	26	22	8	*	0	0	0	*	2	14	25	125
ESCANABA U	52	31	28	28	17	3	*	0	0	1	6	20	29	163
FLINT	19	30	27	25	13	2	0	0	0	1	6	18	27	149
GRAND RAPIDS	22	30	27	25	12	2	*	0	0	1	6	19	28	150
LANSING	43	29	27	26	12	2	0	0	0	*	6	16	27	145
MARQUETTE	24	31	28	28	17	4	0	0	0	*	4	20	29	161
MUSKEGON	21	30	27	26	12	1	0	0	0	*	4	17	27	144
SAULT STE. MARIE	20	31	28	30	22	7	*	0	*	1	7	22	30	178
MINN. DULUTH	19	31	28	30	22	7	*	0	0	3	12	27	31	191
INTERNATIONAL FALLS	22	31	28	30	23	8	*	0	*	5	15	28	31	199
MINNEAPOLIS	22	31	28	27	10	1	0	0	0	1	5	23	30	156
ROCHESTER	10	30	28	26	16	2	*	0	0	2	9	23	30	166
ST. CLOUD	22	31	28	29	17	3	0	0	0	2	11	26	31	178
MISS. JACKSON U	59	11	7	3	*	0	0	0	0	0	*	5	10	36
MERIDIAN	16	13	8	5	1	0	0	0	0	0	1	7	13	47
VICKSBURG	24	8	4	1	0	0	0	0	0	0	*	2	5	21
MO. COLUMBIA	22	27	23	17	3	0	0	0	0	*	1	15	24	110
KANSAS CITY	27	26	21	15	2	*	0	0	0	0	1	13	23	101
ST. JOSEPH	13	30	25	20	5	*	0	0	0	0	2	18	27	127
ST. LOUIS U	23	23	18	12	2	0	0	0	0	0	*	9	19	83
SPRINGFIELD	15	25	20	16	3	*	0	0	0	0	2	15	22	103
MONT. BILLINGS	26	29	25	25	12	2	*	0	0	1	8	21	27	149
BUTTE	29	31	28	30	26	15	3	*	1	11	24	29	30	228
GLASGOW	18	31	28	29	17	3	*	0	0	2	14	28	31	184
GREAT FALLS	24	26	25	25	13	3	*	0	0	2	8	20	24	146
MONT. HAVRE U	57	30	27	27	15	3	*	0	*	3	14	25	29	172
HELENA	21	30	27	28	17	5	*	0	*	4	16	26	29	182
KALISPELL	11	29	27	28	18	4	*	0	*	5	18	25	29	182
MILES CITY	17	31	28	27	16	3	*	0	0	1	11	27	30	174
MISSOULA	16	30	27	28	18	4	*	0	*	3	16	26	30	181
NEBR. GRAND ISLAND	22	31	27	24	9	*	0	0	0	0	6	23	30	150
LINCOLN U	68	29	25	20	5	*	0	0	0	*	4	18	28	129
NORFOLK	16	31	27	25	11	1	0	0	0	1	7	25	30	158
NORTH PLATTE	10	31	28	28	15	2	0	0	0	1	14	28	31	178
OMAHA	25	30	26	22	6	*	0	0	0	*	3	20	29	136
SCOTTSBLUFF	18	31	28	29	17	3	*	0	0	1	12	28	31	180
VALENTINE U	67	31	27	26	12	2	0	0	0	1	10	26	30	164
NEV. ELKO	31	30	27	29	22	11	3	*	1	9	23	28	30	214
ELY	23	31	28	30	24	13	4	*	1	8	22	28	30	218
LAS VEGAS	19	20	11	3	*	0	0	0	0	0	*	8	17	59
RENO	19	29	26	27	20	7	1	0	*	4	18	27	29	188
WINNEMUCCA	12	28	25	27	20	9	2	*	*	7	21	26	28	195
N. H. CONCORD	20	30	28	27	15	4	0	0	0	2	12	21	29	168
MT. WASHINGTON	29	31	28	31	29	19	6	1	3	12	23	28	31	242
N. J. ATLANTIC CITY U	85	20	18	12	2	0	0	0	0	0	*	6	17	75
NEWARK	20	25	22	15	2	0	0	0	0	0	*	9	22	94
TRENTON	29	24	21	15	2	0	0	0	0	0	*	7	21	89
N. MEX. ALBUQUERQUE	21	26	20	15	3	0	0	0	0	0	*	17	26	107
CLAYTON	16	29	25	24	10	1	0	0	0	0	5	22	28	144
RATON	16	31	28	29	18	3	0	0	0	1	16	29	31	185
ROSWELL	51	25	19	11	2	*	0	0	0	0	2	15	25	99
N. Y. ALBANY U	82	28	26	23	8	*	0	0	0	*	3	15	26	129
BINGHAMTON U	65	28	25	23	10	1	*	0	0	*	5	16	26	134
BUFFALO	17	28	26	24	10	1	0	0	0	*	2	14	26	131
NEW YORK U	89	23	22	16	4	0	0	0	0	0	*	6	20	91
NEW YORK	21	23	19	12	1	0	0	0	0	0	0	4	18	77
ROCHESTER	21	29	26	25	10	1	0	0	0	*	3	15	26	136
SYRACUSE	12	29	26	26	8	1	0	0	0	0	3	15	27	135
N. C. ASHEVILLE	31	18	16	12	2	0	0	0	0	0	2	13	19	83
CAPE HATTERAS R	81	5	4	1	*	0	0	0	0	0	0	*	3	13
CHARLOTTE	21	17	13	9	1	0	0	0	0	0	*	8	17	65
GREENSBORO	33	20	17	12	2	0	0	0	0	0	1	12	21	85
RALEIGH	17	18	15	10	1	*	0	0	0	0	1	9	19	73
WILMINGTON U	81	8	7	3	*	0	0	0	0	0	*	2	7	27
WINSTON-SALEM U	56	23	18	11	3	*	0	0	0	0	2	12	21	90
N. DAK. BISMARCK	19	31	28	30	18	4	0	0	0	2	14	28	31	186
DEVILS LAKE U	57	31	28	29	19	6	*	0	*	3	15	28	30	189
FARGO	19	31	28	29	17	5	0	0	0	2	12	27	31	182
WILLISTON U	45	31	28	28	15	3	*	0	0	2	13	27	31	178
OHIO AKRON CANTON	13	28	25	24	11	1	0	0	0	*	3	18	26	136
CINCINNATI (ABBE)	46	23	20	16	3	*	0	0	0	0	2	11	22	98
CLEVELAND	19	27	24	21	8	*	0	0	0	*	1	13	25	119
COLUMBUS	21	26	23	20	6	*	0	0	0	*	3	15	24	117
DAYTON	18	27	23	19	6	*	0	0	0	0	2	15	24	116
SANDUSKY	34	26	24	20	5	0	0	0	0	0	1	12	23	111
TOLEDO U	82	27	25	21	7	*	0	0	0	*	2	14	24	121
YOUNGSTOWN	18	28	26	23	11	1	0	0	0	*	4	17	26	136
OKLA. OKLAHOMA CITY U	63	20	16	8	1	0	0	0	0	0	1	7	17	70
TULSA	22	21	16	10	1	0	0	0	0	0	1	9	18	76
OREG. ASTORIA	8	8	6	8	2	*	0	0	0	0	*	6	6	36
BURNS	11	30	26	28	18	6	1	0	*	3	16	27	30	184
EUGENE	19	15	10	8	3	*	0	0	0	*	2	8	12	59
MEACHAM	17	28	26	27	20	7	1	*	0	10	10	22	27	170
MEDFORD	31	19	15	10	3	*	*	0	0	*	3	13	16	79
PENDLETON	26	22	17	10	2	*	0	0	0	0	2	15	19	86
PORTLAND U	59	9	4	1	*	0	0	0	0	0	*	1	5	22
ROSEBURG U	77	10	7	5	1	*	0	0	0	*	1	4	7	35
SALEM	24	15	11	9	3	*	0	0	0	*	1	9	11	60
SEXTON-SUMMIT	17	20	17	22	14	6	*	0	0	*	4	10	18	112
PA. ALLENTOWN	18	28	26	22	7	*	0	0	0	*	4	16	27	129
ERIE U	84	27	25	23	8	*	0	0	0	0	1	10	23	117
HARRISBURG	23	26	22	18	3	*	0	0	0	*	1	12	23	106
PHILADELPHIA	22	26	23	18	3	*	0	0	0	*	1	12	23	106
PITTSBURGH U	81	24	23	18	5	0	0	0	0	0	1	11	22	103
READING	20	24	21	14	2	0	0	0	0	0	1	8	22	92
WILLIAMSPORT	17	28	25	22	7	1	0	0	0	*	4	16	27	130
R. I. BLOCK ISLAND U	75	22	22	16	2	0	0	0	0	0	*	5	18	85
PROVIDENCE U	51	26	24	18	4	*	0	0	0	0	1	10	23	106
S. C. CHARLESTON	19	10	7	3	*	0	0	0	0	0	*	4	9	33
CHARLESTON U	85	3	2	*	0	0	0	0	0	0	0	1	3	9
COLUMBIA U	68	9	7	3	*	0	0	0	0	0	*	3	9	31
FLORENCE U	56	11	11	4	*	0	0	0	0	0	*	2	11	39
GREENVILLE U	38	12	9	5	*	0	0	0	0	0	*	5	11	42
SPARTANBURG	18	14	10	6	1	0	0	0	0	0	*	5	16	52
S. DAK. HURON U	74	31	28	27	13	2	*	0	0	1	11	26	30	169
RAPID CITY U	56	29	26	26	13	2	*	0	0	1	9	23	29	158
SIOUX FALLS	16	31	28	28	14	2	0	0	0	1	9	26	30	169
TENN. BRISTOL	16	20	17	14	3	*	0	0	0	0	3	14	22	94
CHATTANOOGA U	77	14	11	6	1	0	0	0	0	0	*	6	13	51
KNOXVILLE U	85	17	13	8	1	0	0	0	0	0	1	8	16	64
MEMPHIS	20	17	11	6	*	0	0	0	0	0	1	8	15	57
NASHVILLE	20	18	15	10	1	0	0	0	0	0	1	10	18	74
OAK RIDGE	13	17	14	12	2	0	0	0	0	0	1	14	19	80
TEX. ABILENE	21	15	8	5	*	0	0	0	0	0	*	5	12	45
AMARILLO	20	27	21	17	4	*	0	0	0	0	1	15	26	111
AUSTIN	19	7	4	2	0	0	0	0	0	0	*	2	4	19
BROWNSVILLE	19	1	*	*	0	0	0	0	0	0	0	0	*	1
CORPUS CHRISTI	22	3	1	*	0	0	0	0	0	0	0	*	1	5
DALLAS	20	12	6	3	*	0	0	0	0	0	*	4	8	33
EL PASO	21	16	9	4	*	0	0	0	0	0	*	7	16	52
FT. WORTH U	57	11	8	3	*	0	0	0	0	0	*	2	8	32
GALVESTON U	84	2	1	*	0	0	0	0	0	0	0	*	1	4
HOUSTON	23	4	1	1	0	0	0	0	0	0	0	*	1	7
LAREDO	18	2	1	*	0	0	0	0	0	0	0	*	1	4
LUBBOCK	15	26	19	13	3	0	0	0	0	0	1	15	25	102
MIDLAND	26	18	11	6	1	0	0	0	0	0	*	8	20	64
PORT ARTHUR	11	4	2	1	0	0	0	0	0	0	0	1	3	11
SAN ANGELO	13	13	8	4	*	0	0	0	0	0	*	5	12	42
SAN ANTONIO U	70	5	3	1	0	0	0	0	0	0	*	1	3	13
VICTORIA U	50	4	2	1	0	0	0	0	0	0	*	1	3	11
WACO	25	9	5	2	*	0	0	0	0	0	0	3	7	26
WICHITA FALLS	17	18	12	7	1	0	0	0	0	0	*	7	16	61
UTAH MILFORD	13	30	26	27	17	4	*	0	0	2	15	27	30	179
SALT LAKE CITY	32	28	23	20	7	1	0	0	0	*	5	22	27	134
WENDOVER	12	29	24	18	4	*	0	0	0	0	3	22	29	129
VT. BURLINGTON	64	29	27	26	13	1	0	0	0	*	6	18	28	148
VA. LYNCHBURG U	81	20	17	10	2	*	0	0	0	0	1	9	19	78
NORFOLK U	85	13	11	5	*	0	0	0	0	0	*	2	10	41
RICHMOND U	58	19	17	9	1	0	0	0	0	0	*	7	18	71
ROANOKE	14	20	17	14	2	0	0	0	0	0	1	12	20	86
WASH. OLYMPIA	19	18	14	15	7	1	0	0	0	*	4	10	14	84
SEATTLE U	64	8	4	2	*	0	0	0	0	0	*	1	5	20
SPOKANE U	74	26	22	17	5	*	0	0	0	*	6	16	24	116
STAMPEDE PASS	18	31	27	30	23	11	2	*	0	1	11	26	30	191
TATOOSH ISLAND R	59	4	1	1	0	0	0	0	0	0	0	*	1	7
WALLA WALLA U	47	20	12	5	1	*	0	0	0	*	1	10	17	66
YAKIMA U	46	28	22	17	6	1	*	0	0	1	7	21	27	130
W.I. SAN JUAN P. R.	62	0	0	0	0	0	0	0	0	0	0	0	0	0
W. VA. CHARLESTON	14	22	18	16	5	*	0	0	0	0	3	14	21	97
HUNTINGTON U	12	21	17	14	3	0	0	0	0	0	1	14	20	91
PARKERSBURG	73	22	21	15	4	*	0	0	0	*	2	12	21	97
WIS. GREEN BAY U	69	30	28	26	12	1	0	0	0	*	6	20	29	152
LA CROSSE	11	31	27	28	10	*	0	0	0	*	5	22	30	153
MADISON U	77	30	27	25	9	*	0	0	0	*	4	19	29	143
MILWAUKEE U	85	29	26	22	7	*	0	0	0	*	3	16	26	129
WYO. CASPER	22	30	27	27	18	5	*	0	0	3	13	24	28	174
CHEYENNE U	83	29	27	27	18	6	*	0	*	2	13	24	28	174
LANDER U	64	31	28	29	18	6	*	*	*	4	17	28	31	192
SHERIDAN	21	31	28	29	17	3	*	0	0	2	14	27	30	182
YELLOWSTONE MAM.H.S.	30	31	27	30	23	12	3	*	*	7	19	28	30	210

* LESS THAN ONCE IN 2 YEARS.

DATA FROM AIRPORT, EXCEPT THOSE MARKED WITH U FOR URBAN AND R FOR RURAL.

THESE CHARTS AND TABULATIONS WERE DERIVED FROM "NORMALS, MEANS, AND EXTREMES" TABLE IN U. S. WEATHER BUREAU PUBLICATION LOCAL CLIMATOLOGICAL DATA (THROUGH 1961 USUALLY).

DUE TO ROUNDING TO WHOLE NUMBERS THE SUM OF THE MONTHLY VALUES MAY NOT EQUAL THE ANNUAL.

(Source: Climatic Atlas of the U.S.)

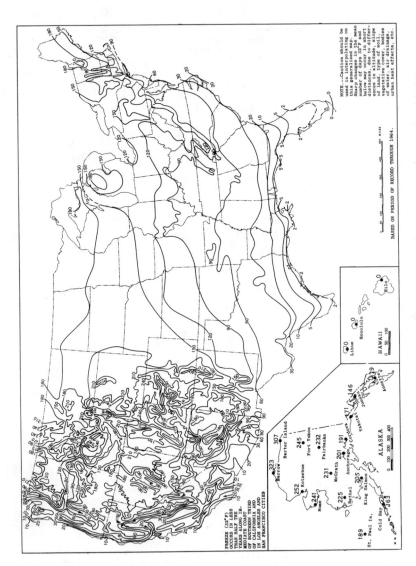

FIGURE 2-9. Mean annual number of days minimum temperature 32 degrees F and below *(from the Climatic Atlas of the U.S.).*

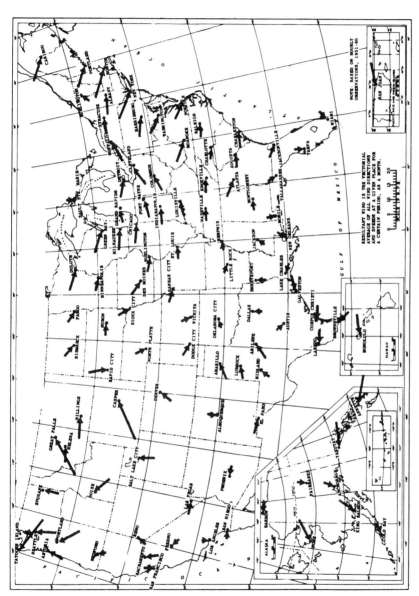

FIGURE 2-10. Mean resultant surface wind direction and speed midwinter month (January) *(from the Climatic Atlas of the U.S.).*

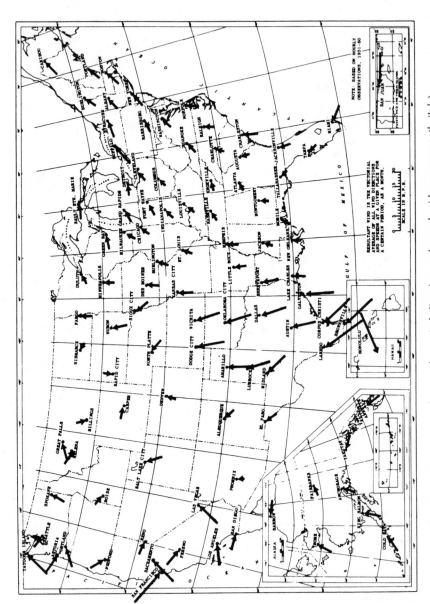

FIGURE 2-11. Mean resultant surface wind direction and speed midsummer month (July) (from the Climatic Atlas of the U.S.).

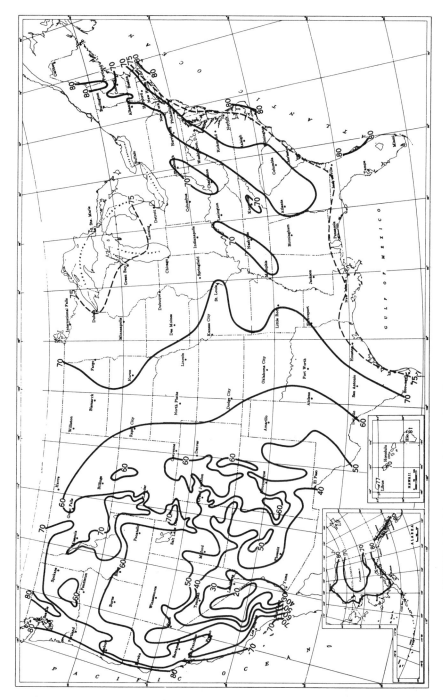

FIGURE 2-12. Mean annual relative humidity in percentages *(from the Climatic Atlas of the U.S.).*

Quick Reference Chart 2-6
MEAN RELATIVE HUMIDITY IN PERCENT

STATE AND STATION	YEAR	JANUARY 1 A.M.	7 A.M.	1 P.M.	7 P.M.	APRIL 1 A.M.	7 A.M.	1 P.M.	7 P.M.	JULY 1 A.M.	7 A.M.	1 P.M.	7 P.M.	OCTOBER 1 A.M.	7 A.M.	1 P.M.	7 P.M.	ANNUAL 1 A.M.	7 A.M.	1 P.M.	7 P.M.
ALA,BIRMINGHAM	54	80	81	61	67	77	77	50	54	86	84	56	68	84	83	49	64	81	81	54	63
MONTGOMERY	68	84	83	61	67	81	80	52	56	87	85	58	68	85	84	49	60	83	83	55	73
ALASKA,ANCHORAGE	20	74	74	72	74	75	67	53	67	84	74	62	72	81	81	68	79	79	74	63	73
FAIRBANKS	30	68	69	68	70	74	63	47	61	88	72	52	66	83	83	69	81	78	71	58	69
JUNEAU	16	78	79	76	79	85	72	64	80	88	78	68	79	88	85	79	87	85	79	72	82
NOME	25	77	80	76	79	82	82	77	81	91	87	83	85	83	83	76	81	83	83	78	81
ARIZ,PHOENIX	21	67	74	47	39	42	53	28	21	42	52	31	23	52	60	32	27	50	59	33	27
YUMA	68	51	57	32	32	38	55	21	19	41	60	28	25	45	58	27	28	43	58	27	26
ARK,LITTLE ROCK	18	76	80	66	68	73	80	55	56	81	86	56	59	79	85	51	60	77	83	57	61
CALIF,FRESNO	65	88	90	73	67	67	80	44	35	44	52	28	16	65	74	42	35	66	74	47	39
LOS ANGELES	20	71	65	46	52	74	78	50	55	79	84	49	53	74	76	47	57	73	75	47	53
RED BLUFF	15	77	81	71	59	58	68	44	34	34	49	29	16	52	62	42	31	56	66	47	36
SACRAMENTO	20	87	90	82	70	75	85	57	45	60	77	47	29	67	78	54	39	73	83	60	46
SAN DIEGO	19	77	76	58	60	78	81	62	61	83	86	68	64	80	82	62	65	79	80	61	62
SAN FRANCISCO	20	83	87	77	69	83	87	69	68	87	90	70	67	82	87	69	65	83	87	71	67
COLO,DENVER	20	60	60	44	49	60	67	40	39	57	67	32	34	56	62	34	35	59	64	38	40
GRAND JUNCTION	18	74	78	62	60	48	57	35	30	37	48	28	23	54	57	37	35	53	61	41	37
CONN,HARTFORD	43	75	73	61	67	79	71	52	61	88	78	55	70	86	82	54	70	82	76	56	68
D.C,WASHINGTON	65	70	73	56	64	69	68	45	55	84	79	53	68	82	80	50	70	76	75	52	65
FLA,APALACHICOLA	25	85	87	69	80	85	85	66	74	86	85	72	76	84	86	64	75	85	86	68	76
JACKSONVILLE	23	85	88	56	74	83	84	47	66	88	86	57	75	88	90	58	78	86	87	55	73
KEY WEST	19	82	84	69	79	79	79	66	75	79	77	68	74	83	84	70	78	81	81	68	76
MIAMI	17	83	86	55	74	81	81	56	70	87	84	65	76	87	88	64	78	85	85	60	75
TAMPA	64	85	87	60	74	82	80	53	68	88	83	62	79	87	66	58	74	86	84	58	74
GA,ATLANTA	71	76	80	64	69	72	75	51	56	85	83	57	68	78	79	53	63	77	79	56	64
HAWAII,HILO	10	85	81	68	85	87	82	69	84	89	82	68	82	87	81	68	86	87	82	69	84
HONOLULU	13	80	80	62	74	75	71	57	71	74	69	56	70	76	72	58	69	76	73	58	71
LIHUE	10	84	84	68	81	81	77	66	76	81	77	65	78	82	79	67	79	82	79	67	78
IDAHO,BOISE	20	81	82	75	74	63	73	48	38	39	54	34	23	61	68	49	43	63	71	53	46
POCATELLO	21	80	82	76	74	59	72	46	39	38	57	30	21	59	72	48	41	61	72	52	46
ILL,CAIRO	68	--	80	68	--	--	75	55	--	--	82	56	--	--	84	54	--	--	81	59	--
CHICAGO	36	79	80	71	75	73	75	58	64	77	76	56	61	74	79	56	65	77	78	61	67
SPRINGFIELD	17	84	85	73	79	78	82	57	61	81	85	53	57	79	85	52	63	81	85	60	66
IND,EVANSVILLE	56	81	82	69	73	75	75	53	57	84	78	52	57	80	82	51	61	80	80	58	63
FT.WAYNE	44	83	85	74	80	79	78	57	63	82	78	51	57	83	85	56	67	82	82	61	68
INDIANAPOLIS	20	81	83	72	78	76	80	57	62	83	84	55	61	80	87	53	65	81	84	60	68
IOWA,DES MOINES	19	78	79	71	73	71	79	54	54	78	85	56	56	71	80	52	55	77	82	61	62
DUBUQUE	79	--	81	70	76	--	75	53	58	--	78	53	60	--	82	55	66	--	80	59	67
SIOUX CITY	66	76	77	68	71	74	77	52	52	79	80	52	51	74	79	52	56	77	79	58	60
KANS,CONCORDIA	72	73	78	63	68	68	77	50	50	68	77	47	48	67	78	48	56	70	78	54	57
DODGE CITY	67	74	79	56	62	69	76	47	47	67	76	41	44	69	78	46	52	71	78	48	52
TOPEKA	18	75	78	64	67	72	79	53	53	76	81	51	50	73	80	50	54	75	81	56	58
WICHITA	66	76	78	63	66	71	75	52	51	72	78	48	49	70	76	51	54	73	78	54	56
KY,LOUISVILLE	71	78	78	68	69	74	72	54	56	83	77	52	57	80	81	52	59	79	77	57	61
LA,NEW ORLEANS	37	83	85	67	73	82	84	59	66	84	84	63	71	79	82	59	68	82	84	62	69
SHREVEPORT	66	82	83	67	67	79	82	56	57	83	86	56	62	82	85	54	59	81	84	58	62
MAINE,CARIBOU	15	75	73	69	73	79	77	59	69	88	82	58	71	85	86	61	76	82	79	62	73
EASTPORT	63	--	74	69	73	--	77	69	76	--	85	75	84	--	81	70	79	--	80	71	78
MASS,BOSTON	62	66	72	59	67	68	67	53	63	79	72	55	70	75	75	53	70	73	71	56	69
NANTUCKET	71	79	81	73	80	86	79	69	84	93	85	74	90	84	81	69	82	85	81	71	84
MICH,ALPENA	45	79	80	70	76	77	77	61	68	82	77	61	68	82	85	65	76	80	80	65	73
DETROIT	26	77	81	70	76	73	75	53	60	77	75	51	56	79	83	55	67	77	79	59	66
GRAND RAPIDS	51	86	85	78	80	77	76	53	60	83	76	50	54	84	84	56	72	83	81	61	68
MARQUETTE	55	74	76	72	74	70	72	65	68	77	74	65	69	76	78	65	73	76	76	68	72
SAULT STE.MARIE	64	78	78	75	78	80	78	61	69	91	84	61	70	88	87	68	79	85	82	67	75
MINN,DULUTH	18	77	78	73	73	75	80	59	60	87	88	61	65	82	88	62	72	81	84	65	69
INTER NAT'L FALLS	17	73	74	68	71	73	78	53	55	85	88	57	59	81	86	60	69	80	82	62	65
MPLS.-ST.PAUL	52	75	76	67	72	70	75	52	55	78	79	51	54	73	79	55	61	76	78	58	62
MISS,VICKSBURG	63	84	82	65	67	87	82	57	60	93	87	61	70	92	84	55	66	89	84	59	66
MO,KANSAS CITY	66	74	66	37	66	68	74	53	54	72	76	49	53	67	76	51	55	71	77	55	59
ST.LOUIS	65	77	77	65	68	72	73	54	58	74	73	50	55	74	76	52	58	75	76	57	61
SPRINGFIELD	68	80	82	68	72	74	77	56	57	84	82	57	60	78	82	54	61	81	81	60	63
MONT,HAVRE	50	76	82	71	78	65	79	47	44	57	74	38	35	67	79	50	54	69	79	54	55
HELENA	66	70	68	64	64	64	69	48	43	56	64	37	33	68	71	53	51	67	69	52	49
KALISPELL	50	80	79	79	74	68	78	54	54	59	74	46	34	77	84	67	61	71	79	63	55
NEBR,LINCOLN	14	78	79	67	72	75	81	54	54	76	82	50	50	72	78	47	56	77	81	57	61
NORTH PLATTE	15	80	83	61	62	73	82	50	47	74	84	50	47	75	84	47	51	77	84	54	54
NEV,ELY	7	72	75	60	57	54	66	36	30	37	50	22	21	53	62	34	29	55	64	38	34
LAS VEGAS	11	52	59	41	34	26	35	18	14	23	31	19	14	28	35	21	16	31	38	23	18
RENO	49	76	77	67	56	58	69	37	32	43	62	25	20	66	71	42	32	62	70	44	35
WINNEMUCCA	65	80	84	66	65	56	70	37	31	32	48	20	17	58	66	36	43	58	68	40	38
N.H,CONCORD	18	77	79	61	71	78	73	48	61	89	79	50	68	87	86	52	73	83	80	53	69
N.J,ATLANTIC CITY	70	--	79	68	74	--	76	65	75	--	81	72	82	--	80	64	75	--	79	67	77
N.MEX,ALBUQUERQUE	24	63	70	49	46	40	52	27	23	49	61	33	30	50	62	35	33	50	60	36	32
ROSWELL	51	61	70	45	41	42	58	28	24	57	73	38	35	61	74	41	40	54	68	38	35
N.Y,ALBANY	16	73	75	63	71	74	73	50	60	84	79	52	64	83	86	54	72	79	78	56	68
BINGHAMTON	11	81	82	68	77	79	79	54	64	88	86	56	71	85	88	55	75	84	84	59	73
BUFFALO	20	78	79	72	77	77	77	58	67	82	79	53	63	79	82	58	73	80	79	61	71
CANTON	43	82	80	71	80	76	75	60	66	82	75	56	65	81	82	62	74	81	78	63	72
NEW YORK	71	67	72	61	66	69	69	54	62	72	76	57	67	72	75	58	66	70	73	58	65
SYRACUSE	55	77	76	70	76	77	71	56	65	82	74	54	65	81	79	59	74	80	74	60	71
N.C,ASHEVILLE	45	--	82	59	69	--	77	47	57	--	88	56	73	--	87	50	67	--	84	54	67
RALEIGH	18	78	82	57	69	75	78	46	57	89	86	55	72	86	89	53	74	82	84	53	68
N.DAK,BISMARCK	55	76	74	67	70	76	79	50	51	79	81	48	48	74	81	51	57	77	79	56	59
DEVILS LAKE	22	--	76	72	75	--	82	56	54	--	86	53	53	--	83	57	60	--	82	62	63

City																					
FARGO	68	75	75	70	75	77	82	56	60	81	84	51	57	75	83	56	65	78	81	60	66
WILLISTON	43	74	74	67	69	68	76	49	47	71	77	45	41	70	79	53	53	73	77	55	55
OHIO,CINCINNATI	75	81	82	69	74	75	79	53	59	84	84	51	59	80	87	51	65	80	83	57	65
CLEVELAND	66	82	80	73	76	76	74	54	65	81	74	52	62	78	77	54	68	80	77	59	68
COLUMBUS	68	82	83	72	76	77	75	55	62	84	77	51	59	80	82	52	64	81	79	58	66
OKLA.OKLAHOMA CITY	63	75	79	62	65	71	76	50	52	76	80	49	51	71	79	58	57	74	79	54	57
OREG.BAKER	12	71	83	81	78	52	76	71	46	42	69	68	36	56	78	74	51	59	78	75	55
PORTLAND	19	83	87	82	78	76	87	66	54	71	85	63	47	86	91	79	67	79	87	73	61
ROSEBURG	71	90	91	87	79	77	89	63	53	62	85	53	38	86	93	77	61	78	90	70	58
PA,HARRISBURG	67	71	73	60	68	69	67	50	67	81	75	50	63	80	79	53	67	75	74	54	64
PHILADELPHIA	19	74	76	60	68	74	73	49	59	83	79	52	65	82	83	53	70	78	78	54	66
PITTSBURGH	65	76	77	67	73	69	72	50	59	80	77	53	60	75	80	52	63	75	77	56	65
R.I.PROVIDENCE	19	71	73	60	67	75	71	51	65	87	79	57	72	83	83	55	74	79	77	56	70
S.C.CHARLESTON	17	84	87	56	76	85	84	49	70	93	88	64	81	90	90	56	84	88	87	56	77
COLUMBIA	57	79	81	57	65	78	77	47	55	86	83	55	68	86	84	50	65	83	81	53	63
S.DAK.HURON	67	78	76	72	72	76	80	54	51	79	81	52	49	75	81	52	56	79	80	60	59
RAPID CITY	9	71	71	60	66	66	71	48	47	64	70	42	40	60	64	42	46	67	70	50	52
TENN.KNOXVILLE	67	80	83	65	69	71	75	49	54	84	83	55	66	82	86	51	63	79	82	56	63
MEMPHIS	64	80	79	68	70	74	74	56	58	82	81	58	63	81	80	54	62	79	79	59	64
NASHVILLE	69	80	84	66	72	74	76	51	55	82	81	53	61	81	84	51	58	79	81	56	62
TEX.ABILENE	68	71	73	50	55	65	72	40	48	61	74	39	40	69	77	47	52	66	75	45	48
AMARILLO	64	70	73	51	54	62	70	40	38	66	76	41	42	68	76	45	50	66	74	45	46
AUSTIN	29	78	83	63	64	80	84	55	54	79	88	50	51	78	85	50	54	78	85	55	56
BROWNSVILLE	32	87	88	66	76	87	79	59	82	86	91	57	68	87	90	57	73	87	89	61	72
DEL RIO	49	70	81	59	53	59	77	48	40	58	80	50	41	68	83	55	54	64	80	53	47
EL PASO	67	36	61	40	35	31	40	22	16	48	60	34	30	48	59	35	36	43	54	32	28
FT.WORTH	57	73	78	61	61	70	78	53	52	66	78	48	45	67	77	51	53	69	77	53	53
GALVESTON	69	86	87	77	83	86	86	73	80	83	82	69	74	79	82	66	72	83	84	71	77
HOUSTON	46	84	85	66	73	86	87	59	67	90	90	58	66	85	86	54	68	86	87	59	68
SAN ANTONIO	17	76	81	61	59	76	82	54	51	75	88	48	43	76	83	51	53	76	83	54	52
UTAH,SALT LAKE CITY	26	80	80	70	72	63	69	42	39	45	55	27	23	65	69	41	42	64	69	46	44
VT.BURLINGTON	18	78	80	69	77	77	74	54	64	84	76	54	66	82	82	59	74	81	78	60	71
VA.NORFOLK	66	77	79	61	71	77	74	52	67	88	81	61	76	85	81	60	74	81	79	58	72
RICHMOND	25	79	82	58	71	75	76	46	59	89	84	55	73	87	89	53	77	83	82	53	70
ROANOKE	12	69	72	54	62	67	70	44	53	83	82	52	65	79	82	49	63	74	77	51	61
WASH.NORTH HEAD	62	82	87	79	84	85	88	79	81	91	92	87	84	89	90	85	85	86	89	83	84
SEATTLE	20	82	85	80	73	74	85	63	52	70	85	62	46	86	91	79	68	78	87	72	60
SPOKANE	66	84	85	80	76	67	75	53	35	49	64	39	24	76	81	65	50	70	77	61	49
TATOOSH ISLAND	59	82	85	83	83	83	87	80	80	93	94	89	87	88	91	86	87	86	90	84	85
WALLA WALLA	52	--	82	73	80	--	69	44	43	--	54	29	24	--	72	50	54	--	70	51	51
YAKIMA	11	84	85	75	74	60	74	38	31	55	72	35	26	79	85	55	49	70	79	52	46
W.VA.ELKINS	14	83	82	65	77	80	81	53	62	96	94	61	76	90	93	53	75	88	88	59	73
PARKERSBURG	60	--	82	66	74	--	74	49	58	--	80	52	67	--	84	52	70	--	80	56	68
WIS.GREEN BAY	30	74	76	67	73	73	76	56	61	80	78	55	59	78	82	57	69	77	79	61	66
MADISON	23	77	80	71	76	75	78	54	59	81	79	52	56	79	83	56	66	79	81	61	67
MILWAUKEE	68	76	76	68	74	75	76	63	68	82	75	59	65	77	80	61	70	78	77	64	70
WYO.CHEYENNE	67	61	58	48	52	68	72	47	49	65	70	36	41	64	65	39	47	65	66	44	48
SHERIDAN	18	70	73	61	65	70	77	48	46	63	73	37	35	66	72	44	47	69	75	50	51
P.R.SAN JUAN	56	83	81	71	77	84	76	72	78	86	78	76	82	88	80	74	84	86	79	73	80
V.I.ST.CROIX	6	81	76	66	80	85	74	66	81	86	77	70	84	90	80	72	87	86	77	69	83

TIME IS EASTERN STANDARD (75th MERIDIAN); SUBTRACT 1 HOUR FOR CENTRAL (90th M.), 2 HOURS FOR MOUNTAIN (105th M.), AND 3 HOURS FOR PACIFIC (120th M.) TIME. BASED ON RECORDS THROUGH 1959, EXCEPT IN A FEW INSTANCES. TAKEN FROM "NORMALS, MEANS, AND EXTREMES" TABLE IN U.S. WEATHER BUREAU PUBLICATION, LOCAL CLIMATOLOGICAL DATA,

(Source: Climatic Atlas of the U.S.)

winter (humidification) and removed during the summer (dehumidification). Natural and man-made means of humidification and dehumidification are discussed in Chapter 9.

2-2. TRANSFER OF HEAT

Heat is a form of energy that is always around us. There is no such thing as cold: cold is simply the absence of heat. All heat is said to be absent at the temperature of −460 degrees F, the temperature known as *absolute zero*. At absolute zero, all molecular activity ceases. Absolute zero is a temperature of −460 degrees Fahrenheit (F), −273 degrees Centigrade (C), 0 degrees Kelvin (K), and 0 degrees Rankin (these temperatures are the absolute zero temperature on the various scientific thermometer scales).

When we speak of *cold*, or of *cool*, we really mean that the temperature is low, relative to a warmer temperature. When the outdoor temperature is 15 degrees F, we say that it is cold, but relative to 10, 0, or −17 degrees F, it would be warm. Heat pumps (Chapter 12) make use of this principle and operate in a reversed air-conditioning cycle; heat from outdoors, even when the temperature is 40 degrees F or somewhat lower, is warm relative to the refrigerant in the heat pump system and can thus be used to heat a home.

Heat is transferred from one place to another by three methods: *conduction*, *convection*, and *radiation*. The method of heat transfer called *conduction* is a point-by-point process in which one part of a body is heated by direct contact with a source of heat and neighboring parts become heated successively. Thus, if a metal rod is placed into a heated furnace, the heat travels along the rod by conduction. Metals such as copper and silver are good heat conductors whereas nonmetals such as wood, glass, paper, and water are not good heat conductors. Good conductors of heat are also good conductors of electricity.

The transfer of heat by *convection* refers to the circulation of warmed fluids or air and is brought about by changes in density that accompany changes in temperature. Water in a kettle on the kitchen stove, in a hot water heater boiler, and in a hot water supply heater is heated by convection. The base of the water vessel is heated first. The water touching the bottom in the center is heated, becomes less dense and less heavy, and rises. Cooler, more dense water from the top moves down the sides of the vessel to the base where it is heated and rises. The currents formed are convection currents.

In some types of home heating, heating units are placed in the rooms and the air which comes into contact with them circulates the heat by convection. Steam or hot water radiators transfer some heat by radiation, but far more heat is transferred by convection. The draft in a chimney is produced by convection; the heated gases weigh less than a corresponding column of cold air and consequently, the heated gases rise. A *gravity* hot air furnace works on the same principle; the heated air is less dense and therefore rises through ducts to the rooms and to the room ceilings. Colder, more dense air at the floors falls through return ducts to the heater for reheating.

We are all familiar with the principle that heat rises; we can see it on a clear day coming off of our automobile hoods, in smoke rising from a cigarette, and in similar situations. Perhaps you think then that heat is not lost through a cellar floor or through the side walls of a room. This is incorrect thinking because heat travels *through* a solid material from a warmer area to a colder area. Heat travels in a solid in *all directions* by conduction.

The transfer of heat by radiant energy is an illustration of *radiation*. The process involves the conversion of internal energy into radiant form at the heater, the transmission of the energy by waves through space, and the reversion of radiant energy into internal energy wherever the radiation is absorbed. The electric radiator type heater operates on this principle; it has a heating element mounted at the focus of a parabolic reflector. When the element is heated by an electric current flowing through it, the element emits not only light waves that the eye sees, but also infrared waves that are not seen. A reflector behind the heating element concentrates the waves into a beam. Radiation striking a body may be reflected from it, absorbed by it, or transmitted through it, and these effects may occur singly or in combination. Bodies that absorb radiation become heated, and then internal energy is increased.

A practical example of heat radiation is found in every home where a fireplace is used for heating. Contrary to most beliefs, the heat entering a room from a fireplace is practically all in the form of infrared rays originating in the flames, the coals, and the stone or brick walls. The air that is heated within the fireplace does not enter the room but is carried up the chimney as a convection current. This rising current of air draws fresh air into the room and into the fire, supplying fresh oxygen to the burning wood. The room air that literally goes up the chimney is replaced by cold air coming in and around windows, doors, and walls (more on this and what to do about it in Chapter 14).

2-3. SOME DEFINITIONS

Three definitions used in weatherizing with which you need to become familiar are R, U, and k. R is the measure of a product's ability to decrease heat flow. The higher the value of R, the better its ability to decrease heat flow. For example, four-inch fiberglass insulation batts for attic insulation have a value of approximately R-11 whereas a six-inch batt has a value of approximately R-19. The R values are additive; that is, if R-11 and R-19 batts are used together, the value is R-30. All building materials have R values (some R values are shown in QRC 7-2).

Quick Reference Chart 2-7
WEATHERIZATION INSPECTION

Inspect for	Ref.	Location/ Materials/ Comments
1. Broken or loose windowpanes and sashes	Sec. 3-1	
2. Holes in walls	Sec. 3-2	
3. Holes in floors	Sec. 3-3	
4. Holes in roof	Sec. 3-4	
5. Cracks in foundation	Sec. 3-5	
6. Heat leaks between living and nonliving areas	Sec. 3-6	
7. Loose-fitting fireplace damper	Sec. 3-7	
8. Incorrectly operating thermostat	Ch. 4	
9. Missing or loose caulking	QRC 5-2	
10. Missing or damaged weather stripping	QRC 6-2	
11. Need for insulation or supplementary insulation	QRC 7-1	
12. Need for storm windows and doors	Ch. 8	
13. Excess humidity level	Ch. 9	
14. Excess moisture	Ch. 10	
15. Ineffective use of nature	Ch. 11	
16. Need for furnace service	Ch. 12	
17. Need for air-conditioning service	Ch. 13	
18. Fireplace inefficiency	Ch. 14	

U value is a term used to describe the insulating characteristics of materials; it is the coefficient of the transmission of heat. The R value describes the resistance of a specified material whereas the U value includes all of the components involved such as the siding, sheathing, insulation, and wallboard of a wall. The lower the U value, the better is the resistance to heat.

Thermal conductivity is shown by the k value where k represents the amount of heat that passes through a uniform composition (homogenous) material one square foot in area by one inch thick in one hour with a temperature difference of one degree Fahrenheit between the inner and outer surfaces. The k value is expressed in British thermal units (Btu) per hour. The lower the k value, the higher its insulating ability. One Btu is the energy required to increase the temperature of one pound of water by one degree Fahrenheit.

Definitions of weatherization terms and of home-building terms are defined in the text when used. A comprehensive glossary is located at the back of the book.

2-4. CONSTRUCTING A WEATHERIZATION PLAN

To weatherize most efficiently and economically, you need to plan ahead. The easiest way to do this is to first read through this book paying special attention to material on where to look for problems and what materials are needed. The book is presented in sequential order of weatherizing; in other words, there is no sense in insulating (Chapter 7) if you don't first stop the escapement of heat in winter and infiltration of heat in the summer (Chapter 3), control your thermostat (Chapter 4), caulk (Chapter 5), and weather strip (Chapter 6).

Once you have some idea of what to inspect for, use QRC 2-7 to inspect the inside and outside of your home and to formulate your plan. Use the location/material/comment column to make note of the exact location where work is to be done, the materials you think you'll need, and any other comments pertinent to the accomplishment of the job.

With your planning list and priorities in hand, watch your newspaper for preseason, do-it-yourself, and after-season sales of weatherizing supplies. Buy your supplies during the week; have your tools ready; then, work on the weekends (Figure 2-13).

2-5. SAFETY

In weatherizing your home, as with any home repair, you must always be conscious of safety, not only for yourself but also for any other person within your work area. Be conscious of electrical hazards, mechanical hazards, and of the dangers of falling from a ladder or roof. These hazards and dangers, along with suggestions on how to work safely to reduce the probability of an accident, are discussed in the following subsections.

Electrical Shock

Unless you are an electrician or know how to work with electricity, you should not attempt to make wiring connections to electrical circuits; obtain the services of a skilled electrician familiar with all facets of house wiring and electrical codes.

There are two dangers from electrical shocks: the shock received from the current which in itself can be fatal, and the *after result* of receiving a minor shock. After result means the harm that can occur from the recoil of your hand and arm when you receive a mild shock; you may smash your hand against another object and receive a laceration, bruise, sprain, or even a broken bone.

caulking needed
around flashing

cracked glass needs
replacement

caulking
needed

weather stripping
needed

panes need
reglazing

latch needs
resetting for
tighter fit

masonry damage
requires repair

threshold with
vinyl insert needed

FIGURE 2-13. In addition to needing paint, this door indicates that this home is in need of
weatherization (*courtesy Pease Company, Ever-Strait Division*).

There are three properties to every electrical circuit: *current, voltage,*
and *resistance*. The *current* is the flow of electrons and is the property of
electricity that kills people. *Voltage* is the force or pressure against the current
and *resistance* is the property of a material to resist the flow of current.

Voltage, current, and resistance are interrelated. Voltage is measured in volts, current in amperes, and resistance in ohms.

It is the electrical *current* passing through the body that causes death. The voltage does not necessarily kill the victim. It is said that a person has been killed by 40 volts (direct current), but there have been persons who have survived a shock of 15,000 to 20,000 volts. Survival at the higher voltage is possible because of the violent muscular reaction which often knocks the victim away from the current. A lower voltage of 115 to 450 volts doesn't do this. It is agreed that death will most likely occur to a person receiving more than 0.1 ampere of current. Persons with weaker hearts will die from less current. The average major electrical applicance draws 15 amperes of current. This is not meant to threaten but to caution you. You should not fear electricity, *but you must respect it.*

Shock is felt by different individuals at currents as follows:

	Ampere	Equivalent Milliamperes*
Noticeable shock	0.0005 to 0.005	0.5 to 5.0
Victim's muscles become paralyzed and the victim cannot let go of electricity contact	0.005 to 0.030	5.0 to 30.0
Heart muscles twitch (fibrillation) preventing a rhythmic pumping	0.050 to 0.200	50 to 200
Breathing interrupted	over 0.100	over 100
Death likely to occur	over 0.100	over 100

* A milliampere is 1/1000 of an ampere of electrical current.

In the event of electrical shock:

- Without touching the victim, shut off the electrical power by pulling a plug, turning off a circuit breaker, or by placing a power switch to off.
- Remove the victim from the electrical contact.
- If breathing has stopped, immediately begin artificial respiration and continue until a physician declares the victim dead. Refer to a Red Cross textbook for mouth-to-mouth artificial respiration techniques.
- Electrical safety for protection of yourself and those around you can be practiced most easily and effectively by using your own common sense. Read the suggestions in Quick Reference Chart 2-8 and always keep them in mind.

When you are working within an electrical unit on the mechanical system or if electricity is not needed within the unit, disconnect the unit plug

---

Quick Reference Chart 2-8
ELECTRICAL SAFETY

- Be sure the unit you are working on is electrically grounded by a third wire that connects it to an earth ground.
- Never install, repair, or use an electrical unit in a wet location. If necessary, place dry boards on the floor under your feet.
- Always wear rubber-soled shoes.
- Develop safe, methodical servicing habits.
- Turn the unit off before unplugging it.
- If the power cord plugs into the unit as well as the power receptacle, always connect the cord to the unit before connecting it to the power receptacle.
- When disconnecting the unit, always remove the plug from the power receptacle first.
- Grasp the plug and not the cord to remove the plug from the power receptacle.
- Don't run power cords near hot pipes, radiators, damp or wet areas, etc.
- Concentrate on your efforts; know when electrical power is on and when it is off. Always have the unit plug in sight.
- Never touch a bare wire if the unit is plugged into a power receptacle.
- If the unit is plugged into a power receptacle, use the *one-hand rule*; one of your hands is placed in your pocket and the other hand is used for working. This prevents an electric current from passing from an electrical terminal or wire, through one hand and arm, through your chest and heart, and through the other arm and hand, back to the frame of the unit.

---

from the power receptacle. Keep the plug in sight at all times so you know whether or not there is power flowing to the unit.

Mechanical Dangers

When most people think of installing or repairing something they think of the dangers involved from electrical shock. But, in addition to electrical shock, there are mechanical dangers that can hurt the do-it-yourselfer. Moving gears, pulleys, belts, sharp edges, improperly used tools, and particles from grinding wheels all pose potential injury to the careless do-it-yourselfer. To decrease the possibility of injury from mechanical parts, continually practice the mechanical safety procedures listed in QRC 2-9.

Always use the correct tool for the job at hand. If you are unfamiliar with the proper selection, use, and care of hand tools used for home repairs, automobile repairs, and other do-it-yourself and professional jobs, you should obtain a copy of *Everyone's Book of Hand and Small Power Tools* by G. R. Drake (hardcover No. 0-87909-260-2 or softcover No. 0-87909-217-3, available from Reston Publishing Company, Inc., Reston, Virginia 22090, or order through your local bookstore).

Quick Reference Chart 2-9
MECHANICAL SAFETY PROCEDURES

- Roll up your shirt sleeves.
- Ensure that adequate light is available to see within the unit.
- Keep your tools out of your way.
- Keep your hands away from moving gears, pulleys, shafts, and electrical terminals.
- Inspect for sharp points and edges; trim these as necessary with a file to protect yourself.
- Keep your tools sharp, clean, and operative.
- Use the proper tool for the job.
- Keep a firm grip on the tool being used so that if it is hit by a moving part, it will stay in your hand.
- Don't clean, oil, or do other work on any unit when the unit is operating.
- Wear goggles or a face shield when dusting and grinding.

Ladder Safety

Before you use a ladder, check it carefully to ensure that it is safe to use. Check the ladder for loose, cracked, or broken rungs. Then check the side-rails for cracks or breaks. Finally, check the feet at the base of the ladder. Repair the ladder, as required, before using it or replace it with a new ladder.

Extend the ladder (if possible) and place the ladder against the house so that the top of the ladder extends two feet (about two rungs) above the highest point to be worked on. The base of the ladder should extend from the house a distance equal to one-fourth the length of the ladder. For example, the base of a 20-foot ladder should extend from the house 5 feet. Make sure that the feet of the ladder are flat and well supported. When you are on the ladder, keep your hips between the rails; do not over extend your reach. Instead, get off the ladder and relocate it. Always keep both hands on the ladder when ascending or descending. Always keep one hand on the ladder while working.

You can buy a ladder stabilizer to mount to the top of your extension ladder. It is a cylindrical tubular steel frame that increases the stability of the ladder. Sears, Roebuck and Company is one source.

Roof Safety

If you must work on a roof, pick a warm, dry, windless day. Think safety. Be sure that the ladder to the roof is secure and that someone is holding it. Wear soft-soled shoes; walk in a crouch.

It is safest if you have a device on the roof on which you can walk. You can use a ladder to give you safer footing. Pull a section of an extension ladder up on the roof and secure it with ropes that pass over the top of the house and secure to a tree, a post, or similarly stable object on the other side of the house. Another method is to use a 1- by 8-inch board the height of the roof. Screw 1- by 8-inch crosspieces on the board at 12-inch intervals to fashion a ladder. On one end of the board, bolt on a large steel angle that will fit over the roof piece and down the other side about 18 inches. Also tie a piece of rope from the top of the board (around the top "step") and anchor the rope over the top of the house and down the other side to a tree or post.

chapter 3

STOPPING WINTER HEAT ESCAPEMENT AND SUMMER HEAT INFILTRATION

Before you spend money and time to perform sophisticated energy–saving home improvements, you need to stop the major places of winter heat escapement and summer heat infiltration into the home. This involves locating and repairing cracked and broken windowpanes and sashes, holes in exterior walls, holes in floors, holes in the roof, cracks in the foundation, heat leaks between living and nonliving areas, and the secure closing of fireplace dampers. Repairing each of these items is covered in the following sections.

3-1. WINDOWPANES AND SASHES

Windows account for large heat losses in winter and for large heat gains in summer. Heat passes directly through the glass because glass has very little heat conduction resistance. If the glass is cracked, pieces missing, or if a small pane out of a multiple-pane window is missing, the window's thermal resistance is about as effective as an open door.

In addition to cracked or broken glass, air quickly leaks past cracked, crumbling, or missing glazing compound (used to hold the pane of glass in place in the window frame). Loose-fitting sash (frames) at the top, bottom, and at the junction of double-hung windows accounts for additional heat losses. Finally, air also leaks around the outside of the window where the *casing* meets the siding on the outside and where it meets the walls on the inside. This section describes reduction of heat losses at windows by replacing cracked or broken glass, securing sash locks, and tightening the casing. Reducing heat losses around windows by *caulking* and *weather stripping* are covered in detail in Chapters 5 and 6, respectively. If you want to consider replacing an entire window with a new one, refer to Chapter 8.

When replacing glass, especially in storm doors and in windows subject to breakage, such as cellar windows that are vulnerable to baseballs, use safety glazing materials approved by the safety councils. These glazing materials include acrylic plastic sheet, tempered glass, extruded sheets of polyvinyl chloride, and extruded acrylic sheets. Thicknesses are $1/10$, $1/8$, and $1/4$ inch. The $1/10$ or $1/8$ inch materials can be placed in the old frames. The $1/4$ inch materials can be installed in storm doors without the frame; cut the material to the size of the original frame and hold it in place with sash holding clips.

The plastic materials are cut by scoring a straight line with several strokes of a sharp knife along a metal straightedge. The material is then placed on a flat surface with the scored line facing upward and just over the edge of the surface. Wearing heavy protective gloves, snap the material downward along the line to separate the pieces of the material. As an alternative method of cutting the material, use a small-toothed blade with few teeth per inch in a saw (seven teeth or less per inch). Run the saw at as slow a speed as possible to keep from melting the material (causing it to join back together just behind the blade as you cut). Sand the material edges smooth and clean the material with mild soap and water. Do not use cleaners containing silicon. You can use a paste wax to fill minor scratches and restore the luster.

If you prefer to use tempered glass because of its durability, scratch resistance, good appearance, and high strength, you have a slight problem; you cannot cut it. You can install it, however, thanks to PPG Industries' replacement panel that is adaptable to most storm doors. First measure the width and length of the storm door frame. Then purchase the panel with the largest glass size that is just smaller than the needed dimensions. A wide plastic-surrounding frame comes already attached to the glass: the plastic frame is scored in $1/8$ inch increments. Cut through the proper size scored line to produce the overall size of the needed replacement pane (the frame can be trimmed up to 3 inches in both width and length).

Inspect each windowpane for cracks and inspect the glazing compound. If the compound is brittle and has fallen or is falling out, it's time to replace it because heat passage will be quite easy, especially when the wind blows. Replace a pane of glass in a wooden frame as described in DIY 3-1; glass is replaced in a metal frame in the same manner as in a wooden frame with the following exceptions. When the glass is removed from a metal casement window frame, you will find metal spring clips instead of glazing points. Reinstall the spring clips to hold the new glass in place. In aluminum-framed windows, the glass is held in place by a rubber or plastic gasket. Remove this gasket along with the broken glass; ensure that all glass slivers are removed. Measure the glass opening and then subtract $1/32$ inch (instead of $1/8$ inch) for material expansion. Lay the new glass (or acrylic plastic) in the frame and

Do-It-Yourself 3-1
REPLACING GLASS

Materials: glazing compound, linseed oil, glazing points, house paint, gloves, glass or acrylic plastic

Tools: ruler, stiff putty knife, safety glasses, hammer, propane torch or heavy duty electric soldering iron, pliers

Procedure:

1. Use the putty knife to remove the old loose glazing compound (this isn't necessary with aluminum-framed windows). The compound can be softened by carefully warming the putty with a low-flame propane torch or soldering iron, but take care not to get the surrounding glass hot or to burn the wood frame. Be sure to wear safety glasses.
2. When the glazing compound is out, or as it is coming out, remove the glass. Be sure to wear gloves to handle the glass.
3. When the glass is removed, use the putty knife to clean out any remaining glazing compound and to pry out old glazing points. You can also pull the points out with pliers. In an aluminum window frame, pull the gasket out.
4. Measure the width and length between the frame; then subtract ⅛ inch from each dimension to allow for expansion of the materials. Take the measurements to your hardware store (for aluminum-framed windows, subtract $^1/_{32}$ inch instead of ⅛ inch; proceed to step 9).
5. Using a rag, paintbrush, or a toothbrush, coat the raw wood framework with linseed oil or house paint; let it dry before proceeding. The oil or paint prevents the frame from drawing the oil out of the new glazin compound.
6. Place a bead of glazing compound against the frame behind where the glass will be placed (Fig. 3-1). This bead acts as a shock absorber between the glass and the frame. If the glazing compound is stiff, roll it in your palms to soften it.
7. Place the glass in position and press it lightly into the bead of glazing compound. Install glazing points as required to hold the glass in place (Fig. 3-2). On a window 8 by 10 inches, you only need one glazing point in the center of each side. Place the glazing point flat against the glass with the tip toward the wood. Hold the edge of the putty knife in position against the back of the point. Tap the putty knife lightly until the point is secured in the frame.
8. Apply glazing compound to the glass and frame to form approximately a 45–degree bevel, as on the other window panes (Fig. 3-3). Smooth it with the knife edge and your finger. Following the glazing compound directions, paint over the compound to give it a protective coating.
9. For aluminum-framed windows, place the glass in position in the frame. Replace the gasket by starting at one corner and working around the frame.

carefully replace the gasket by pressing it under the frame lip. Start at one corner and work around the frame. (In replacing glass in a storm door, use safety glass or a plastic material.)

FIGURE 3-1. A bead of compound about 1/8 inch in diameter is rolled in the palms and then placed against the frame (behind where the glass will be placed) to act as a cushion.

In an emergency, small holes or cracks can be patched by criss-crossing duct tape, freezer tape, or masking tape across the hole or crack and the glass. If a large area of glass is missing, cut a piece of heavy cardboard (as from a carton) slightly larger than the size of the windowpane. Tack the cardboard securely to the inside of the window frame with thumb tacks.

A lot of air can pass over the top of the upper window sash, under the bottom of the lower sash, and at the center where the top and bottom sashes of double-hung windows meet. The sash lock must fit tightly and be kept latched in the winter to prevent heat loss and in the summer to reduce heat gains. If the lock does not latch securely, relocate one or both pieces of the lock. Fill the old screw-holes with a wood filler; drill new holes and screw the lock piece(s) securely into place.

As a temporary quick-fix for air passage over and under window sashes, pull down the top sash and stuff cloth or newspaper in place; do the same at the bottom and then place cloth where the windows meet along the sash lock.

Another place where heat infiltrates around windows is at the casing (the decorative molding which covers the inside edge of the jamb and the rough opening between the window unit and the wall). Check for drafts along the sides of the window next to the wall. If there are drafts, screw or nail the casing

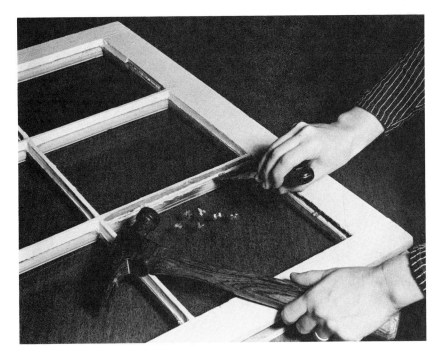

FIGURE 3-2. The glazing points are tapped into place with a putty knife and hammer. Angle the knife so that the tapping force is against the wood frame, not against the glass.

to the wall. If necessary, apply a thin bead of caulking along the casing/wall junction (caulking is discussed in detail in Chapter 5).

3-2. HOLES IN WALLS

Holes in the outer walls of your rooms can allow a lot of heat to escape, especially if the walls are not insulated. Holes can be temporarily filled with a piece of cloth or crumpled newspaper, or fixed permanently (DIY 3-2).

Do-It-Yourself 3-2
PATCHING A HOLE IN A WALL

Materials: a. for plaster or plaster-board walls: patching plaster
   b. for wood walls: wood filler
   c. for masonry walls (inside): cement or plaster
   d. general: piece of old window screen, piece of 12-inch, thin wire or string

Tools: flexible putty knife

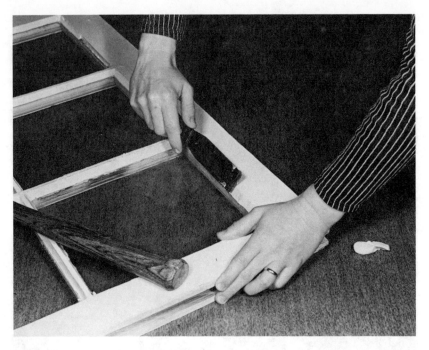

FIGURE 3-3. A bead of compound about 3/8 inch in diameter is beveled against the glass and frame.

Do-It-Yourself (continued)
PATCHING A HOLE IN A WALL

Procedure:

1. Remove all loose plaster and dust from around the hole.
2. Cut a piece of flexible window screening slightly bigger than the hole; the screen will be placed behind the hole to prevent filler material from passing through and also to anchor the filler. Tie the center of a piece of 12-inch wire or string around a couple of strands of the screen wire at the center of the screen.
3. Holding onto the wire or string ends, carefully place the screen through the hole. Then gently pull on the ends until the screen is flush with the back of the hole. Place a wooden pencil across the hole and tie the wire or string ends to the pencil so that the pencil holds the screen securely to the back of the hole (Fig. 3-4).
4. Prepare the patching material according to the manufacturer's instructions.
5. Using a cloth or brush and water, wet the edges of the plaster or plaster board; this keeps the plaster from drawing water out of the patching material. Fill the hole to about ⅛ inch from the surface.
6. After the material in the hole is dry, cut the wire or string and remove the pencil. Cut the string ends back, flush with the filler material.
7. Apply a second coat of patching material. When this is dry, it can be sanded and then finished to match the existing wall (with paint, varnish, or similar material).

FIGURE 3-4. A pencil and wire hold the screen in place behind the hole. The screen provides an anchor for the patching material.

3-3. HOLES IN FLOORS

Sometimes you find access holes drilled through your floors for such things as television antenna lead-ins or for some unknown reason. Or perhaps, the molding along the floor and wall junction is loose. If the space under the floor at these locations is unheated, then a significant amount of heat can be lost through the opening. You can feel a draft at these areas when there is a temperature difference of 10 degrees F or more.

Drafts through holes in the floor through which wires pass can be blocked easily with a little material such as cotton, foam, or fiberglass insulation pushed into the hole. This can easily be removed if the wiring needs to be moved. For empty holes, pack as described above but leave about a ⅛ inch space near the surface. Fill the remainder of the hole with a wood putty that matches the color of the floor. After the putty dries, sand it and apply a finishing material such as varnish, shellac, lacquer, or paint to the floor.

If drafts occur around loose molding, perform the steps in DIY 3-3.

Do-It-Yourself 3-3
TIGHTENING FLOOR MOLDING TO ELIMINATE DRAFTS

Materials: 1-½ inch finishing nails
Tools: drill and drill bit (diameter slightly smaller than diameter of nail), hammer, nail set (or a larger nail to use as a nail set), awl or ice pick

Procedure:

1. Locate loose molding. Start nailholes with an awl or ice pick.
2. Drill hole(s) through the molding into the floor.
3. Drive the nail(s). Set the nail head(s) slightly below the surface of the molding using another nail or a nail set.
4. Fill the nail hole(s) with a colored putty stick that matches the molding, or fill the holes with wood putty and apply the applicable finish.

3-4. HOLES IN THE ROOF

Holes in the roof let in water as well as letting hot air transfer. Holes require immediate repair before water severely damages roof sheathing, ceilings, walls, and even floors if sufficient water leaks in. The presence of water marks on ceilings or walls should trigger you into action.

The easiest way to find the leak is to go into the attic above the area where the leak is coming through the ceiling or wall when it is raining. The place where the water is coming through the roof *may not be* (and probably isn't) directly above the location where the water is coming through the ceiling or wall, but is further up the roof. The water may come through the roof and then run on the underside of the sheathing or along the rafters until it drops. Follow the trail up to the source and mark the location with a crayon. If you have insulation installed, you will have to remove it along the line until you find the source of the leak.

When the weather clears, you can proceed to fix the roof. If the roofing material is soft such as tarpaper, asphalt shingle, wood shingle, aluminum, or similar material, pound a long nail from the attic at your crayon marking up through the sheathing and roofing materials so that you can easily locate the source of the leak on the roof exterior. If the roof is slate or other hard brittle material, you'll need to locate the location on top of the roof by measuring from a reference. First, establish a reference indoors such as the chimney, soil pipe, or roof edge. Measure inside. Then, using the same reference on the outside, measure to locate the area of the leak. With the leak located on the outside of the roof, inspect further for cracked or broken roofing material and damaged flashing around the chimney or soil pipe. Then proceed to repair the roof as directed in DIY 3-4.

Do-It-Yourself 3-4
REPAIRING ROOFS

Materials: broad-headed galvanized roofing nails, patching materials (see text below for applicable repairs)
Tools: hammer, stiff broom, wide putty knife

Procedure:

NOTE
This DIY is divided into five areas: flashing, flat roofs, asphalt shingle roofs, slate roofs, and wood shingle or shake roofs. Follow the applicable procedure(s) for your roof. Aluminum, clay tile and concrete tile roofs are difficult for the homeowner to repair; consult a professional roofer.

Do-It-Yourself 3-4 (continued)
REPAIRING ROOFS

Flashing

1. Inspect the flashing around the chimney, soil pipe, and roof junctions.
2. If the flashing is loose, nail it secure with roofing nails. Cover the nail heads with roofing cement. Also apply roofing cement where the flashing meets with other materials.

Flat Roof

1. Leaks generally occur on flat roofs around the chimney, soil pipe, and low spots where water collects. If there is water, remove it. Roof repair work should only be performed on a warm day.
2. Use a stiff broom to remove grit or pebbles from the area where the leak is occurring.
3. If there is a blister in the roofing material where it leaks, cut the blister with a knife; make a single, long, straight cut. (For larger patches, refer to step 7.)
4. Work roofing cement under the blister with a wide putty knife. Then nail along both sides of the cut with roofing nails. Spread roofing cement over the cut and nail heads.
5. Cut a piece of tarpaper or shingle larger than the patched area and with about 2 inches of excess material on each side.
6. Nail the patch in place and then spread roofing cement over the patch, nails, and edges.
7. For larger patches, cut the damaged material away in a rectangular pattern. Apply roofing cement to the entire area and under the edges of the remaining material. Cut a new piece of material to fit the cut out area plus a 2-inch overlap on each side. Nail the patch in place and cover with roofing cement; overlap the cement 2 inches on all sides. Add another piece of material 2 inches larger on all sides than the previous piece. Nail the piece in place. Coat the edges and the nail heads with roofing cement.

Asphalt Shingle Roof

1. Missing, broken, or otherwise damaged shingles can be replaced without reroofing. A shingle roof is constructed, starting with a bottom row first; subsequent rows of shingles are laid up the roof. Therefore, to replace a damaged shingle, you have to carefully lift up the good shingle(s) on the row above and slip in new shingles under them. Because the shingles become more pliable as they become warm, replacement should be done on a warm day to prevent the breaking of good shingles.
2. Since the shingles are nailed on, you'll need to remove the nails. Carefully lift the shingle(s) on the row above the damaged shingle.
3. Remove the nails with a pry bar or chisel.
4. Slide the new shingle into place. If it won't go all the way in, cut off the corners.

5. Nail the shingle in place and apply roofing cement over the nail heads.

6. If the new tile or the tiles above it persist in curling up, place a dab of roofing cement under the corners and press the corners into place. The heat of the sun will cause the tiles to flatten out and adhere at the corners.

7. If you do not have a replacement shingle, slide a sheet of aluminum up under the row of good shingles. Coat the bottom of the aluminum with roofing cement.

Slate Roof

1. A damaged slate is removed and replaced in a manner similar to asbestos tiles; however, the row above the damaged slate cannot be bent up for access to the nails.

2. Make a nail-puller from a piece of strap iron (Fig. 3-5).

3. Slide the nail-puller up under the tile and hook it onto the nail. Then bang the end of the puller toward the bottom of the roof, pulling the nail out.

4. Place the new slate in position up under the next higher row as far as possible. If the slate needs to be cut to size, score along the line with a chisel. Wear gloves. Hold the slate in position over a sharp edge with the line along the edge. Snap the slate to break it on the scored line. Again, place the tile in position. Mark the centers for two holes where nails will be placed to hold the tile in position. Remove the tile again.

5. Drill the two holes through the slate with a masonry bit; the diameter of the hole should be slightly larger than the roofing nail diameter.

6. Apply roofing cement on the area where the slate will be placed. Put the slate in place and nail it through the drilled holes with galvanized roofing nails. Cover the nail heads with roofing cement.

Wood Shingles and Shakes

1. A wood shingle or shake roof can be repaired as described for slate roofs.

2. In lieu of replacing the shingle or shake, you can cut a piece of aluminum to shape and then drive it up under the leaking shingle or shake, with a hammer and a block of wood.

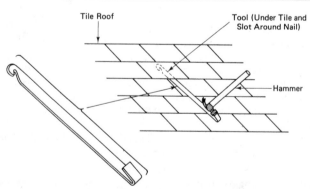

FIGURE 3-5. Make the slate shingle nail puller from strap iron. Cut a notch in one end and bend the other end around so it can be struck with a hammer.

3-5. *CRACKS IN THE FOUNDATION*

Inspect the foundation walls, joints, and floors for cracks and holes. If the openings are above grade, they'll let air readily pass through; if below grade, water may be seeping in. Repair cracks and holes as described in DIY 3-5.

Do-It-Yourself 3-5
REPAIRING FOUNDATIONS

Materials: mortar mix and fine, sharp sand *or* two-part epoxy *or* quick-setting hydraulic cement (see below)

Tools: cold chisel, hammer, brush (wire or stiff)

Procedure:

Dry Cracks and Holes

1. If the crack is more than ⅛ inch wide, use a chisel and hammer to shape the crack to a wedge shape, wider at the rear than at the front. The wedge shape will anchor the patching material.
2. Brush the crack out and then vacuum out all dust.
3. Mix either a mortar mix (A) or a two-part epoxy mix (B):
   A. Mortar mix: mix one part mortar mix and three parts fine, sharp sand. Wet the crack with water; this keeps the crack from drawing the water out of the mix during curing. Force the mortar into the crack. Smooth off the patch and keep the patch damp for several days during curing.
   B. Two-part epoxy mix: mix the two parts (two colors) together until the mix is one color. Trowel the mix into the crack. When 30 minutes have passed, wet the trowel or your finger with water and smooth the patch.

Leaking Holes

1. Undercut the hole as described in step 1 above. Remove loose material.
2. Mix the quick-setting hydraulic cement and mold it into a tapered plug.
3. As the plug starts to set, push it into the hole like a cork and hold it in place until the plug sets. Smooth the surface with a trowel before the plug hardens completely.

3-6. *HEAT LEAKS BETWEEN LIVING AREAS AND NONLIVING AREAS*

Heat can also easily leak through openings within the home between living and nonliving areas such as through openings around loose-fitting attic doors, doors to unused rooms, folding stairs and trapdoors in the ceiling to attic areas, around light fixtures to the attic, and around pipes passing from one floor to another. These air leaks can be partially stopped by cloth, small rugs, or newspapers placed or pushed into the openings. *Do not* put any flammable materials (cloth, rugs, paper, and so on) near a chimney or flue; you *must* use mortar or plaster to block infiltrating or exiting air around chimneys and flues.

Loose-fitting doors to the attic and to unused rooms should be weather-stripped. Refer to Chapter 6 for a description of different types of weather stripping and installation procedures.

Folding stairs, trapdoors, and lights can be sealed with a closed cell foam tape (Fig. 3-6). Wipe the area where the tape is to be applied with turpentine or alcohol to get rid of dirt and grease. Measure the required length, cut the tape from the roll, peel the paper off the back, and place the strip in position. Press on the strip to affix it. Other types of weather stripping may be used in place of the foam tape (refer to Chapter 6). Be sure to use a fireproof material around lights.

Material can be placed around pipes and the metal trim rings placed back in position to cover the material. For a more permanent fix, caulk the opening and then reposition the trim rings (refer to Chapter 5).

FIGURE 3-6. Foam tape can be used to stop heat transfer around openings in folding stairs, trap doors, and around lighting fixtures (*courtesy Macklanburg-Duncan Co.*).

3-7. FIREPLACE DAMPERS

A fireplace damper can let a lot of heat up the chimney if it remains open after the fire is out. At the same time, you can't close the damper until the fire is out or else the smoke will back into the room.

Your best solution is to close the damper as much as possible without allowing the fire to smoke. If you have a tempered glass screen in front of the fire, you can close that off too. If you don't have a glass front, you can place a sheet of metal across the complete fireplace opening.

CAUTION

Place only a nonflammable material such as tempered glass or metal in front of the opening. Do not use a flammable material.

If the fireplace damper does not fully close, you can work on it as soon as the fireplace is cold. Failure to close is probably caused by accumulated dirt, ash, and pitch. Reach up through the fireplace throat to the damper. By moving the damper lever and the damper, you'll probably be able to determine the trouble. Reach over the damper and clean off the smoke shelf behind the damper too. Clean out the accumulations with a brush. It will help if you have a shop vacuum. (Refer to Chapter 14 for additional information on fireplaces.)

chapter 4

CONTROLLING
YOUR
THERMOSTAT

Want a *no-cost, no-effort* way to lower your heating and cooling costs? Lower your thermostat a few degrees in the winter and raise it a few degrees in the summer. You can get used to the change in temperature gradually if you begin at the start of the season, or if you're in the middle of the season now, change it a couple of degrees at a time over a period of several days.

The National Bureau of Standards estimates that for each degree Fahrenheit you set your thermostat back in the winter, you'll save two percent on your fuel bill in cold climates and three percent in moderate climates. If you set the thermostat back 10 degrees F at night for eight hours, you'll be able to save 10 to 15 percent on your fuel bill. The Federal Energy Administration estimates that by keeping your thermostat of your air-conditioned home at 78 degrees F instead of 72 degrees F, you'll save from 12 to 47 percent on your air-conditioning bill, depending upon where you live. Heating setback fuel savings published in a Housing and Urban Development publication are shown in Fig. 4-1. Locate the zone in which you live; then refer to the chart which lists the approximate percentage you can save by setting your thermostat back five or eight degrees F.

In cold climates, it is recommended that the thermostat not be set back more than seven degrees F from the daytime temperature. Otherwise the extra fuel required to raise the temperature in the morning would be sufficient to nullify the fuel saved the night before, or even cause the use of additional fuel.

If no one is home during the day, keep your thermostat down in winter and up in summer. When you are away for a day or two, set the thermostat to 60 degrees F in the winter and turn your air conditioner off in the summer.

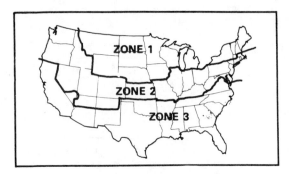

	ZONE 1	ZONE 2	ZONE 3
5⁰ turn-down	14%	17%	25%
8⁰ turn-down	19%	24%	35%

FIGURE 4-1. Setting your thermostat down in the winter can save a lot of money; similar savings can be realized by setting it up in the summer. Reasonable settings are, winter, 68 degrees F daytime and 60 degrees F nighttime; summer, 78 degrees F, day and night (*Courtesy U.S. Department of Housing and Urban Development*).

You can become accustomed to temperatures of 65 degrees F or lower, although this isn't recommended if there are elderly people in the home or if small children are playing on the floor. Dress warmly for the lower temperatures; wear loose–fitting wool or cotton clothes which trap the body heat. Several layers of light clothing are warmer because of the air space between the layers, and you can shed one layer at a time as you warm up. The same goes for bedtime: wear wool or cotton pajamas, long underwear, socks, and perhaps even a night cap for your head. Sleep under several layers of blankets, quilts, or comforters. They should keep you warmer as they trap your body heat between the layers. Covers can be pinned in place to keep them from coming off; this is particularly important with small children.

4-1. THERMOSTAT CONTROLS

If you hate to get out of bed in the morning to a cold house, you can invest in a thermostat timer or similar device. You can set these to automatically turn the heat down to a predetermined temperature late in the evening and then reset the temperature to the daytime level early in the morning. The result is that you cut down on the use of fuel during the night, but have a warm house again by the time the alarm clock goes off. There are several types of timers and other devices to choose from.

One type is the electric automatic thermostat-timer (Fig. 4-2). Two finger-set adjustments are used to set the high and the low temperatures desired. A clock, similar to an alarm clock, is set to the time when the temperature is to be lowered and another time is set for the temperature to be returned to the higher level. Some models are powered by house current, others by a battery. A time-delay thermostat that has a wind-up timer rather than an electric timer wired into a thermostat is also available. The thermostat-timer shown in Fig. 4-2 is for a combination heating and cooling system; thermostat-timers are also available only for heating systems or cooling systems.

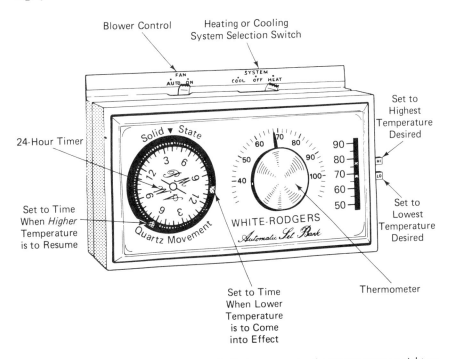

FIGURE 4-2. Automatic thermostats automatically lower or raise the temperatures at night or day, winter and summer (*courtesy White-Rodgers Div., Emerson Electric Co.*).

Another device for automatically controlling the thermostat is a timer-triggered resistance heater (Fig. 4-3) that keeps your home temperature 6 to 13 degrees F lower for up to 12 hours as set on the dial control mounted one inch underneath the thermostat. The device connects to any 120-volt ac convenience outlet. When you go to bed, you set the timer to the number of hours you want the temperature lower minus 1½ hours, and set a slide switch to the temperature differential desired (6 to 13 degrees F differential). When

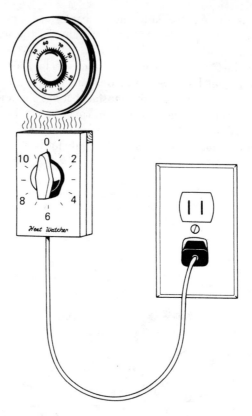

FIGURE 4-3. A clock-triggered heating device "fools" the thermostat during the night, allowing the house temperature to drop 6 to 13 degrees F (adjustable) below the daytime temperature to save energy. The device allows the home heating system to return to the daytime temperature before you wake up (*courtesy M.H. Rhodes Inc.*).

the timer is set, a small electrical current flows through a resistance that produces heat which rises to the thermostat and "fools" the thermostat into thinking that the home is warmer than it really is. At the end of the time period, the timer interrupts the electrical power to the resistance heater; the thermostat detects the loss of heat and turns the home heating system on. (For example, if you go to bed at 11 p.m. and expect to rise at 7:30 a.m., you want the house cool for 8½ hours minus 1½ hours equals seven hours to be set on the timer. The heating system needs about 1½ hours to heat the home back to the daytime temperature.)

A similar device to the one just discussed uses an automatic 24-hour timer that is plugged into a 120-volt ac convenience outlet. A cable connects from the timer to the resistance heating device. The turn-on and turn-off time of the heating device is set one time and doesn't have to be set each night; resetting is required only if the operating time is to be changed.

Another type of thermostat control (THERM O GUARD by Dynamic Electronic Controls Inc.) operates by room light intensity. As long as a certain level of light is available, the control keeps the day temperature control in operation to maintain the selected temperature (the light-sensitivity threshold for switch-over to the night or day temperature setting is adjustable by the user). The heating (or cooling) system is turned off by the control when there is a loss of light and back on when light is available. In a home, the heat level can be changed to the night setting at different times automatically when the lights are turned off at bedtime and on again when the lights are turned on in the morning. The disadvantages are that when you wake up in the morning and turn the light on, the house is at the lower nighttime temperature and the heating system may take up to an hour and a half to bring the home to the daytime temperature. You also need to keep the light source on the entire time that you want the daytime temperature. Although not too practical for the homeowner, the control is practical in a building or office which is unoccupied for periods of time such as evenings, weekends, and holidays. And, if one of the employees works overtime, the heating (or cooling) remains on until the employee leaves for the evening and turns the lights off.

4-2. HEATING/COOLING THERMOSTATS

The thermostats in Fig. 4-4 illustrate different types of thermostats used with combination heating and cooling systems (more on these systems in Chapters 12 and 13). While those shown may vary from the type of system in your home, the information applies to most thermostats. Operating controls include a heat/cool/off/auto system mode select, blower control, and temperature selector.

The *heat/cool/off/auto selector* is set to the season of the year: HEAT for winter and cooler days or COOL for summer and warmer days. Some thermostats have an AUTO position; when in AUTO, there is automatic heat/cool changeover as required. For example, the system may heat the home on a cool morning and then cool the home when the afternoon sun is hot. An OFF position shuts down the entire system regardless of the temperature setting of the thermostat.

The *blower control* has an ON or CONT setting for continuous blower operation and INT or AUTO setting for intermittent or automatic blower operation, used *only* when the heating element is hot or the cooling compressor is on. The ON or CONT setting usually gives the best comfort performance because the blower is on continuously which helps prevent the stratification of layers of different temperature air.

The *temperature selector* dial or lever is used to select the home (or the zone-controlled section of the home) temperature desired. Some thermostats have one temperature setting dial or lever for all seasons; others have separate temperature selectors for heating and cooling.

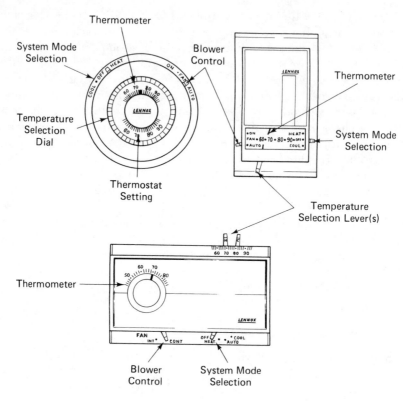

FIGURE 4-4. Thermostats control the temperature of your home during the heating and cooling season, if you have both heating and cooling systems. The text describes the various controls (*courtesy Lennox Industries Inc.*).

chapter 5

CAULKING

Caulking is the process of closing the natural openings that occur at the junction of two different building materials or two different parts of the house such as at the junction of wooden window frames and asbestos shingle siding or where the porch juts out from the house. Because of varying temperatures during the days throughout the years, the different materials expand and contract at different rates. This causes the junction between the materials to open and close as the temperature varies.

It is necessary to have a caulking material that will "bead" across the gap between the materials and that will remain flexible as the temperature changes, causing the materials to move. A caulk that dries out will become brittle quickly, will crumble, and will fall out, making recaulking necessary. Therefore, buy a caulk that will last five to ten years; a little extra money spent at the outset will save you time and recaulking costs later. Caulking is an inexpensive, easy, do-it-yourself task that will cut fuel bills. Experts estimate that caulking *and* weather stripping (Chapter 6) cost between $75 and $105 if you do the work yourself and double that if you have a contractor do it; but you'll save $30 to $75 per year on your heating bill and another $25 to $60 per year on your air-conditioning bill.

Failure to have adequate caulking between the junctions of different materials and different parts of the house can cause several problems: heat escapes in the winter; heat infiltrates in the summer; uncomfortable drafts occur; water may seep in and eventually cause wooden frames and sheathing to rot; water may seep in, freeze, and cause structural damage; dust and dirt blow in; insects are able to enter the home; and paint failures, including peeling and mildew, may occur.

5-1. TYPES OF CAULK

There are five types of caulking compounds (Quick Reference Chart 5-1). Of these, the acrylic latex is the most effective and durable caulk available to the homeowner at a reasonable price; recaulking won't be necessary for up to 10 years. Most caulking is available in white, gray, black, and sometimes other colors and comes in 11-ounce disposable cartridges or in 1- or 5-gallon containers.

Caulking also comes in a caulking cord/weather strip. The strip is a plastic sealant that does not adhere to surfaces. It is intended to be forced

Quick Reference Chart 5-1
TYPES OF CAULK

Type	Use	Comments	Durability (Years)
Oil base	Bonds to most surfaces—wood, masonry, and metal.	Oil tends to ooze out, causing caulk to stain surfaces and to dry out causing shrinkage, brittleness, loss of adhesion, cracking. It eventually falls out. Relatively easy to apply. Inexpensive.	1 to 2
Latex base, water-thinned	Recommended mainly for indoor use. May be used on exterior except at cement/concrete joints and must be painted. Not for use where joints move. Don't use in joints more than ¼ inch wide.	Goes on and cleans up easily. The polyvinyl acetate (PVA) in it fades from light and it disintegrates when used on cement/concrete. Loses adhesive qualities under moderate joint expansion and contraction. Must be painted when used on exterior. Fast drying and takes oil or latex paint well. Also in an ethylene vinyl acetate (EVA). Best latex is the acrylic (see below). Very easy to apply.	2 to 5

Type	Use	Comments	Durability (Years)
Butyl rubber	For metal surfaces. Good for gutters and downspouts. Best for metal to masonry joints. Don't use in joints more than ⅜ inch wide.	Adheres well to metal surfaces. Water resistant. Difficult to apply. Requires a solvent to clean up. Some require 7 days curing before painting. Moderately expensive.	Long–lasting
Silicone	Adheres well to all surfaces and materials except paint.	Known as an elastomer caulk. Extremely flexible with long–service life. Primers are required for some surfaces and they are not too easy to apply. Costs two to three times as much as other caulks. Most cannot be painted.	Longest lasting of all caulks; some are guaranteed up to 20 years against cracking, crumbling, and drying out.
Acrylic latex	Suitable for all applications on all types of surfaces—wood, concrete, metal, glass, stone, brick, stucco, marble, or any other. Can be used indoors as well as outside for any type of caulking job. Seals joints up to ½ by ½ inch.	Adheres well, is flexible, is durable, and is not affected by light. It has excellent resistance to cracking, peeling, and loss of color. Tools clean up easily in water. Cures within an hour leaving a thin flexible surface that can be painted with alkyd or water-based paints. Does not stain. Very easy to apply. Moderately expensive; the higher-priced acrylics give higher levels of performance because there is more acrylic material in them.	Up to 10 years.

between two surfaces that will hold it in place; it should not be applied to joints that move.

5-2. WHERE AND WHEN TO CAULK

Caulking is used to close the junction of two dissimilar materials or two different parts of the house to prevent heat transfer, water entrance, or insect entrance. Quick Reference Chart 5-2 lists the external junctions that are caulked. Caulking is also used internally around bathtubs, bathroom walls and floors, and kitchen counters (the subjects of other texts). Weather stripping (Chapter 6) is applied along the edges of windows and doors where movement occurs during opening and closing.

Visually inspect all of the points where caulking should be located, as listed in QRC 5-2 and shown in Fig. 5-1. If there is no caulking, or if the existing caulking is old, brittle, or cracked, put in new caulking. Do your caulking when the outside temperature is higher than 45 degrees F; this is required so that the caulking can bond properly to the materials and cure properly.

Quick Reference Chart 5-2
WHERE TO CAULK

1. Joints between siding and window frames. Also check the window glazing. If glazing is missing, the glass can be caulked or reglazed (Section 3-1).
2. Joints between siding and door frames.
3. Joints between siding and foundation (also inside the basement where the sill rests on the foundation).
4. Where siding meets at corners.
5. Joints where siding meets a stone or brick wall.
6. Joints where two lengths of siding meet.
7. Joints between wing extensions, porches, and the main body of the house.
8. Joints between siding and trim.
9. Joints where the underside of eaves and gable molding meet.
10. In older homes, open the windows and check for openings in the outer frame; caulk any openings.
11. Seal around plumbing and wire insulating strips that pass through the siding or foundations.
12. Around air conditioners.
13. Cracks or nail holes in siding.
14. Loose flashing around chimney.
15. Around wall caps and vents for fans.
16. All roof flashing.
17. Around window glass, in place of glazing compound.

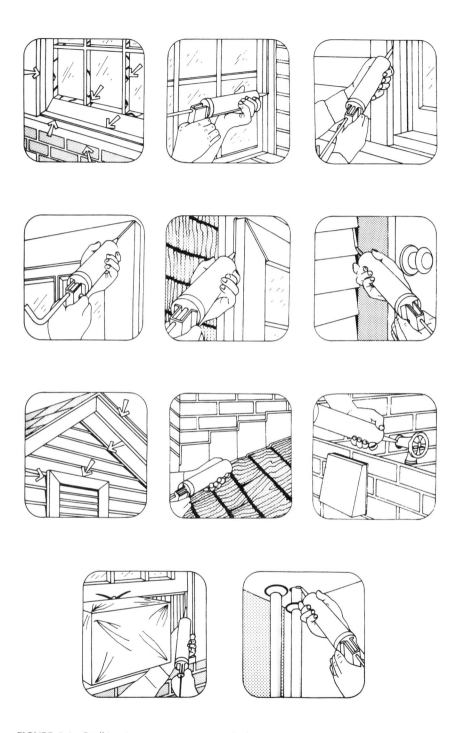

FIGURE 5-1. Caulking is an easy, economical, do-it-yourself task. Caulk at the junctions shown here and discussed in QRC 5-2 (*courtesy Macklanburg-Duncan Co.*).

5-3. HOW TO CAULK

The easiest type of caulking for the homeowner to use is the 11-ounce disposable cartridge that fits into a *half-barrel* caulking gun. Another type, bulk caulking contained in 1- or 5-pound cans, is loaded into a *full-barrel* caulking gun.

It is difficult to estimate the amount of caulking compound necessary to close gaps, but use the following as a guide:

	Ounces	Number of Cartridges
Per window and door	5	½
Foundation sill	40	4
Chimney for two-story house	20	2

A disposable cartridge has a tapered plastic tip that is cut off at a 45-degree bevel to aid in applying the caulking; the closer the cut is made to the cartridge, the wider the bead of caulking will be. After the bevel is cut on a new tube, a stiff wire, a long nail, or a narrow screwdriver blade is inserted into the tapered tip to break the seal. The loading rod of the gun is placed with the handle down and is retracted fully; the handle is then raised which places the notches of the loading rod into the trigger mechanism. The cartridge is placed in the gun (the rear of the cartridge is dropped in before the tip) and the trigger is pulled to advance the loading rod into the caulking tube, making it ready for use. Follow the caulking procedure in DIY 5-1.

To load bulk caulking into a full-barrel caulking gun, unscrew the end from the gun. Retract the gun loading rod fully. Using a putty knife or a stick, fill the gun with bulk caulking. Replace the end onto the gun, raise the handle on the loading rod, and pull on the trigger to advance the loading rod.

Do-It-Yourself 5-1
HOW TO CAULK

Materials: the correct type of caulking compound for the job; solvent such as turpentine for removing any oil or grease; oakum, caulking cotton, urethane, or similar filler for large cracks; primer, if the caulking specifies it

Tools: caulking gun, stiff putty knife, ladder, old screwdriver

---

Do-It-Yourself 5-1 (Continued)
HOW TO CAULK

Procedure:

SAFETY
Be sure to use a ladder safely. Have someone hold the ladder in place. Do not lean out from the ladder; relocate it. Be sure the ladder is level. Use both hands to climb the ladder.

NOTE
Caulking should only be applied when the temperature of the caulk and of the materials to be caulked is above 45 degrees F; some caulking compounds may require application at even higher temperatures (see manufacturer's instructions on the caulking container).

1. Inspect the areas described in QRC 5-2 and shown in Fig. 5-1. If caulking is missing, or if it is old, brittle, or broken up, new caulking is required.
2. Scrape out the old caulking with a stiff putty knife, taking care not to gouge the woodwork.
3. Clean away all old material and dust. Remove grease or oil with a solvent.
4. If the caulking compound used requires a primer, apply the primer per the manufacturer's instructions.
5. If the cracks are larger than ½ inch, push in a filler material to within at least ½ inch from the surface (Fig. 5-2).

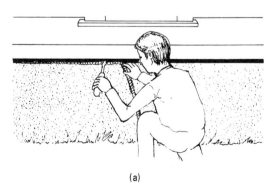

(a)

FIGURE 5-2. Where the cracks are large, fill the crack to 1/2 to 1/4 inch from the surface with oakum, glass fiber insulation strips, caulking cotton, or similar material (*courtesy U.S. Department of Housing and Urban Development*).

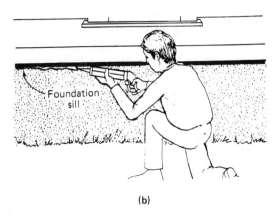

(b)

FIGURE 5-2. (Continued).

6. Holding the caulking gun at a 45-degree angle, apply a continuous bead of caulking from one end of the crack to the other. The bead must overlap both materials along the crack and should be raised slightly above the crack (Figs. 5-3 and 5-4). If cord caulking is used, force it into the cracks, allowing part of the surface to remain above the crack.
7. If necessary, smooth or "dress" the bead by following the manufacturer's instructions. For example, latex caulking can be dressed by rubbing the surface with your finger wet with a soapy water solution. Butyl caulking is dressed with mineral spirits.

FIGURE 5-3. Caulking is applied with a half-barrel gun and disposable caulking cartridge. The gun is angled down and pulled smoothly at a steady rate as the trigger is steadily squeezed (*courtesy Rohm and Haas Company*).

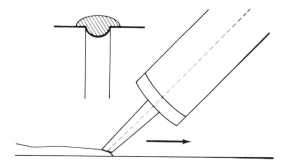

FIGURE 5-4. Make sure the bead of caulk overlaps both sides of the crack for a tight seal (*courtesy U.S. Department of Housing and Urban Development*).

chapter 6

WEATHER STRIPPING

Caulking and weather stripping work as a team to halt heat losses in the winter and heat gains in the summer through openings around doors and windows. Whereas caulking is a creamy soft material applied from a gun generally to the outside around casings of windows and doors (as well as many other joints), weather stripping is in the form of foam strips, rubber strips, and metal strips applied to the sides and/or the insides of windows and doors. Weather stripping, as is caulking, is an easy do-it-yourself job that is inexpensive for the results it produces: it costs about $2 per window and $6 per door to weather-strip it yourself. If you have a contractor do the work, it will cost from two times to four times as much. Weather stripping installed by the homeowner is economical in all climates above 2,000 degree days.

Besides keeping heat inside in the winter and preventing heat infiltration in the summer, weather stripping keeps moisture from getting between the window and the storm window causing frost, and keeps out dust and dirt. Weather stripping also provides vibration and shock-dampening features and can be used to seal out light (as for stripping doors and windows in a photography darkroom).

6-1. TYPES OF WEATHER STRIPPING

Weather stripping comes in a variety of sizes, shapes, methods of attachment, and materials, including hair felt, wool felt, polyurethane foam, vinyl, aluminum, bronze, stainless steel, and combinations of these. You may have need for one or more types of weather stripping; be sure to look at your windows and doors and use this chapter as a guide before you go to a hardware store and buy the wrong type or size. Figures 6-1 and 6-2 and QRC 6-1 illustrate and describe some of the various types of weather stripping available.

Seals for door thresholds may be one of three types: a metal strip with a vinyl "bulb" insert mounted under the door, a flexible strip mounted to the door bottom, or a metal strip with a flexible felt or vinyl insert mounted to the face of the door along the bottom. All types are screwed into place. When the metal strips are placed on the floor, it is usually necessary to trim the door bottom at an angle to assure a tight fit between the door and the strip.

(a) Self-Adhesive Plastic Foam Tape. Felt Looks Similar, but has no Self Adhesive Backing

(b) Flexible Aluminum with Felt Insert

(c) Flexible Aluminum with Vinyl Bulb Insert

(d) Rolled Tubular Vinyl

(e) Vinyl Covered Polyurethane

(f) Wood Strip Edged with Foam

(g) Aluminum Strip with Vinyl Insert

(h) Bronze, Aluminum, or Stainless Steel Roll with Built in Tension

FIGURE 6-1. Weather stripping teams up with caulking to prevent heat losses in the winter and infiltration in the summer around windows and doors (*courtesy Macklanburg-Duncan Co.*).

(i) Metal Fold-Back Automatically Spaces Itself When Properly Placed Against the Door Stop

(j) Double Vinyl or Rubber Strips for Garage Door Bottoms

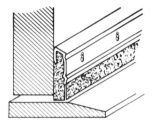
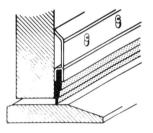

(k) Aluminum with Felt Strip Insert

(l) Aluminum with Vinyl Strip Insert

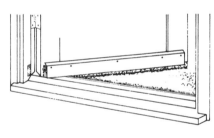

(m) Automatic Door Seal—Vinyl Bottom Raises Up to Clear Rug when Door Opens

(n) Aluminum Threshold with Vinyl Insert for Average Depth Rugs

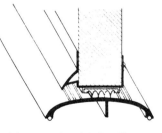

(o) Aluminum Threshold with Vinyl Insert for Thick Pile Rugs

(p) Installs on Either Threshold or Door Bottom

(q) Threshold with One Piece Drip Cap and Door Bottom

FIGURE 6-1. Continued).

81

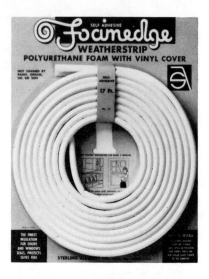

FIGURE 6-2. You need to know what kind of weather stripping, the width, length, and how you will attach the weather stripping before you buy. Vinyl covered polyurethane foam is shown here. One stripping attaches with thumb tacks, the other is self-adhering (*courtesy Teledyne Mono-Thane*).

Quick Reference Chart 6-1
TYPES OF WEATHER STRIPPING

Type	Use	Comments	Installation
Felt (wool pre-shrunk felt is better than hair felt) and felt strip hemmed with aluminum (Fig. 6-1A and 6-1B).	Door jambs. Top and bottom of double-hung windows. Will not stand up against the friction of moving windows.	Inexpensive. Can be used just about anywhere for stop-gap purposes, but will require replacement in 1 to 2 years. The aluminum hem gives more body to the felt. Do not paint the felt as it will harden and lose the flexibility required for sealing.	No. 4 upholstery tacks or ⅜-inch nails with heads. With door closed, press felt against door stop and door to slightly compress it. Tack into place (Fig. 6-4B, 6-4C).
Plastic foam tape (Fig. 6-1A).	Door jambs. Top and bottom of windows. Will not stand up against the friction of moving windows.	Inexpensive. Very easy to install. Can be used just about anywhere for stop-gap purposes, but will require replacement in 1 or 2 years. Do not paint	Self–adhesive. Clean and dry the surface to be sure it is free of dirt. Remove grease or oil with denatured alcohol. Strip backing off of foam and

Type	Use	Comments	Installation
		the foam as the foam will harden and lose the flexibility required for sealing.	press foam into place with fingers. Cut off at desired length. Put in place on a warm day to assure adhesion (Fig. 6-4B).
Wood strips edged with foam (Fig. 6-1F).	Door jambs. Will not stand up against the friction of moving pieces.	Easy to install. Wood strip can be (and should be) painted to match wood trim; however, do not paint the foam as it will harden and lose its flexibility required for sealing. Damaged by hard objects hitting it as they pass through door.	Nails. Measure and cut both sides. Close door and hold strip in place with slight force against door. Nail every 10 inches with small finishing nails. Measure, cut, and install top piece. Paint. Stripping can be attached to metal with contact cement (Fig. 6-4F).
Rolled tubular vinyl (Fig. 6-1D).	Door jambs. Windows. Seals sides, top, bottom, and center of windows.	On double-hung windows, the roll should press lightly against the frame of the sash at sides. On doors and casement windows, install the strips so that the tube is slightly compressed when door or window is closed. Can be bought in 17-foot or continuous lengths. Loses its flexibility at about 20 degrees F.	Tacks or self-adhering. Start at one corner. Place the tubing so that rounded part makes slight contact against door or window when closed. Attach to the door or window frame or stop, not to the moving parts (Fig. 6-4B, 6-4D, 6-4E).
Vinyl or rubber strips (Fig. 6-1J).	Garage door bottoms.	Double seal.	Nails. Raise garage door to convenient height. Starting at one end, nail strip into place. Cut off remainder (Fig. 6-3K, 6-3L).

Type	Use	Comments	Installation
Vinyl-covered polyurethane bulb (Figs. 6-1E, 6-2).	Door jambs. Windows. Seals sides, top, bottom, and center of windows.	This is a good weather stripping having good recovery in cold weather. Loses its flexibility at about 20 degrees F. Can be bought in 17-foot or continuous lengths.	Tacks, small nails, or self–adhering. Start at one corner. Place the tubing so that rounded part makes slight contact against door or window when closed. Attach to the door or window frame or stop, not to the moving parts (Fig. 6-4B, 6-4D, 6-4E).
Extruded aluminum strip with vinyl bulb (Fig. 6-1G).	Doors.	Similar to the wood strip with foam but this stripping is more durable and does not require painting.	Nails. Measure and cut strips for both sides. Close door and hold strip in place with slight force against door. Nail into place through holes provided. Measure, cut, and install top piece (Fig. 6-4G).
Coil spring of bronze, aluminum, or stainless steel (Fig. 6-1H).	Doors and windows.	Built-in tension. Very durable. Invisible when installed and is protected from damage by its location.	Nails. Refer to DIY 6-2, 6-4 and Figs. 6-5, 6-7, 6-4H.
Fold-back bronze or aluminum (Fig. 6-1I).	Doors.	The fold-back design automatically spaces itself when it is properly placed against the door stop. It is invisible when installed.	Nails or screws. Installs similarly to the coil stripping, but because of its design, is automatically spaced. Strip is nailed or screwed every 3 inches. If the strip does not have holes, punch them with an awl or ice pick (Fig. 6-4H, DIY 6-2).

Type	Use	Comments	Installation
Extruded aluminum strip with vinyl insert. Also comes with felt strip (Fig. 6-1K, 6-1L).	Door bottoms.	The vinyl insert is more durable than the felt insert. Can be used alone or to supplement threshold strips.	Screws. Refer to DIY 6-5 and Fig. 6-8A, 6-8B.
Automatic door bottoms (Fig. 6-1M).	Door bottoms.	A flexible vinyl bottom raises automatically to clear the carpet when the door opens and lowers again on closing. This keeps the strip from dragging on the carpet.	Screws. Refer to DIY 6-5 and Fig. 6-1M, 6-3H.
Extruded aluminum threshold with vinyl insert. One design also includes a rain-drip cap for exposed exterior doors (Figs. 6-1N, 6-1O, 6-1P, 6-1Q).	Thresholds of doors.	Ideal way to seal gaps under doors. Different shapes and sizes available for average rugs, thick pile rugs, shag rugs. Vinyl insert is replaceable.	Screws. Refer to DIY 6-6 and Fig. 6-8C, 6-8D.

6-2. WHERE TO WEATHER-STRIP

To determine the extent of weather stripping needed around your home, inspect each of the windows and doors. Look for strips of vinyl, metal, or foam rubber around the edges. If the weather stripping is broken, badly damaged, deteriorated, or missing, you'll need to apply new weather stripping. If metal stripping is already installed, but is letting drafts through because it is damaged, you can replace it or *supplement* it with one of the other types of weather stripping. Don't take storm doors and combination storm windows for granted either; check the weather stripping seals between the panels and the seal between the door frame and the door. Some combination storm windows use the spring tension of a piece of metal to hold the window against the frame. If necessary, bend the metal to increase the pressure forcing the window tighter against the frame.

Fig. 6-3 and QRC 6-2 tell you where weather stripping is necessary.

Windows

(a) Seal the Inside Edges

(b) Seal the Top, Bottom and Center of the Window

Doors to the Outside

(c) Gasket Type Weather Stripping Against the Stop

(d) Metal Stripping Along the Frame

(e) Weather Stripping Storm Doors

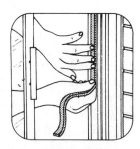  

(f) Weather Stripping Sliding Glass Doors

(g) Threshold Stripping Seals Under the Door

(h) Door Stripping Automatically Raises to Clear the Rug as the Door Opens

FIGURE 6-3. Weather stripping is used on all doors to the outside, doors to unheated areas, and windows (*courtesy Macklanburg-Duncan Co.*).

Interior Doors

(i) Weather Strip Doors to Basement, if it is Unheated

(j) Weather Strip Interior Doors to Unheated Areas

Garages

(k) Weatherstrip Garage Door Bottom

(l) Seal Garage Door Bottom and Sides

(m) Weather Strip Attic Hatches

Miscellaneous

(n) Weather Strip Around Air Conditioners to Prevent Heat Infiltration During the Air Conditioning Season

(o) Weather Strip Around Lighting Fixtures to Unheated Areas

FIGURE 6-3. (Continued).

1. Doors to the exterior.
2. Garage doors, especially if garage is heated.
3. Interior doors to *unheated* (or *unair-conditioned*) rooms such as laundry, attic, unused rooms, storage, utility, garage, basements and workshops.
4. Trapdoor to attic.
5. Folding stairway to attic.
6. All windows (weather stripping of windows in unheated areas is recommended in cold climates).
7. Electrical light fixtures in ceilings or walls that are backed by unheated areas.

Foam weather stripping with a self-sticking adhesive is effective as a "stop-gap" against drafts around windows. However, when time and weather permit, it should be replaced with a more durable strip. It is best to apply the self-sticking foam on a warm day because it helps the strip adhere better (blow hot air on the back of the strip with a hair dryer to make it adhere better on cool days).

Other stop-gap procedures include the stuffing of material or newspaper in the top, center, and bottom of windows. A piece of rug or similar material can be folded and placed along the crack at the bottom of a door to prevent air infiltration.

When you buy weather stripping (as well as caulking), you get just about what you pay for. The least expensive weather stripping is the least durable (lasts one or two years) and the easiest to install. Higher priced weather stripping lasts considerably longer; however, the *difference in price* between the least expensive and the higher priced material is not much money. The differences to be considered are the time involved in installation and the long range durability. If you plan to stay in your home for at least several years, you certainly don't want to weather-strip every year to two. The inexpensive, less durable materials can be quickly installed. The more durable weather strip may take you an hour per installation to measure, cut, and nail or screw into place. In summary, spend a little more money and invest a greater amount of time for installation; you'll have a more effective weather stripping job that will last about five times longer.

Weather stripping is applied to doors either along the stops, if you're using gasket type weather stripping (Fig. 6-4), or on the framework along the outside edges of the door, if you're using metal tension stripping (Fig. 6-4H, 6-5). The material is shaped so that when it is nailed into place, it protrudes slightly along the edge away from the door. As the door closes on it, its spring tension holds it against the door providing a weather seal. If more tension is needed, a screwdriver can be run down the underside of the strip carefully

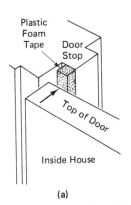

(a)

Attachment of Plastic Foam Tape. Felt is Similarly Attached, but is Tacked Rather Than Self Adhering

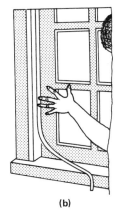

(b)

Attachment of Foam Plastic Tape or Felt by Tacking. In this Application, these Types of Stripping Provide Only a Temporary "Stop Gap". Installation of Rolled Tubular Vinyl or Vinyl Covered Polyurethane Would be as Shown Here Also

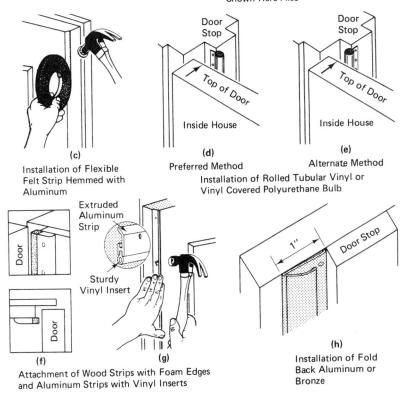

(c)

Installation of Flexible Felt Strip Hemmed with Aluminum

(d)

Preferred Method

(e)

Alternate Method

Installation of Rolled Tubular Vinyl or Vinyl Covered Polyurethane Bulb

(f) **(g)**

Attachment of Wood Strips with Foam Edges and Aluminum Strips with Vinyl Inserts

(h)

Installation of Fold Back Aluminum or Bronze

FIGURE 6-4. Methods of applying different types of gasket-type weather stripping are shown (A to G). Metal fold-back stripping is shown in H (*courtesy Macklanburg-Duncan Co.*).

Installation of Built-In Tension Weather Stripping

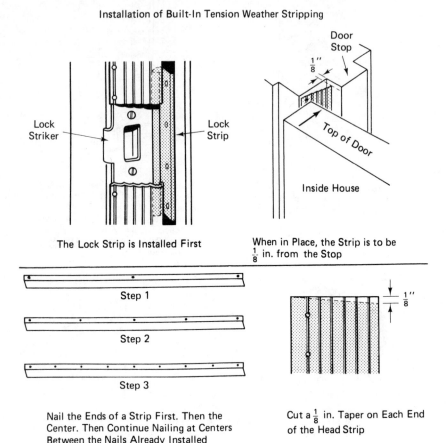

The Lock Strip is Installed First

When in Place, the Strip is to be $\frac{1}{8}$ in. from the Stop

Step 1

Step 2

Step 3

Nail the Ends of a Strip First. Then the Center. Then Continue Nailing at Centers Between the Nails Already Installed

Cut a $\frac{1}{8}$ in. Taper on Each End of the Head Strip

FIGURE 6-5. The procedures for installing metal built-in tension weather stripping around a door are shown.

bending the metal out slightly more; be careful *not* to put dimples (slight bends or dents) in the strip.

On windows, gasket type weather stripping is applied along the stops, to the top of the upper sash, the bottom of the lower sash, and to the bottom of the upper sash where it meets the top of the lower sash (Fig. 6-4, 6-6, 6-3B). Metal spring-tensioned weather stripping is placed in the channels behind the sash frame, under the sash cords (Fig. 6-6). On double-hung windows, a piece is also nailed to the frame at the top of the upper sash, the frame at the bottom of the lower sash, and to the outside of the bottom of the upper window so it seals against the lower sash.

If your garage is heated, the doors and windows should be weather-stripped (and caulked as well); even if it's not heated, weather strip-

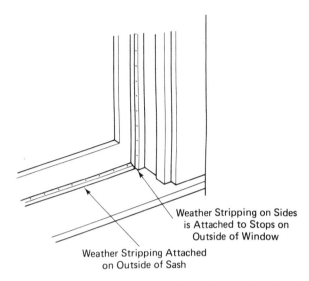

Weather Stripping on Sides
is Attached to Stops on
Outside of Window

Weather Stripping Attached
on Outside of Sash

FIGURE 6-6. Gasket-type weather stripping is attached to the outsides of windows, but may also be attached inside.

ping will help to keep out the bitter cold and blowing rain, snow and dust. The garage doors should have a specially designed double rubber or plastic strip on the bottom edge. On the sides and top, use a weather stripping consisting of metal strips with vinyl flaps that bear against the door (Fig. 6-3K, 6-3L).

Other places where weather stripping should be installed are on interior doors (and hatches) to unheated attics, basements, laundry areas, storerooms, and workshops. These doors to unheated areas should be sealed as carefully as exterior doors. (A seal on the workshop door keeps a lot of dust from spreading throughout the house too.)

6-3. HOW TO WEATHER-STRIP

There are two general types of weather strip installations: concealed and visible. The concealed weather stripping is completely protected by the closed door or window and is partially protected when open. Concealed weather stripping is metal with a built-in spring tension. On doors, concealed weather stripping is mounted around the door frame adjacent to the door; on windows, it is placed in the channels behind the sash frame under the sash cords.

Visible weather stripping is mounted on door stops and window stops. Visible stripping slightly decreases the size of the openings of doors and windows and it may be damaged when objects are moved through the openings.

All types of weather stripping are installed so that the stripping material is slightly *compressed* when the door or window is closed. Gasket types of weather stripping are installed on the *stops* of doors and windows outside or inside to assure a good seal; be sure you don't mount the strip on a stop so that it is *in the path* of one of the windows of a double-hung window when it is moved up or down. This may necessitate the installation of some of the stripping on the window interior and some on the exterior.

Since there are so many different types of weather stripping, be sure to read and follow the manufacturer's instructions carefully. In their absence and as supplementary information, do-it-yourself charts 6-1 through 6-6 describe the installation procedures for various types of weather stripping:

6–1. Doors with gasket type weather stripping

6-2. Doors with built-in tension metal weather stripping

6-3. Windows with gasket type weather stripping

6-4. Windows with built-in tension metal weather stripping

6-5. Door bottom strips

6-6. Thresholds

Do-It-Yourself 6-1
HOW TO WEATHER-STRIP DOORS WITH GASKET TYPE WEATHER STRIPPING

Materials: the correct type and size of weather strip for the job (see QRC 6-1); tacks, nails, or screws, as applicable (usually provided)

Tools: hammer or conventional screwdriver to fit screw (3/16 or 1/4 inch), hacksaw or tin snips for cutting metal, knife or small saw to cut wood strips; file for removing burrs from metal

Procedure:

NOTE

The general procedures for installing the particular type of gasket weather stripping such as a self-adhesive, nail-on, or screw-on stripping are described in QRC 6-1.

Doors—Gasket Type Weather Stripping

1. Gasket type weather stripping is applied on the door stop (Fig. 6-4).
2. Clean the surface and prepare the stripping materials as required.
3. Cut the vertical side pieces of stripping to length. If a metal piece is cut, file off burrs.
4. Fasten stripping in place so that when the door is closed, it will compress the stripping slightly.
5. Measure, cut, and fasten the piece to the top of the door frame against the stop in similar manner to steps 3 and 4. Be sure there is continuous contact around the door. Add threshold weather stripping (DIY 6-6).

HOW TO WEATHER-STRIP DOORS WITH BUILT-IN TENSION METAL WEATHER STRIPPING

Materials: weather stripping, nails (usually supplied)

Tools: tin snips to cut the stripping, hammer, ruler, nail set (or a large nail to use as a nail set)

Procedure:

1. Tension metal type weather stripping is installed on the frame against the door stop (Figs. 6-5 and 6-4H). The lock strip is the first piece nailed into place.
2. The piece above the lock is installed next; measure, cut both ends square and locate it so that there is a ⅛-inch gap between the free edge of the strip and the door stop.
3. *Don't* try to drive the nails completely flush with the metal with a hammer, as the strip might be damaged by the hammer. Set the last ⅛ inch of each nail with a nail set. Place the strip in position, put one nail at each end, then one in the center of the strip. Next, place nails in the center of the remaining spaces, and so on, until all nails are in place.
4. Measure, cut, and nail a piece of stripping to fit between the threshold and the lock-striker, overlapping the lock strip already installed. Leave a ⅛-inch gap between the free edge of the strip and the door stop.
5. Measure, cut, and nail stripping for the other side of the door frame. Leave a ⅛-inch gap between the free edge of the strip and the door stop.
6. Measure and cut stripping for the head strip above the door. Cut a ⅛-inch taper on each end so that it mates with the vertical side strips. Nail as previously described. Add threshold weather stripping, as subsequently described.

HOW TO WEATHER-STRIP WINDOWS WITH GASKET TYPE WEATHER STRIPPING

Materials: the correct type of weather stripping for the job (see QRC 6-1); tacks, nails, or screws, as applicable (usually provided)

Tools: hammer or conventional screwdriver (³/₁₆ or ¼ inch), hacksaw or tin snips for cutting metal, file for removing burrs from metal

Procedure:

NOTE

The general procedures for installing the particular type of gasket weather stripping such as self-adhesive, nail-on, or screw-on stripping are described in QRC 6-1.

1. Gasket type weather stripping is applied to window stops or frames, and to the top of the upper sash, bottom of the lower sash, and to the underside of the bottom of the top sash to seal the center of double-hung windows.

HOW TO WEATHER-STRIP WINDOWS WITH GASKET TYPE WEATHER STRIPPING

2. Obtain the proper material for the installation (QRC 6-1).

3. Remove dirt from the area where stripping is to be applied. Use sandpaper or a wood chisel to smooth off any paint runs. If a self-adhesive stripping is used, clean off oil or grease with alcohol.

4. For double-hung windows measure, cut, and install strips (the material should be slightly compressed when correctly in position) to the top of the upper sash, the bottom of the lower sash, the underside of the upper sash, and the stops or window frame along the window when the window is closed. Fasten the stripping according to the manufacturer's recommended directions printed on the weather stripping package and in accordance with the installation instructions of QRC 6-1 and Figs. 6-3A, 6-3B, and 6-6.

5. On casement windows, install the stripping on the window edge that overlaps the frame.

Do-It-Yourself 6-4

HOW TO WEATHER-STRIP WINDOWS WITH BUILT-IN TENSION METAL WEATHER STRIPPING

Materials: weather stripping (nails are usually provided)

Tools: tin snips to cut the stripping, hammer, ruler, nail set (or a large nail to use as a nail set), sandpaper or wood chisel to remove dried paint runs in channels)

Procedure:

1. Metal spring-tensioned weather stripping is placed in the channels behind the sash frame under the sash cords, at the top of the upper sash, the bottom of the lower sash, and on the lower outside edge of the upper sash to seal the window (Fig. 6-7).

2. Move the upper sash to its open position. Use sandpaper or a wood chisel to clean out any paint runs that have accumulated.

3. Cut weather strips for the sides of the window, allowing about ½ inch extra to fit below the sash when it is closed. Slide the strips into place and nail, keeping the free edge ⅛ inch from the stop. Drive the nails to within ⅛ inch of the weather strip; use a nail set to drive the nails the final ⅛ inch.

4. In similar manner, attach a strip to the frame channel above the upper sash (as an alternate, the strip may be attached to the sash). Nip off ⅛ inch of metal from each end at a bevel so that the strip fits correctly.

5. In similar manner, attach a piece of stripping to the outside edge of the upper sash.

6. Close the upper sash and open the lower sash. Install the side strips and bottom strip as described in steps 3 and 4. The bottom strip should be fastened to the frame for the best protection.

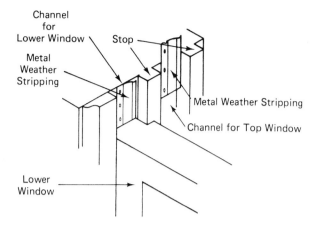

Channel for Lower Window

Stop

Metal Weather Stripping

Metal Weather Stripping

Channel for Top Window

Lower Window

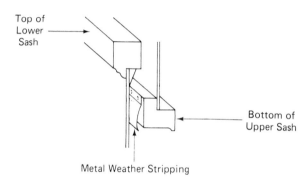

Top of Lower Sash

Bottom of Upper Sash

Metal Weather Stripping

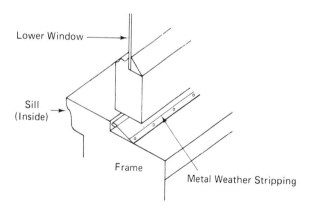

Lower Window

Sill (Inside)

Frame

Metal Weather Stripping

FIGURE 6-7. Built-in tension weather stripping is attached in the window channels, top of the upper sash, outside bottom of the upper sash, and to the frame under the bottom of the lower sash.

Do-It-Yourself 6-5
HOW TO INSTALL DOOR BOTTOM STRIPS

Materials: door strip and screws (usually supplied)

Tools: hacksaw to cut strip to length, screwdriver (conventional, ¼ inch), awl or ice pick to start screw holes, drill with diameter slightly smaller than screw

Procedure:

1. Door bottom strips are located on the bottom of the door and prevent air from passing under the door (QRC 6-1 and Fig. 6-1K, 6-1L, 6-1M).
2. With the door closed, measure and cut the strip to fit the width of the door (Fig. 6-8A, 6-8B).
3. Note that the screw slots are elongated. This is to help you set the strip at the proper height during installation and it permits the strip to be lowered to compensate for wear. Hold the strip in position with the bottom edge touching the floor. Using the strip as a pattern, mark screw hole locations at the *bottom* of the elongated slots. Start the hole with an awl or ice pick.
4. Using a drill, drill the holes for the *end* screws.
5. Attach the strip to the door with two screws. Check the strip for proper sealing and proper clearance over the floor when the door is opened; adjust the height as necessary.
6. Drill the other holes and drive the screws; again, check for proper sealing and clearance and then tighten the screws.
7. If the door bottom strip is an automatic door seal, install the actuating plate to the door frame (Fig. 6-3H).

Do-It-Yourself 6-6
HOW TO INSTALL THRESHOLDS

Materials: threshold and screws (usually supplied)

Tools: hacksaw to fit threshold, screwdriver (conventional, ⅜ inch), old screwdriver and hammer to remove door-hinge pins (if necessary), saw to cut off door bottom (if necessary)

Procedure:

1. The threshold provides a weather seal between the floor and the door. Select the proper type and size for the door and rug height (QRC 6-1 and Fig. 6-1N, 6-1O, 6-1P, 6-1Q).
2. With the door open, measure the door opening width. You may need to make a cardboard pattern of the molding along the door frames so that the threshold can be cut to fit correctly. If necessary, for attaching the threshold to the floor, remove the vinyl or rubber strip and then saw the metal with a hacksaw.
3. If a part of the threshold stripping attaches to the bottom of the door or if the door bottom needs to be cut to fit the strip, remove the door by driving the pins up out of the hinges. Cut the door with a saw as directed by the manufacturer (a bevel cut helps the door operate easier over the strip and seals it better).
4. Install the strip with screws. Replace the door, if removed.

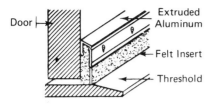

(a) Installation of Extruded Aluminum Strip with Felt Insert

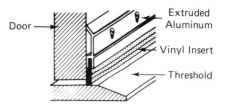

(b) Installation of Extruded Aluminum Strip with Vinyl Insert

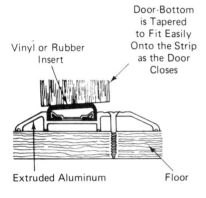

(c) Installation of Threshold

(d) On Some Models, the Threshold is Screwed into Place First; then the Vinyl or Rubber Insert is Installed

FIGURE 6-8. Door bottom strips and thresholds with replaceable vinyl or rubber strips close off any remaining gaps between the threshold and the bottom of the door to stop under door drafts (*courtesy Macklanburg-Duncan Co.*).

6-4. DOOR AND WINDOW SEAL TESTS

After the caulking and weather stripping are completed, check each door and window for a tight seal. If there is quite a temperature difference between the inside and the outside, and if the wind is blowing, you can move the back of your hand along the frames; if any air leaks remain, you'll feel a draft at the air leak. If there is not a temperature difference, wait for a windy day and then make a candle test. Doors can be checked by looking for light coming through openings that are not properly sealed (Fig. 6-9). If you are checking internal doors, stand back on the inside of the door as someone on the outside shines a flashlight beam along the door frame.

FIGURE 6-9. After all caulking and weather stripping is completed, check for openings with the back of your hand, a flashlight, or a candle. Be sure to remove curtains and any other flammable materials from the area when making a candle test (*courtesy Macklanburg-Duncan Co.*).

WARNING

The candle test uses an open-flame candle to check for drafts indicating insufficient or incorrect placement of weather stripping and caulking. Remove all curtains, shades, and all other flammable objects from around the door or window to be checked. Do *not* use the candle test if plastic storm windows are used indoors on the window being tested.

The candle test is used to determine if all drafts have been stopped around doors and windows by effective caulking and weather stripping (Fig. 6-9). Read the warning in the preceding paragraph; do not allow the candle near any flammable material. Do not bring the candle close to the woodwork, rug, or ceiling. Place the candle in a candleholder so that hot wax does not drip down on your hand. Now, light the candle and slowly move it around the door or window being tested. Any draft around the casing, the frame, or the glass panes will cause the flame to flicker indicating the location of the draft. (Be sure the draft is coming from the door or window and not from a hot air blower, people moving, or similar items that could produce air currents.)

chapter 7

INSULATING

Presumably at this point you have completed the first jobs in weatherizing your home—stopping the escapement and infiltration of heat through openings, resetting your thermostat, caulking, and weather stripping. If you don't complete these weatherizing measures first, you are simply wasting your money buying insulation. With these priority weatherization procedures completed, however, the addition of insulation (or supplementing of existing insulation) is a fairly inexpensive and, for the most part, easy, do-it-yourself task that can save you another 20 to 30 percent of your heating and cooling costs.

The installation of insulation is a *permanent* way to reduce costs. You need make only *one* purchase, and pay for it *one* time. Once the purchasing cost of the insulation is recouped, you *never* again have to pay out for reinsulating, but you continue to save every year. In fact, you'll save *more money every year* because you'll use about the same amount of energy, but the cost of the energy that you save will have spiraled upward yearly. In a period of three to seven years, depending upon your house structure, the extent of the insulation, the climate, and whether you simply heat or heat and air-condition, you'll have the insulation paid for by the savings you've made in the cost of energy that was *not* used. Insulation is a good investment in geographic areas that require heating and/or a lot of air conditioning (even if the heating requirements are low). The National Bureau of Standards and the Federal Energy Administration sum it up by saying, "Even though utility bills rise as energy prices increase, the rise will be much less than it would have been without increased insulation. In fact, you might think of energy conservation improvements as a hedge against inflation."

In addition to less use of energy resulting in lower heating and cooling costs, insulation increases your comfort by reducing drafts. Walls, ceilings, and floors are warmer because the flow of air in convection currents is decreased. Insulation also cuts down on noise by absorbing sound waves rather than reflecting them.

7-1. INSULATION FORMS AND MATERIALS

Insulation is available in the form of blankets, loose fill, rigid foam, (spray) foam, and tape (Fig. 7-1). Some forms may be used in more than one type of application; others are used in only one type of application. Materials include mineral wool, cellulose fiber, and plastic. The various forms and the materials from which they are made are discussed in the following paragraphs.

Remember that you need to insulate only once for the life of your home. Therefore, it is important to select the best insulation you can find and afford

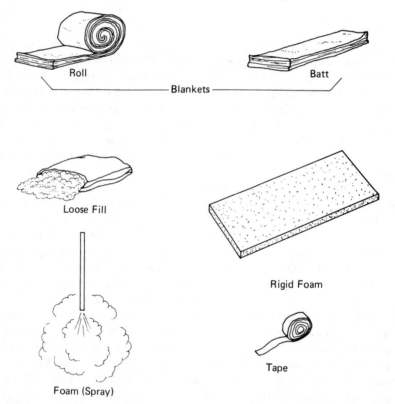

FIGURE 7-1. Insulation is available in different forms for different applications such as ceilings, walls, floors, ducts, hot water heaters, and pipes.

for the particular area to be insulated. Ideally, the insulation should possess the following characteristics:

1. An efficient barrier to the transfer of heat
2. Immune to water vapor
3. Nonsettling
4. Nonburning
5. Rotproof and permanent
6. Vermin resistant
7. Lightweight
8. Odorless

Blanket Insulation

Blankets of insulation are made of countless hair-like mineral wool fibers that are interlaced into a cotton-like pad. The mineral wool fibers are made from molten rock, slag, or fiberglass (with fiberglass being the most prevalent) which are then formed into blankets. The blankets are either unfaced (no vapor barrier) or are faced with an attached single, double, paper, or aluminum vapor barrier containing flanges for stapling the blankets into place on joists, studs, or rafters. Some blankets have a *breather* paper covering the nonvapor barrier side to protect the blankets during shipping and installation; the breather paper is stripped off after the insulation is in place. Blankets may be in the form of 40-foot rolls or 4- or 8-foot batts; widths are 15 inches or 23 inches for joists, studs, and rafters with 16-inch or 24-inch on-center spacing, respectively, and are easy for the do-it-yourselfer to install. Approximate R values for various thicknesses of blanket insulation are:

R-7   2 to 2.5 inches
R-11   3.5 inches
R-19   6 inches
R-22   7 inches
R-30   10 inches
R-38   12 inches

The vapor barrier on a blanket prohibits moisture within the house from passing through or condensing in the insulation. Moisture, also called water vapor or condensation, decreases the effectiveness of insulation and can cause serious structural problems such as wood rotting, paint peeling, and plaster cracking. The barrier is treated paper or foil and covers one side of the blanket. Blanket insulation is installed with the vapor barrier facing the living quarters (unless a competent air-conditioning specialist in your locality advises you differently).

When blankets are combined or are placed over existing loose fill insulation to increase the R value, only the blanket facing the living quarters should have a vapor barrier; the second layer should *not* have a vapor barrier. If you do not have blankets without a barrier, you can remove the barrier by peeling it off. Instead of peeling the barrier off, you may also cut a series of long slashes through the barrier to destroy its function.

Loose Fill Insulation

Loose fill insulation is made from mineral wool (rock, slag, expanded mica, or glass) or cellulose fiber (often reclaimed paper). Mineral wool is easily poured by the do-it-yourselfer into place from a bag, fluffed, and is leveled to the number of inches required for the R value desired. Mineral wool is in the form of nodules and in addition to being poured, can be blown into place with a pneumatic machine, although this is usually done by a professional. It is difficult to effectively fill exterior wall cavities of existing homes by pouring or blowing loose fill insulation because of fire breaks, cross bracing, pipes, and electrical wires and boxes within the walls.

Cellulose fiber insulation is derived from wood pulp (paper-cellulosic wood fiber) that has been highly refined. The insulating qualities are provided by the cellulose fibers and by the air spaces between them and range from about R-3.7 to R-5.2 per inch. Added chemicals make the cellulose fiber insulation fire retardant and vermin resistant.

Cellulose fiber insulation can be poured, blown or sprayed into place in ceilings and walls. Pouring is an easy do-it-yourself task that involves pouring, fluffing, and leveling off. Cellulose fiber can be blown into place by the do-it-yourselfer using a rented machine, but this procedure is usually done by a professional. When blown correctly into walls under air pressure, the material fills voids around pipes, wires, and structural cross bracing. Professionals can also spray the cellulose fibers along with an adhesive binder and fire retardant chemicals onto ceilings and other clean solid surfaces. Cellulose fibers provide a reduction in the transfer of heat, condensation control, and noise reduction, and do not irritate the skin.

When vapor enters an area filled with cellulose fibers, the vapor is held in suspension within the natural cellulose fibers, preventing the vapor from collecting at the base of the area where damage could occur. The moisture dissipates naturally during temperature changes similar to the evaporation of perspiration from the skin.

Rigid Foam Insulation

Rigid foam insulation is available as a nonflexible board having the highest insulation value for insulators; it is made by a number of manufacturers from substances such as polystyrene, polyurethane and polyethylene. Some of these products are combustible while others are said by the manufacturers to be flame retardant, nonburning, noncombustible, nonflammable, nontoxic, and self-extinguishing. These terminologies as applied to manufacturer's foam insulation products are being investigated for truth in the statements. Therefore, you are urged to look for certified statements on the products; you are also advised to contact your local government building codes representative for information on the products approved in your area.

Rigid foam insulation is easy to apply and is especially useful where blanket or fiberous materials aren't practical; just cut it with a utility knife and glue it to walls and ceilings with the appropriate mastic. Any product that is not fireproof must be covered on both sides with fire retardant material ½ inch or thicker such as plasterboard, cinder block, or brick.

Foam (a Spray Applied Via a Nozzle) Insulation

Foam insulation of polystyrene, polyurethane, or urea formaldehyde resins applied through a pressurized nozzle are used in walls between studs and in crowded spaces in attic ceilings such as under flat roofs where there is no access to pour insulation or lay blankets. Installation of foam insulation is a job for the professional who has the equipment and the technical know-how to accomplish the job. It can be readily applied to new structures before closing the cavities or to existing structures by making small openings either externally or internally. You can save some money by making the openings and closings yourself.

Foam is inserted via a pressurized nozzle through small holes in the exterior or interior walls of the house. One manufacturer uses liquid urea formaldehyde resin, a hardener containing a foaming agent, and air, which are mixed at the nozzle producing liquid foam with millions of tiny air cells. The foam flows up and down and around obstacles filling the stud cavities (Fig. 7-2). The foaming material provides not only insulation, but also a good vapor barrier, a hostile environment to rodents and insects, and a sound barrier.

To apply the pressurized foam, sections of siding or shingles are removed from the exterior of the house and holes of 1 to 1¼ inches are drilled into the sheathing on each side of each stud. In brick or stucco homes, holes are drilled through the mortar or some bricks are removed; holes are then drilled through the sheathing. The walls are plumbed to locate obstructions and the foam is applied via a nozzle. The foam sets up in about 45 seconds, so the opening of the house, the foaming, and the reclosing can be accomplished in a day or so by a contractor. Complete curing takes about 30 days. As an alternative to opening the exterior, the foam can be applied through holes in the plaster walls which can then be replastered (Section 3-2), paneled, or covered with a chair rail. You can save from 25 to 50 percent of the cost of contractor-installed foam insulation if you make the openings and closures yourself. But, be sure to discuss the exact procedures with the insulation contractor before you start any work.

FIGURE 7-2. Foam can be sprayed from the exterior or the interior into wall cavities to insulate them. Be sure to hire an experienced, reputable contractor as the quality of application of foam insulation is inconsistent (*courtesy Rapperswill Corp.*).

In general, the plastic foam insulators present some problems. Since it is injected in a wet form, there is a need to properly dissipate the initial moisture as the insulation cures. It is also subject to shrinkage (about two percent) and the initial odor requires some consideration. Sometimes cavities are left unfilled and there have been reports of the foams bursting through walls through loss of control or through swelling. Some of the foams present fire hazards and toxic fumes may be emitted under certain conditions; therefore, it must be used only as permitted by the appropriate building code for the specific application. The quality of application has been very inconsistent (contrary to advertising, many of the cavities are not adequately filled). Ask your local government building codes personnel to recommend a qualified contractor.

Insulating Tape and Duct Tape

Insulating tape is used to keep hot water hot inside of pipes and to prevent the sweating and dripping of pipes. It can also be used to aid in preventing pipes from freezing. Insulating tape is available in fiberglass, cork, and vinyl foam with aluminum in various widths and lengths. Some tapes are self-adhering; others, such as the fiberglass, have a tape included with them to hold the fiberglass securely in place.

Duct tape is a polyethylene-coated, self-adhering tape used to seal seams and holes in hot and cold air ducts. It's useful for other applications around the home too such as sealing aluminum gutters (the tape is waterproof), patching hoses, sealing cracks in windows, repairing tears in plastic, and wrapping tool handles.

7-2. WHERE AND WHAT TO INSULATE

By insulating your living area, you are effectively trying to place your living area in an envelope through which heat and moisture cannot escape in the winter and through which heat and moisture cannot infiltrate in the summer.

Ideally, insulation is to be placed above your *living area* in ceilings or attic roofs; around your living area in the walls; and under your living area below the floors. Note that the words *living area* are emphasized. If no one lives in the attic, then insulation is placed in the ceiling (the attic floor joists) at the top level of the living area; if the attic is finished off for living, then the insulation is placed in the roof rafters, collar beams, and knee walls around the living area. If the basement is unheated, then insulation is placed in the joists of the first floor; if the basement is heated, then there should be insulation part way under the perimeter of the floor slab (this is only possible during new home construction; if you don't have it now, it's too late). If you have an unheated crawl space, then the floor joists above the crawl space are to be insulated. If you have an unheated garage, then the wall of the house joining the garage is to be insulated. Typical insulating "envelopes" are shown in Fig. 7-3; QRC 7-1 describes where to insulate.

FIGURE 7-3. Ideally your living area should be surrounded by insulation (*courtesy National Mineral Wool Insulation Association, Inc.*).

Quick Reference Chart 7-1
WHERE TO INSULATE

There are quite a number of places where insulation can be used to reduce costly energy bills. Inspect these areas for adequate insulation; add insulation, as required. It is better to get some insulation in all areas rather than a lot of insulation in only one or two areas.

1. Attics—the top of the living space
2. Floors over unheated areas (basements, crawl spaces)
3. Exterior walls above ground level

Quick Reference Chart 7-1 (continued)
WHERE TO INSULATE

---

4. Walls along unheated areas (as along garages, crawl spaces)
5. Around heating ducts (also air-conditioning ducts), through unheated (or uncooled) areas
6. Basement ceiling perimeter (where the foundation meets the structure)
7. Foundation walls to a point below the frost line (about 24 inches) or to the cellar floor, if possible
8. Hot water heaters
9. Hot water pipes

7-3. R-RESISTANCE VALUES

Energy savings for heating and cooling are the result of the *resistance* of the insulating and construction materials to the flow of heat, not to the thickness of the material. There are a number of different insulating materials available (as is subsequently discussed) and for the same physical thickness, the resistance to the flow of heat of one material may be much higher than the resistance to the flow of heat of the other material. This resistance to the flow of heat is measured in *R units*. The resistance (R) of a material is a measure of its ability to decrease heat flow; the higher the R value, the more effective is its resistance to the flow of heat. For example, if two similar insulating materials each 4 inches thick had R ratings of R-10 and R-11, the R-11 material would resist the transfer of heat more effectively. Perhaps the R-11 material is more expensive; it would most likely be to your benefit to pay a little extra money *one time* to save more energy (and hence money) *all of the time* after correct installation.

Resistance values for structural materials, finish materials, insulation, air spaces, and surface films are known (QRC 7-2 and Fig. 7-4). With all structural

---

Quick Reference Chart 7-2
RESISTANCE VALUES OF STRUCTURAL MATERIALS, FINISH MATE- RIALS, GLASS, INSULATION, AIR SPACES, AND SURFACE FILMS

Structural Materials:

Wood

Bevel siding, ½ by 8 inches, lapped	R-0.81
Siding shingles, 16 inches, 7½-inch exposure	R-0.87
Plywood, ¼ inch	R-0.31
Plywood, ⅜ inch	R-0.47
Plywood, ½ inch	R-0.62
Plywood, ⅝ inch	R-0.78

RESISTANCE VALUES OF STRUCTURAL MATERIALS, FINISH MATERIALS, GLASS, INSULATION, AIR SPACES, AND SURFACE FILMS

Hardboard, ¼ inch	R-0.18
Softwood, per inch	R-1.25
Softwood, ¾ inch	R-0.94
Sheathing	
Nail-base insulating board, ½ inch	R-1.14
Insulating board, regular density, ½ inch	R-1.32
Insulating board, regular density, ²⁵/₃₂ inch	R-2.04
Concrete blocks, two rectangular cores	
Sand and gravel aggregate, 8 inches	R-1.04
Lightweight aggregate, 8 inches	R-2.18
Concrete blocks, three oval cores	
Cinder aggregate, 4 inches	R-1.11
Cinder aggregate, 8 inches	R-1.72
Cinder aggregate, 12 inches	R-1.89
Sand and gravel aggregate, 8 inches	R-1.11
Lightweight aggregate, (expanded clay, shale, pumice, slag, etc.), 8 inches	R-2.00
Concrete, sand and gravel, per inch	R-0.08
Brick	
Common, per inch	R-0.20
Face, per inch	R-0.11
Building paper, vapor, permeable felt	R-0.06
Plasterboard, ½ inch	RX-0.45
Plasterboard, ⅝ inch	R-0.56
Gypsum plaster, lightweight aggregate, ½ inch	R-0.32
Gypsum plaster, sand aggregate, per inch	R-0.18

Finish Materials:

Flooring, hardwood, ²⁵/₃₂ inch	R-0.68
Tile, cork, ⅛ inch	R-0.28
Tile, floor, asphalt, vinyl, linoleum, or rubber	R-0.05
Terrazzo, floor	R-0.08
Carpet and fibrous pad	R-2.08
Carpet and foam rubber pad	R-1.23
Shingles	
Asbestos-cement	R-0.03
Asphalt, roof	R-0.44
Wood, roof	R-0.94
Asphalt roll roofing	R-0.15
Stucco, per inch	R-0.20
Hardwoods, maple, oak, etc., per inch	R-0.91
Softwoods, fir, pine, etc., per inch	R-1.25
Siding, aluminum over sheathing, average	R-1.8

Glass:

Single (winter) (U=1.13)	R-0.88
Single (summer) (U = 1.06)	R-0.94
Insulating (winter) ¼-inch air space (U = 0.65)	R-1.54
Insulating (summer) ¼-inch air space (U = 0.61)	R-1.64
Insulating (winter) ½-inch air space (U = 0.58)	R-1.72
Insulating (summer) ½-inch air space (U = 0.56)	R-1.79
Storm window (winter) 1- to 4-inch air space (U = 0.56)	R-1.79
Storm window (summer) 1- to 4-inch air space (U = 0.54)	R-1.85
Glass block, 6 by 6 by 4 inches (winter) (U = 0.60)	R-1.67
Glass block, 6 by 6 by 4 inches (summer) (U = 0.57)	R-1.75
Glass block, 12 by 12 by 4 inches with cavity divider (winter) (U = 0.44)	R-2.27
Glass block, 12 by 12 by 4 inches with cavity divider (summer) (U = 0.42)	R-2.38
Plastic sheet, single (U = 1.09)	R-0.92

Insulation:

Loose fill fiber mineral	
glass fiber, per inch	R-2.3
rock wool, vermiculite, perlite, per inch	R-2.7
cellulose fiber, per inch	R-3.7
Blankets	
glass fiber, per inch	R-3.0
rock wool, per inch	R-3.6
Vermiculite	R-2.1
Fiberboard insulation, per inch	R-2.2 to 2.6
Polystyrene (rigid), per inch	R-3.5 to 5.0
Urethane (rigid), per inch	R-5.5 to 6.5
Urea formaldehyde foam, per inch	R-5.0
Ducting insulation—glass fiber duct tape, per inch	R-4.0

Air Spaces (¾ inch):

Heat flow, up	
Nonreflective	R-0.87
Reflective, one surface	R-2.23
Heat flow, down	
Nonreflective	R-1.02
Reflective, one surface	R-3.55

RESISTANCE VALUES OF STRUCTURAL MATERIALS, FINISH MATERIALS, GLASS, INSULATION, AIR SPACES, AND SURFACE FILMS

Heat flow, horizontal
  Nonreflective   R-1.01
  Reflective, one surface   R-3.48

NOTE
The addition of a second reflective surface facing the first reflective surface increases thermal resistance values of an air space only four to seven percent.

Surface Air Films:

Inside (still air)
  Heat flow, up (through horizontal surface)
    Nonreflective   R-0.61
    Reflective   R-1.32
  Heat flow, down (through horizontal surface)
    Nonreflective   R-0.92
    Reflective   R-4.55
  Heat flow, horizontal (through vertical surface)
    Nonreflective   R-0.68
Outside
  Heat flow any direction, surface any position
    15 mph wind (winter)   R-0.17
    7.5 mph wind (summer)   R-0.25

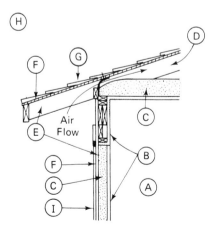

A. Still Air Inside
B. Plasterboard, Plaster, or Paneling
C. Insulation
D. Air Space Above Insulation
E. Sheathing
F. Building Felt
G. Shingles
H. Outside Air
I. Siding

FIGURE 7-4. Structural materials, finish materials, insulation, air spaces, and surface films provide resistance to the transfer of heat.

materials, air spaces, and surface films accounted for, the average uninsulated wall has a thermal resistance of R-3.7 to R-4.8. A sealed air space that has a moisture barrier provides some insulating effect.

Wood is a good natural insulator because of the countless number of tiny dead air spaces in the wood cell structure. The air spaces retard the passage of heat and cold. Thus, wood siding, paneling, and wood window frames provide additional resistance to the transfer of heat.

Resistance units are *additive*. If your unfinished attic floor already contains 6 inches of loose fill mineral fiber (you can measure its depth with a ruler) of let's say R-3 per inch, you already have R-18 insulation. To upgrade the resistance to heat transfer to R-29, you could add 4-inch batts of R-11 insulation. In trying to reach the recommended R value, get as close as possible; a little higher or a little lower won't matter that much.

7-4. HOW MUCH INSULATION DO YOU NEED?

The three most critical areas where insulation are required in the order of priority are ceilings, floors over unheated spaces, and walls. Most industry and government authorities agree that these areas require the following *minimum* R values:

	minimum
ceilings	R-19 to R-22
floors over unheated spaces	R-11 to R-13
walls	R-11 to R-13

Where the climates are extremely cold and/or the costs of energy are higher or air conditioning is used extensively, these resistance values should be extensively increased such as to R-30 ceilings where gas or oil are used for heating and R-38 for electrically heated homes. New homes being built to reduce the use of energy have R-38 ceilings and R-19 walls. Figure 7-5 presents recommendations for the amount of insulation required with today's high residential heating and cooling energy costs and projected increases in energy costs.

Each additional inch of insulation added in the attic saves energy and money, but the incremental savings become progressively smaller. You approach a point of diminishing returns; i.e., the more you add, the less effective the additional insulation becomes. Therefore, it is wise to put a little insulation *everywhere* before you put a lot of insulation anywhere.

To find the best combination of energy conservation measures for your climate and fuel prices, use Quick Reference Charts 7-3 through 7-8. This best combination gives you the largest, long-run net savings on your heating and

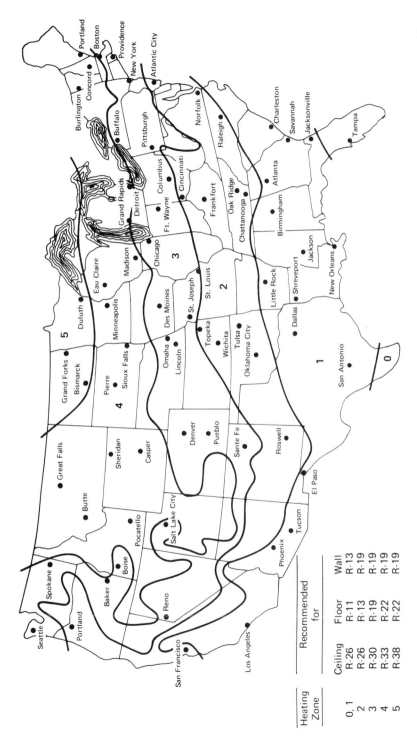

Heating Zone	Recommended for		
	Ceiling	Floor	Wall
0, 1	R-26	R-11	R-13
2	R-26	R-13	R-19
3	R-30	R-19	R-19
4	R-33	R-22	R-19
5	R-38	R-22	R-19

FIGURE 7-5. This map represents recommendations on the amount of insulation required with today's high residential heating and cooling energy costs and projected increases in energy.

Your needs depend on climate and existing insulation. Consult a reputable insulation dealer, local building inspector, or county agent for specific recommendations for your locality.

cooling costs for your investment. By comparing this best combination with what already exists in your house, you can figure out how much more needs to be added to bring your house to the recommended levels. The recommended improvements apply to most houses to the extent they can be installed without structurally modifying the house.

Follow these steps and fill in the information for your house on the worksheet. Information for a typical house located in Indianapolis, Ind. has been filled in as a sample.

1. Locate your city on the heating zone map in Fig. 7-6; write the heating zone number on line 1 of the worksheet.

2. Locate your city on the cooling zone map in Fig. 7-7; write the cooling zone letter on line 2 of the worksheet.

3. The sample house currently uses fuel oil at a cost of 34¢ a gallon to heat. It uses electricity at 4¢ a kilowatt hour to cool. Obtain your unit heating and cooling costs from the utility companies as follows: tell your company how many therms (for gas) or kilowatt hours (for electricity) you use in a typical winter month and summer month (if you have air conditioning). The number of therms or kilowatt hours is on your monthly fuel bill. Ask for the cost of the last therm or kilowatt hour used, including all taxes, surcharges, and fuel adjustments. For oil heating, the unit fuel cost is simply your average cost per gallon plus taxes, surcharges, and fuel adjustments.

4. Locate your *heating index* from QRC 7-3 by finding the number at the intersection of your heating zone row and heating fuel cost column (to the nearest cost shown). (The sample house has a heating index of 20.) If your house is air conditioned, or if you plan to add air conditioning, find your *cooling indexes* from steps 5 and 6. If your house is not air conditioned and it is not planned, your cooling indexes are zero.

NOTE

If your fuel costs shown in QRC 7-3 to 7-5 fall midway between two fuel costs listed, you can interpolate. For example, if your fuel oil costs were 38¢ a gallon, your heating index would be 22.5

5. Locate your cooling index for attics from QRC 7-4 by finding your cooling zone and cooling cost to the nearest cost shown. (The sample house has a cooling index for attics of 5.)

6. Locate your cooling index for walls from QRC 7-5 by finding your cooling zone and cooling cost to the nearest cost shown in the table. (The sample house has a cooling index for walls of 2.)

7. Find the sum of your heating index and cooling index for attics. (The sample house sum is 25.)

8. Find the sum of your heating index and cooling index for walls. (The sample house sum is 22.)

9. Find the resistance value (R) of insulation recommended for your attic and around attic ducts from QRC 7-6. (The recommended resistance value is R-30 for attic floors and R-16 for ducts in the sample house.)

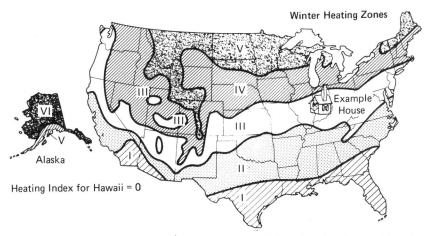

FIGURE 7-6. Locate your home and note the number of the winter heating zone (*courtesy U.S. Department of Commerce, National Bureau of Standards*).

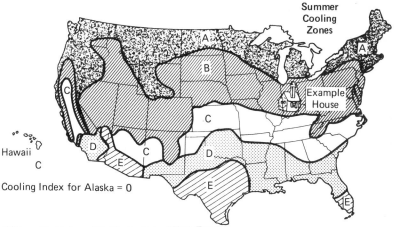

If Your House is on the Borderline of Two Zones,
Select the Zone in which the Climate is More
Typical of Your Area.

FIGURE 7-7. Locate your home and note the letter of the summer cooling zone (*courtesy U.S. Department of Commerce, National Bureau of Standards*).

10. Find the recommended level of insulation for floors over unheated areas from QRC 7-7. (The sample house is R-19.) Using QRC 7-7, check to see whether storm doors are economical for your home. Storm doors listed as optional may be economical if the doorway is heavily used during the heating season (refer to Chapter 8 on storm doors).

11. Find the recommended level of insulation for your walls and ducts in unheated areas from QRC 7-8. (The sample house should have full-wall insulation if none existed previously and R-16 insulation around ducts.) QRC 7-8 also shows the minimum economical storm window size in square feet for triple-track storm windows. The sample house should have storm windows on all windows 9 square feet in size or larger where storm windows can be used (refer to Chapter 9 on storm windows).

WORKSHEET

Example:		Your Calculations:
Climate:		*Climate:*
1. Heating zone	_III_	Heating zone
2. Cooling zone	_B_	Cooling zone
Fuel costs:		*Fuel costs:*
3. Heating energy	_oil_	Heating energy
4. Cost per unit	_34¢/gal_	Cost per unit
5. Cooling energy	_electric_	Cooling energy
6. Cost per unit	_4¢/KWH_	Cost per unit
Indexes:		*Indexes:*
7. Heating	_20_	Heating
8. Cooling (attic)	_5_	Cooling (attic)
9. Cooling (wall)	_2_	Cooling (wall)
10. Heating + cooling (attic)	_25_	Heating + cooling (attic)
11. Heating + cooling (wall)	_22_	Heating + cooling (wall)
Best Combination		*Best Combination*
12. Attic insulation (batt)	_R-30 (10 inches)_	Attic insulation
13. Duct insulation (in attics)	_R-16 (4 inches)_	Duct insulation (in attics)
14. Insulation under floors	_R-19 (6 inches)_	Insulation under floors
15. Storm doors	_optional_	Storm doors
16. Wall insulation (blown in)	_R-14 full-wall (3½ inches)_	Wall insulation (blown in)
17. Duct insulation (in unheated crawl spaces, etc.)	_R-16 (4 inches)_	Duct insulation (in unheated crawl spaces, etc.)
18. Storm windows (minimum size)	_9 sq. ft._	Storm windows (minimum size)
19. Weather-strip and caulk windows and door frames	_all_	Weather-strip and caulk windows and door frames

FROM TABLE 7-6 (items 12–13)

FROM TABLE 7-7 (items 14–16)

FROM TABLE 7-8 (items 17–18)

Quick Reference Chart 7-3
HEATING INDEX

Cost per unit*

Type of fuel:										
Gas (therm)	9¢	12¢	15¢	18¢	24¢	30¢	36¢	54¢	72¢	90¢
Oil (gallon)	13¢	17¢	21¢	25¢	34¢	42¢	50¢	75¢	$1.00	$1.25
Electric (kWh)				1¢	1.3¢	1.6¢	2¢	3¢	4¢	5¢
Heat pump (kWh)	0.9¢	1.1¢	1.5¢	1.8¢	2.3¢	2.9¢	3.5¢	5.3¢	7¢	8.8¢

H E A T I N G ZONE										
I	2	2	3	3	4	5	6	9	12	15
II	5	6	8	9	12	15	18	27	36	45
III	8	10	13	15	20	25	30	45	60	75
IV	11	14	18	21	28	35	42	63	84	105
V	14	18	23	27	36	45	54	81	108	135
VI	22	28	36	42	56	70	84	126	168	210

*Cost of last unit used (for heating and cooling purposes) including all taxes, surcharges, and fuel adjustments.
(Courtesy U.S. Department of Commerce, National Bureau of Standards)

Quick Reference Chart 7-4
COOLING INDEX FOR ATTICS

Type of air conditioner:					*Cost per unit**				
Gas (therm)			9¢	12¢	15¢	18¢	24¢	30¢	36¢
Electric (kWh)			1.5¢	2¢	2.5¢	3¢	4¢	5¢	6¢
C		A	0	0	0	0	0	0	0
O	Z	B	2	2	3	4	5	6	7
O	O	C	3	5	6	7	9	11	13
L	N	D	5	6	8	9	12	15	18
I	E	E	7	9	11	14	18	23	27
N									
G									

* Cost of last unit used (for heating and cooling purposes) including all taxes, surcharges, and fuel adjustments.
(Courtesy U.S. Department of Commerce, National Bureau of Standards)

Quick Reference Chart 7-5
COOLING INDEX FOR WALLS

Type of air conditioner:					*Cost per unit**				
Gas (therm)			9¢	12¢	15¢	18¢	24¢	30¢	36¢
Electric (kWh)			1.5¢	2¢	2.5¢	3¢	4¢	5¢	6¢
C		A	0	0	0	0	0	0	0
O		B	1	1	2	2	2	3	4
O	Z	C	2	2	3	4	5	6	7
L	O	D	3	3	4	5	7	8	10
I	N	E	4	5	6	8	10	13	15
N	E								
G									

* Cost of last unit used (for heating and cooling purposes) including all taxes, surcharges, and fuel adjustments.
(Courtesy U.S. Department of Commerce, National Bureau of Standards)

Quick Reference Chart 7-6
ATTIC FLOOR INSULATION AND ATTIC DUCT INSULATION

Index— Heating Index Plus Cooling Index for Attics	Attic Insulation				Duct Insulation*	
		Approximate Thickness				
	R-Value	Mineral Fiber Batt/Blanket	Mineral Fiber Loose-Fill**	Celluose Loose-Fill**	R-Value	Approximate Thickness
1-3	R-0	0"	0"	0"	R-8	2"
4-9	R-11	4"	4-6"	2-4"	R-8	2"
10-15	R-19	6"	8-10"	4-6"	R-8	2"
16-27	R-30***	10"	13-15"	7-9"	R-16	4"
28-35	R-33	11"	14-16"	8-10"	R-16	4"
36-45	R-38	12"	17-19"	9-11"	R-24	6"
46-60	R-44	14"	19-21"	11-13"	R-24	6"
61-85	R-49	16"	22-24"	12-14"	R-32	8"
86-105	R-57	18"	25-27"	14-16"	R-32	8"
106-130	R-60	19"	27-29"	15-17"	R-32	8"
131-	R-66	21"	29-31"	17-19"	R-40	10"

* Use Heating Index only if ducts are not used for air conditioning.
** High levels of loose-fill insulation may not be feasible in many attics.
*** Assumes that joists are covered; otherwise use R-22.
(Courtesy U.S. Department of Commerce, National Bureau of Standards)

Quick Reference Chart 7-7
INSULATION UNDER FLOORS AND STORM DOORS

Index– Heating Index Only	Insulation Under Floors* R Value	Mineral Fiber Batt Thickness	Storm Doors
0-7	0**	0"**	None
8-15	11**	4"**	None
16-30	19	6"	Optional
31-65	22	7"	Optional
66–	22	7"	On all doors

* If your furnace and hot water heater are located in an otherwise unheated basement, cut your Heating Index in half to find the level of floor insulation.
** In Zone I and II R-11 insulation is usually economical under floors over open crawl spaces and over garages; in Zone I insulation is not usually economical if crawl space is closed off.
(Courtesy U.S. Department of Commerce, National Bureau of Standards)

Figure 7-8 shows the recommended minimum thickness of styrofoam foundation insulation (**DOW STYROFOAM®**). Such insulation is installed during construction, but can be used as a guide to the homeowner interested in insulating basement walls with rigid foam. The studded foam-filled walls are covered with ½-inch or thicker fire retardant plasterboard.

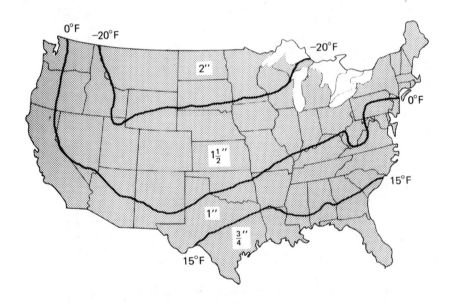

FIGURE 7-8. The minimum thickness of STYROFOAM® brand insulation recommended for foundation perimeter applications is shown (*courtesy of and © The Dow Chemical Company, 1976*).

Quick Reference Chart 7-8
WALL INSULATION, DUCT INSULATION, AND STORM WINDOWS

Index— Heating Index Plus Cooling Index for Walls	Wall Insulation (blown in)	Insulation Around Ducts In Crawlspaces and in Other Unheated Areas (Except Attics)* Resistance and Approximate Thickness	Storm Windows (Triple-Track) Minimum Economical Window Size
0-10	None	R- 8 (2")	None
11-12		R- 8 (2")	20 sq. ft.
13-15		R-16 (4")	15 sq. ft.
16-19	Full-	R-16 (4")	12 sq. ft.
20-28	Wall	R-16 (4")	9 sq. ft.
29-35	Insulation	R-16 (4")	6 sq. ft.
36-45	Approximately	R-24 (6")	4 sq. ft.
46-65	R-14	R-24 (6")	All windows**
66-		R-32 (8")	All windows**

* Use Heating Index only if ducts are not used for air conditioning.
** Windows too small for triple-track windows can be fitted with one-piece windows.
(Courtesy U.S. Department of Commerce, National Bureau of Standards)

You now know your best combination of energy conservation improvements. Of course, the size of your investment depends on your existing insulation and the size of your house.

Some of the recommended improvements in this book are not appropriate for all houses. For instance, insulation cannot be added under floors in houses built on concrete slabs. In such cases, the other recommended improvements should still be added to the extent indicated. Similarly, R-30 insulation may be recommended for your attic although only R-19 may fit at the eaves or in areas where the attic is floored. In this case, you should still put R-30 insulation wherever it fits.

Now you need to calculate the quantity of material you need to insulate. You need to know if you already have existing insulation in place, and if so, how many inches, what is the material, and what is the R value of the material already installed. If blankets are in place, you can read the R value on the backing. Otherwise, determine the material (you may need an insulating expert to look at it) and measure its thickness. Then determine the R value per inch of the material from QRC 7-2 and multiply it by the number of inches. For example, you inspect your unheated attic and find that there are four inches of loose fill rock wool between the joists. The R value of the installed insulation is:

$$4 \text{ inches} \times 2.7 = 10.8$$
$$R = 10.8$$

Let's say you would like to increase the insulation value to the minimum level—between R-19 and R-22. You have several choices (let's say the joists are 8 inches). You can add 4 inches more of loose fill material to bring the level to the top of the joists or you can add 4 inches in blankets. Investigate the materials available in your area and decide whether you'd prefer to add loose fill material or blankets. As far as cost is concerned, select insulating material on the basis of its *cost per resistance unit* (instead of cost per inch). Also consider its durability and its resistance to flame-spread and vermin. Most insulation materials are protected from these hazards.

In our previous example with an R-10.8, you'd get the following results with different materials:

Material	R per in.	Times 4 in. to fill joist	Plus original 10.4 equals total R value
Blanket rock wool	3.6	14.4	24.8
Blanket glass fiber	3.0	12.0	22.4
Loose fill, glass fiber	2.3	9.2	19.6
Loose fill, rock wool	2.7	10.8	21.6
Loose fill, cellulose fiber	3.7	14.8	25.6

Since some of the R value combinations exceed the R-19 to R-22 you desired, you can either use less material, say 3 inches, or be content with the higher thermal resistance which will save you additional energy dollars.

Another question that must be answered is how much insulation (how many rolls, bags, etc.) must be bought for the area to be insulated. The quantity can be roughly calculated by multiplying the length and width of each area to be covered; then add the areas together. Take this *square feet* of area measurement, along with the depth of insulation that you need, to your dealer who will be able to tell you how many rolls, batts, or bags you'll need. You should probably buy just a little bit more so that you'll be sure to have enough for the job; if you have some left over, you can usually use it some other place or you can return the unopened materials to the store.

You may find that the amount of money needed to finance the best combination of energy conservation improvements is more than you can afford all at once. If this is the case, consider taking a low-cost, long-term home improvement loan. Whether it is to your advantage to borrow money depends to some extent on the existing condition of your house. A house that is poorly insulated compared to the levels recommended in this book requires a greater investment in energy conservation than a house which is close to these recommended levels. However, the poorly insulated house will yield much greater savings on fuel bills after the improvements are made. This means that your investment will generally pay back fast enough to cover the monthly payments on a long-term home improvement loan. Once the loan is paid off, the additional savings are free and clear.

If you feel you just can't afford to invest in the best combination of energy conservation improvements for your house, you can still make the most of a limited energy conservation budget. Keep in mind the idea of a balanced combination: not spending too much on one improvement in relation to the other improvements.

To find this less costly, but still balanced combination of improvements, decrease each of the index numbers you used in QRC 7-6, 7-7, and 7-8 by the same percentage, say 20 percent. Use the new index numbers to find a new combination of improvements in these tables. Keep reducing your index numbers by the same percentage until you reach a balanced combination you can afford.

In the sample house, index numbers of 25 in QRC 7-6, 20 in QRC 7-7, and 22 in QRC 7-8 were used. Reducing these by 40 percent, for instance, gives us new index numbers of 15, 12, and 13, respectively. Using these numbers gives us the following balanced combination shown in the table on page 122.

Based on the sample level of insulation, this new combination would cost about 37 percent less for the best combination. If this is still more than can be afforded, the index numbers can be reduced by 50 percent or even more.

	Less Costly Combination	Best Combination
Attic insulation (batts)	R-19	R-30
Duct insulation (attics)	R-8	R-16
Floor insulation	R-11	R-19
Storm doors	none	optional
Wall insulation (blown in)	full-wall	full-wall
Duct insulation (other areas)	R-8	R-16
Storm windows (minimum size)	12 square feet	9 square feet

On the other hand, if you think that you will eventually add all of the recommended improvements in your best combination, but that it will take a year or two to get them all in, it is usually best to start with those which provide the first level of protection such as insulation in places where none exists. Add additional insulation as you can afford it.

7-5. VAPOR BARRIERS AND VENTILATION

Moisture is present in the air and is known as humidity. Humidity is at a high level (50 to 100 percent) in the summer (in many parts of the U.S.) and is lower in the winter (10 to 50 percent) particularly in hot-air heated homes. With increased moisture in the air in the winter, however, our bodies perspire less and we feel warmer. We sometimes purposely add moisture into the air (Chapter 9) in the winter to increase our comfort level; we can do this with a humidifier, by placing pans of water on radiators, and by having live plants in the home. Moisture is also added to the air from the steam of cooking in the kitchen and from the tiny water droplets created when showers are taken. It is desirable to keep some of this moisture in the living area to increase comfort (excessive amounts of moisture are not desirable however, because they cause windows and walls to sweat; therefore, exhaust fans are used in kitchens and bathrooms to draw out excessive moisture). In addition to keeping the moisture within for comfort, it is also desirable to keep the moisture from passing out of the warm living area where it can meet cold building materials causing the vapor (gas) to cool and condense into water. Water greatly reduces the effect of insulation and it causes wood to rot, metal to rust, plaster to crack, and paint to bubble and peel. Vapor barriers not only keep water vapor within the living space, but also are an extremely effective barrier against air infiltration. For these reasons, then, a *vapor barrier* is used next to the warm living space side of insulation to prevent moisture from passing through to the insulation itself and to the other building materials.

Some insulations, such as blankets, may have a vapor barrier attached and others, such as some rigid foams, have their own vapor barriers. However, for loose fill and other insulation without a vapor barrier, you will need to position a separate barrier. A barrier of 2-mil (0.002-inch) thickness (or greater) is used on walls, floors, and ceilings; a barrier of at least 4 mil (0.004 inch) is used for a ground cover under a slab of concrete or over the earth in a crawl space. A vapor barrier of polyethylene sheeting is more effective than a vapor barrier attached to the insulation because the sheeting covers the entire stud area and does not require that the insulation be compressed at the edges to permit the stapling of flanges to the studs. A *foil-backed plasterboard* may also be used instead of vapor-backed blankets or sheeting; the foil side of the plasterboard is faced against the studding.

Plastic faced panels and vinyl wallcoverings are not only decorative, but also are vapor barriers. Other vapor barrier faces on some insulating products are asphalt coated papers and plastic films.

Where insulation has been blown into an existing wall or ceiling, vapor-barrier protection is necessary. This can be provided by painting the interior walls and ceilings with two coats of a good vapor-resistant paint: paint with a low permeability—this can be a high gloss oil based enamel or other finish. Refer to your local paint store for a brand recommendation.

If you have closed cavities that you want to insulate, have an insulation expert advise you on how to insulate the cavity so that moisture will not be a problem. If water vapor (moisture) is allowed to condense in the closed, insulated cavity, it will cause the insulation to lose its effectiveness and it could damage the structure. Moisture problems can be eliminated with the proper installation of a vapor barrier, adequate ventilation, and the avoidance of high humidity settings.

No space should be enclosed by a vapor barrier on each side (inside and outside). Caution should be exercised when plywood exterior sheathing or foam sheathing that has a better vapor resistance than the interior vapor barrier is used. The exterior skin should be four to five times more vapor *porous* than the interior skin.

Anytime that a material is exposed to heat on one side and cold on the other, and water vapor is present, the vapor will condense forming water droplets. As previously mentioned, water can cause insulation to lose its effectiveness and it can damage structural materials. Therefore, a means must be available to prevent the condensation of vapor into water. This is accomplished by providing adequate ventilation so that air can circulate through the confined area to prevent condensation (and to rid the area of any condensation accumulation). The added ventilation also increases summer comfort by cooling off the area (this is especially important if the area is the attic).

Refer to Fig. 7-9; if you live in zone I and you have an insulated attic, install a vapor barrier (unless the insulation is blown into a finished attic). If

FIGURE 7-9. Locate your home in one of the zones and use this as a guide to vapor barrier and ventilation requirements (*courtesy Department of Housing and Urban Development*).

you can't install a vapor barrier or if the underside of the roof shows signs of condensation, add ventilation area equal to 1/300 of the attic floor level. This area should be divided evenly between ventilation louvers installed along the eaves and in the gables (louvers are discussed in Chapter 10). At least two unblocked ventilation ports (louvers) should be provided on opposite ends of the insulated area for cross ventilation.

If you live in zone II (Fig. 7-9) and you've insulated your attic and you *do not* have air conditioning, install a vapor barrier toward the living space if you'll be living in the finished attic. Add ventilation with a louver area of 1/300 of the attic floor area. If the house *is* air-conditioned, add ventilation with an area of 1/150 (instead of 1/300) of the attic floor area. At least two unblocked ventilation ports (louvers) should be provided on opposite ends of the insulated area for cross ventilation.

Crawl spaces need to be vented also. Even with a plastic vapor barrier on the ground, the crawl space becomes too damp unless adequate ventilation is provided. The dampness decreases the insulation effectiveness and causes structural damage.

If your crawl space is part of your hot-air furnace system, seal the crawl space as tightly as possible; the air moving through the space in the winter to the heating system will be sufficient to keep moisture out. You can close off the vents in the winter (check about once a month; if there is condensation, open the vents slightly), but keep them open in the summer.

All other crawl spaces should be vented; keep the vents open in the summer to vent out the dampness and closed tightly in the winter to make your newly added insulation effective. You can cover the vents in the winter to make them airtight (check about once a month to be sure no moisture is accumulating; if it is, open the vents slightly and let the space dry out).

Methods of adding or increasing ventilation by the use of louvers and fans are discussed in Chapter 10.

7-6. ATTIC INSULATION

The attic may be a nonliving space or a living space, depending upon your needs. If it is a nonliving space, you have the following options available to you on how insulation can be installed. (If you ever plan to use the attic for a living space, then insulate it as if it were a living space.)

1. If there *isn't any* flooring at all in the attic, you can
   a. Install blanket insulation between the floor joists (DIY 7-1).
   b. Install loose fill insulation between the joists (DIY 7-2).
   c. Supplement existing insulation between the joists with additional loose fill or blanket insulation (DIY 7-1, 7-2).
2. If there *is* flooring in the attic, you can
   a. Remove the flooring, perform any of the procedures in item 1, and then replace the flooring.
   b. Have loose fill insulation blown under the flooring (DIY 7-3).
   c. Install blanket insulation between the rafters and collar beams (this option should be used if you plan on finishing the attic later [DIY 7-1]). The gable end walls are also insulated between studs.
3. If the attic is finished off, you can
   a. Have a contractor blow insulation into the ceiling, sloping·rafters, gable end walls, and outer attic floors. Insulation blankets are installed in the knee walls (Section 7-1, DIY 7-1).
   b. Install blankets and/or loose fill in all attic spaces surrounding the living space that you can reach. You should be able to do the floor space outside of the living area (and over the ceiling area of the floor below) and the knee walls quite easily. From behind the knee walls, you may be able to slide blankets on top of the ceiling, up toward the peak of the roof between the rafters. Use a long pole to slide the blankets up (DIY 7-1, 7-2).
   c. Remove part of the finished ceiling and install loose fill or/and blanket insulation (DIY 7-1, 7-2).

7-7. EXTERIOR WALL INSULATION

Exterior walls are insulated in the following ways:
1. Have a contractor blow loose fill or foam insulation into the walls (Section 7-1).
2. Pour loose fill insulation from the attic into the walls between the studs. This may not be too effective because of voids in the insulation caused by braces, fire blocks, pipes, and electrical wires and boxes which snag the material and prevent it from falling into all cavities (DIY 7-2).
3. Have the exterior insulated when new siding is placed on the house.

When the old exterior is replaced or re-sided, rigid foam insulation can be installed. On new structures, rigid foam can be used instead of sheathing.

4. If interior walls are damaged (or you simply want to insulate), you can remove the old inside wall surface, add blanket or rigid foam insulation between the studs, and resurface the wall with plasterboard (required over rigid foam) or paneling. As an alternative, you can add studding or furring over the walls, add insulation blankets or rigid foam, and face the studs (DIY 7-1).

Having insulation blown (or foamed) into existing walls that are not currently insulated appears to be economical in many climates and in regions and homes where fuel prices are high. Wall insulation reduces heat transfer, causes a higher mean radiant temperature, and increases the comfort of the occupants. However, this technique of insulating must be carefully considered because of the possibility of moisture problems from water vapor (you can't add a vapor barrier) and because there may be voids in the insulation cavities because of the blockage of insulating material by structure braces, fire stops, electrical wires and pipes. Since the process must be done commercially, it may be too expensive for payback in a reasonable time.

In new construction, a really heavily insulated wall can be attained by double insulation. On the outside of the walls, styrofoam is used in place of sheathing. This is covered with siding. Full blankets of insulation with a vapor barrier are placed between the studding. Finally, drywall or paneling is attached to the studding.

7-8. "INTERIOR" WALL (NEXT TO UNHEATED AREAS) INSULATION

Close inspection of the interior walls of your home that join unheated areas such as garages and porches may reveal that the builder saved some money by not insulating the wall. If you're lucky, the studs are still open. Add insulation by one of the following methods:

1. If the studs are open, as in a garage, install blanket insulation with the vapor barrier toward the living (heated) area. The blankets "force fit" between the studs because the blankets are a little wider than the opening. You can place a couple of headed common nails at the top and bottom and at a couple of other strategic points along the insulation as necessary. Drive the nails in to slightly compress the material, then pull the insulation around the nail back out to "fluff" it again (DIY 7-1).

2. If the wall is closed, look for an opening at the top where you can pour in

loose fill insulation. If there is no opening, perhaps you can remove a top board to create an opening (DIY 7-2).

3. Have a contractor blow insulation into the wall (Section 7-1).

7-9. FLOOR INSULATION

Floors above unheated areas are insulated easily by installing blanket insulation and fastening it in one of two methods. The vapor barrier (and hence the mounting flange) is placed between the joists facing the living area (not the basement):

1. Wire, such as chicken wire, is stapled to the joists and the insulation blankets are placed on the wire (DIY 7-1).
2. The insulation blankets are held between the joists with pointed wires that act as a spring force against the blankets (DIY 7-1).

7-10. BASEMENT WALL INSULATION

You can easily insulate your basement walls on the interior; it's more difficult to insulate from the exterior because you have to dig up earth around the foundation (external insulation around the exterior foundation wall is usually a new construction procedure). Interior basement wall insulation should be applied at least two feet below the level of the earth fill against the foundation (frost usually reaches no deeper than two feet). However, if you can afford the slight additional cost, run the insulation from the ceiling to the floor of the basement. Insulate by one of the two methods below:

CAUTION
Residents of Alaska, Minnesota, northern Maine and other areas exposed to extreme frost penetration should contact local insulation contractors or local housing authorities regarding the procedure for insulating basements. Use of these methods could cause heaving of the foundation in extreme frost penetration conditions.

1. Stud the wall with wood to give sufficient depth for the insulation of the R value desired (2 by 3 inches is usually sufficient). Install blankets at least from the ceiling to 2 feet below the earth line, but preferably all the way to the floor with the vapor barrier toward the living area. Cover the studding with plasterboard or paneling (DIY 7-1).
2. Stud the wall with wood to allow insertion of the desired thickness of insulation. Install rigid foam (1-inch thickness is usually sufficient) between the studs. Face the studding with fire resistant plasterboard (DIY 7-4). (You can also glue rigid foam to the inside foundation wall first followed by studding on the outside of the foam.)

7-11. CRAWL SPACE INSULATION

Crawl spaces are insulated by running blankets from headers to the ground for 2 feet (DIY 7-1). A vapor barrier is placed on the ground and tucked under the insulation (DIY 7-5). Rough 2 by 4 inch lumber, old bricks, or stones are then placed on the blankets to hold the blankets in place. The crawl space must be adequately ventilated.

CAUTION

Residents of Alaska, Minnesota, northern Maine and other areas exposed to extreme frost penetration should contact local insulation contractors or local housing authorities regarding the procedure for insulating crawl spaces. Use of these methods could cause heaving of the foundation in extreme frost penetration conditions.

7-12. HOW TO INSTALL INSULATION IN CEILINGS, WALLS, FLOORS, AND EXTERIOR WALLS

This section describes *how* to install blanket, loose fill, and rigid insulation:

DIY 7-1 • How to Install Blanket Insulation

DIY 7-2 • How to Install Poured Loose Fill Insulation

DIY 7-3 • How to Install Blown-in Loose Fill Insulation

DIY 7-4 • How to Install Rigid Foam Insulation

Do-It-Yourself 7-1
HOW TO INSTALL BLANKET INSULATION

Materials: blankets of insulation

Tools: scissors, staple gun, boards (1 by 8 inches in lengths of 6 to 8 feet should be adequate) for placing across joists in unfinished attics to walk on when installing insulation

Procedure:

CAUTION

Place boards across the joists to walk on to support your weight while installing insulation in an unfloored attic. Do not place any weight down between the joists or you may partially fall through the ceiling of the floor below.

You can use blanket insulation in unfinished attic floors, roof rafters, walls, under floors, and around crawl spaces. Blankets are available in a number of different thicknesses and R values, and in rolls or 4- or 8-foot batts; blankets also come with or without an attached vapor barrier.

Do-It-Yourself 7-1 (continued)
HOW TO INSTALL BLANKET INSULATION

NOTE

In each of the following groups of procedures, read and understand *all* of the procedure before work is begun. Check for and repair any roof leaks before insulating (Section 3-4).

Unfinished Attic Floors

1. Lay blankets between the joists with the vapor barrier toward the living area (Fig. 7-10). Press the blankets to the bottom of the joists carefully so that the insulation is not compressed; the insulation is wide enough to make a snug fit between the joists.

2. Cut long runs first to conserve material; the remaining short pieces can be used as needed to fill the shorter places. Insert the blankets under electrical wires. Do not allow the insulation to block vents; but it should cover the wall top plate.

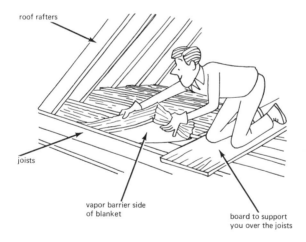

roof rafters

joists

vapor barrier side of blanket

board to support you over the joists

FIGURE 7-10. Install blankets in unfinished attic floors with the vapor barrier down (*courtesy National Mineral Wood Insulation Association, Inc.*).

3. Cut the runs to fit snugly around bracing (Fig. 7-11). Cut the adjacent piece to fit snugly against the first piece.

4. Do not allow blankets to come any closer than 3 inches to any lighting fixtures; this is a National Electrical Code requirement. Place a piece of metal around the fixture (3 inches from it) to keep the insulation away (Fig. 7-12).

5. The space between the chimney and the wood framing is to be filled with a noncombustible material (Fig. 7-13). Unfaced noncombustible blanket materials can be used (you can simply pull the face off blankets).

6. If you need more than one layer of blanket insulation to reach the required R

FIGURE 7-11. Cut the blankets to fit snugly at joist braces (*courtesy Department of Housing and Urban Development*).

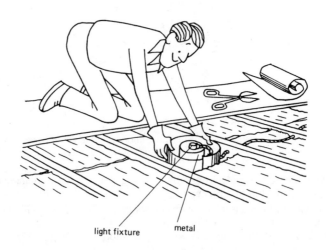

FIGURE 7-12. Place a piece of thin metal a distance of 3 inches from a light fixture to keep insulation away from the fixture (*courtesy National Mineral Wool Insulation Association, Inc.*).

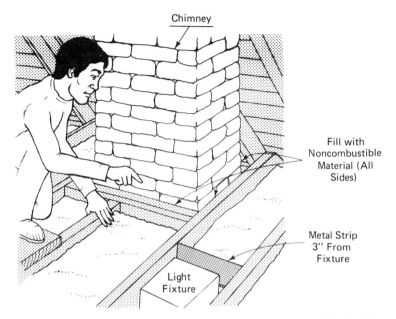

Chimney

Fill with
Noncombustible
Material (All
Sides)

Metal Strip
3″ From
Fixture

Light
Fixture

FIGURE 7-13. The space between the chimney and the wood framing is filled with a non-combustible material such as unfaced blankets (*courtesy Department of Housing and Urban Development*).

Do-It-Yourself 7-1 *(continued)*
HOW TO INSTALL BLANKET INSULATION

---

value, place it at *right angles to the joists* and press the blankets close together. The second layer should *not* have a vapor barrier. Buy insulation without a vapor barrier for the second layer. As an alternative, you may strip the barrier facing from the insulation or make long slashes in the barrier to allow any water vapor to pass through (Fig. 7-14).

Roof Rafters and Collar Beams

1. Blankets having an attached vapor barrier are easy to attach to the roof rafters and collar beams with a staple gun.
2. You need to provide ventilation above the insulation (between the insulation and the sheathing) and at the peak of the house all along the ridge. This will involve the addition of 2 by 4 inch collar beams between each set of roof rafters if they are not already installed. If necessary, cut lengths of 2 by 4 inch wood and nail them in place as collar beams (Fig. 7-15).

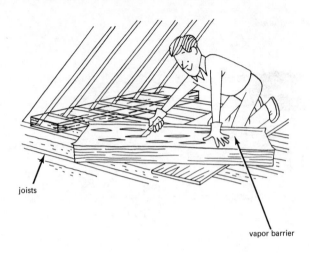

joists

vapor barrier

FIGURE 7-14. A second layer of blanket insulation is placed perpendicularly to the joists. The second layer should *not* have a vapor barrier; if it does, peel it off or slash the barrier to allow any vapor to pass through (*courtesy National Mineral Wool Insulation Association, Inc.*).

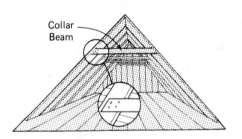

Collar Beam

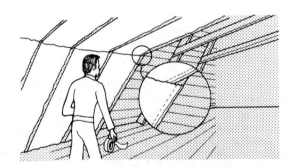

FIGURE 7-15. If the attic is to be lived in, insulation is attached to the roof rafters and collar beams. Install insulation along the rafters first followed by the collar beams (*courtesy Department of Housing and Urban Development*).

132

Do-It-Yourself 7-1 *(continued)*
HOW TO INSTALL BLANKET INSULATION

3. If no insulation already exists, place the insulation blanket between the rafters and the collar beams with the vapor barrier facing the living area. The flange on the vapor barrier is stapled to the edges of the rafters and collar beams (Fig. 7-16). This provides a continuous vapor barrier. If foil-faced insulation is used, the flanges are stapled to the sides of the rafters. Run along all of the rafters first followed by the collar beams (don't try to make a single run up one set of rafters across the collar beams and down the other side; this would make gaps that would be difficult to fill).

4. If insulation already exists on the collar beams, simply pile another layer on *top* of it running it in a direction perpendicular to the first layer. The new layer of insulation should *not* have a vapor barrier; if it does, pull it free from the insulation or slash the barrier with a knife.

5. If insulation already exists on the rafters, but is insufficient to fill the gap (the depth of the rafters minus 1 inch for air circulation), additional insulation can be added, or the existing inadequate material can be replaced. First slash the vapor barrier on the existing insulation. Then cut the old insulation loose and push it into the rafters (but be sure to allow at least 1 inch between the insulation and the roof for air circulation). Then install the new insulation with a vapor barrier as discussed in step 3.

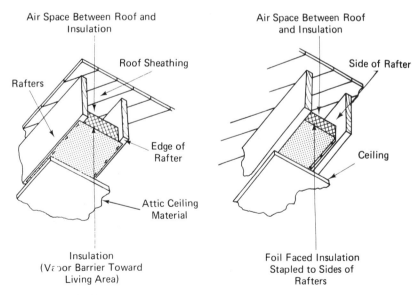

FIGURE 7-16. The flanges of blanket insulation are stapled to the edges of the rafters; if foil-faced insulation is used, the flanges are stapled to the sides of the joists.

Do-It-Yourself 7-1 (continued)
HOW TO INSTALL BLANKET INSULATION

Walls

1. New walls (or walls that are not finished such as in an unfinished room, garage, etc.) are easily insulated with blankets of insulation. You can use blankets with or without vapor barriers. If blankets without vapor barriers are used, one solid piece of plastic vapor barrier can be installed over the entire wall in one operation (Section 7-13, DIY 7-5) or you can use plasterboard that is faced with a vapor barrier. Studding may be 2 by 4 inches in which case R-11, 15-inch wide blankets can be used between studs 16 inches on center; on new construction, 2- by 6-inch studs may be used on 24-inch centers which allow you to use R-19 blankets in 23-inch widths. Blankets cannot be used to insulate existing *finished* walls unless the walls are torn down and rebuilt or unless additional studding and a new wall are added which then slightly decrease the living space.

2. To install blanket insulation to studded walls, simply start at the top between two studs with the vapor barrier facing the living area, lightly press the blanket into place, and cut it off at the bottom so that the complete cavity between two studs from ceiling to floor is filled. Staple the flanges to the studs so that a complete vapor barrier is formed along the wall (Fig. 7-17). If foil-faced blankets are used, the flange is stapled to the sides of the studs so that an air gap exists between the blanket and the wall facing (plasterboard or paneling) that will be attached.

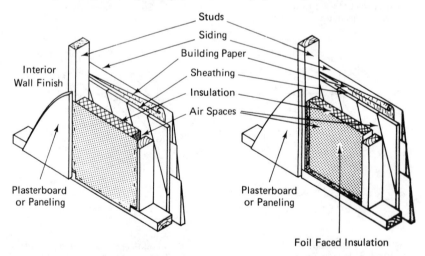

FIGURE 7-17. Blanket insulation is easily attached to studding with staples; the vapor barrier faces the living area. Foil-faced blankets are stapled to the sides of the studs so that an air space exists between the insulation and the finished walls (*courtesy National Forest Products Association*).

Do-It-Yourself 7-1 (continued)
HOW TO INSTALL BLANKET INSULATION

3. Close all nooks and crevices. Stuff small pieces of insulation into spaces around windows. Wear gloves to keep fiberglass particles from causing skin irritations (Fig. 7-18).
4. Add paneling or plasterboard to the wall to finish it off.

pieces of insulation
stuffed into place

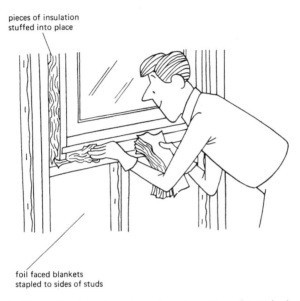

foil faced blankets
stapled to sides of studs

FIGURE 7-18. Stuff insulation into all nooks and crevices. Cover this with plastic strips as a vapor barrier (*courtesy National Mineral Wool Insulation Association, Inc.*).

CAUTION
Residents of Alaska, Minnesota, northern Maine and other areas exposed to extreme frost penetration should contact local insulation contractors or local housing authorities regarding the procedure for insulating basements. Use of the methods in step 5 could cause heaving of the foundation in extreme frost penetration conditions.

5. Studding and blanket insulation are easily installed to basement walls. Eliminate all dampness coming through the walls first (Section 3-5). Install top and bottom plates and studs (on 16- or 24-inch centers). Then install blanket insulation as described in steps 1 to 3; the insulation should extend down to at least 2 feet below the earth line, but extending the blankets to the floor is the most effective means of insulating. Cover the studs with plasterboard or paneling (Fig. 7-19).

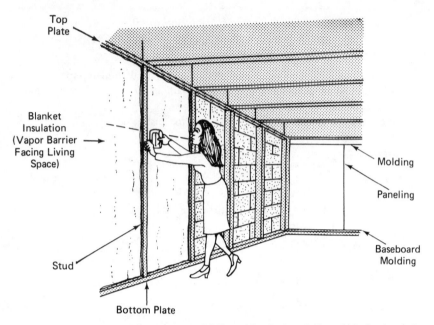

FIGURE 7-19. Basement walls are studded, followed by the installation of blanket insulation. Plasterboard or paneling can be used to finish the wall.

Under Floors

1. Areas such as unheated porches, unheated garages, or unheated basements which are under floors that are living areas need to be insulated. This is an easy do-it-yourself task using blanket insulation fit into the floor joists; a foil-faced vapor barrier is best because an air gap will be left. There is one slight problem: the vapor barrier and hence the mounting flange must be installed (as always) with the vapor barrier toward the living area, and this means that the barrier is up inside the joists and therefore there is no flange to staple to the joists. There are two solutions: you can either put wire across the joists and slide the blankets onto it and into place, or you can secure it in place with pointed wires that are made for this purpose (Fig. 7-20).

2. If you use wire, start at a wall at one end and staple the wire to the joists. Using short sections of blanket insulation, lay the blankets with the vapor barrier facing the living space on the wire between the joists and carefully work the blankets into place. Plan sections to begin and end at obstructions such as cross bracing; cut pieces to fit the cross bracing so that continuous insulating is accomplished (similar to Fig. 7-11 except upside down). Fold and fit the ends of the blankets securely against the floor as shown in Fig. 7-21.

3. If pointed wires are used, carefully press the blankets into the joists until the vapor barrier meets the floor; do not compress the insulation. Position pointed wires to hold the insulation in place.

4. Be sure to vent crawl spaces (Section 7-5).

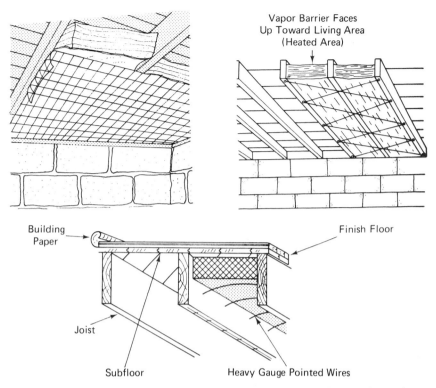

Vapor Barrier Faces
Up Toward Living Area
(Heated Area)

Building Paper

Finish Floor

Joist

Subfloor

Heavy Gauge Pointed Wires

FIGURE 7-20. When insulation is installed under floors, install it with the vapor barrier facing up to the living area. Three alternate methods of holding the insulation in place are shown.

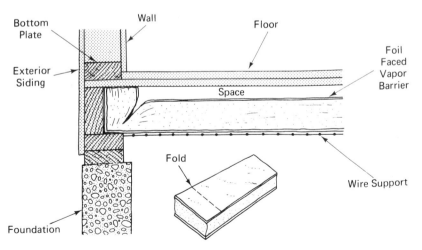

Bottom Plate

Wall

Floor

Foil Faced Vapor Barrier

Exterior Siding

Space

Fold

Wire Support

Foundation

FIGURE 7-21. Foil-faced insulation aids the air space in insulating. Be sure the ends of the insulation fit snugly against the bottom of the floor to prevent loss of heat (*courtesy Department of Housing and Urban Development*).

Do-It-Yourself 7-1 (continued)
HOW TO INSTALL BLANKET INSULATION

Crawl Spaces

CAUTION

Residents of Alaska, Minnesota, northern Maine and other areas exposed to extreme frost penetration should contact local insulation contractors or local housing authorities regarding the procedure for insulating crawl spaces. Use of these methods could cause heaving of the foundation in extreme frost penetration conditions.

1. Crawl spaces are insulated by placing blanket insulation from the headers and band joists down the walls to the ground and then extending inward toward the center of the crawl space for 2 feet. A plastic vapor barrier covers the ground.
2. On the walls that run parallel to the joists (Fig. 7-22), press the blankets against the subfloor and the band joists with the vapor barrier facing the inside of the crawl space. Nail the blankets to the band joist using ½- by 1½-inch wooden nailing strips. Drop the blankets down the side wall, fold them, and extend them 2 feet toward the center of the crawl space; cut them off.
3. On the walls that run perpendicular to the joists, first press short pieces of blanket insulation against the headers (Fig. 7-23); they should fit snugly. Then run blankets from the sill down the wall, fold, and extend outward two feet; cut them off. Hold the blankets in place against the sill with ½- by 1½-inch wooden nailing strips. Be sure to butt the blankets closely together.
4. Lay down a plastic vapor barrier (DIY 7-5).

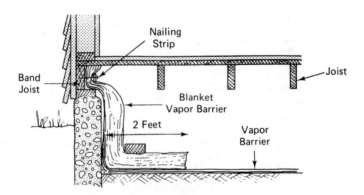

FIGURE 7-22. On the walls of crawl spaces parallel to the joists use nailing strips to attach blanket insulation to the band joists (*courtesy Department of Housing and Urban Development*).

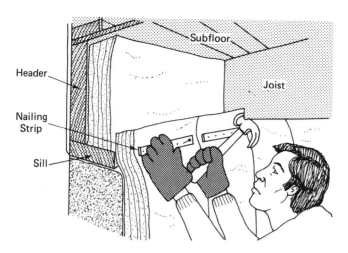

Header

Nailing
Strip

Sill

Subfloor

Joist

FIGURE 7-23. On the walls of crawl spaces perpendicular to the joists, install short pieces of blanket insulation between the joists against the header. Attach blankets to the sill with nailing strips (*courtesy Department of Housing and Urban Development*).

Do-It-Yourself 7-2
HOW TO INSTALL POURED LOOSE FILL INSULATION

Materials: bags of loose fill insulation, vapor barrier (as required), cardboard for baffles (as required)

Tools: boards (1 by 8 inches in lengths of 6 to 8 feet should be adequate) for placing across joists in unfinished attics to walk on when pouring insulation, board used to level the insulation, rake to fluff insulation, staple gun for fastening vapor barrier and baffles

Procedure:

CAUTION

Place boards across the joists to walk on to support your weight while installing vapor barriers and insulation in an unfloored attic. Do not place any weight down between the joists or you may partially fall through the ceiling of the floor below.

You can use loose fill insulation as an easy do-it-yourself task to insulate your unfinished unfloored attic or exterior walls. Check for and repair any roof leaks before insulating (Section 3-4). Proceed as follows:

Unfinished Unfloored Attic

1. Install a separate vapor barrier, if needed (DIY 7-5).
2. Cut and place cardboard baffles (staple into place, if required) to prevent the insulation from blocking eave ventilation (Fig. 7-24) and exhaust fans. Place

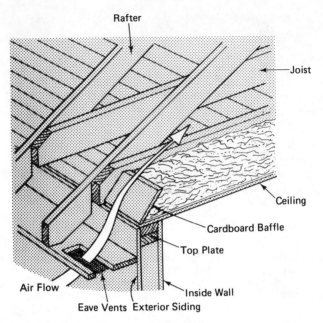

Rafter

Joist

Ceiling

Cardboard Baffle

Top Plate

Air Flow

Inside Wall

Eave Vents Exterior Siding

FIGURE 7-24. Place cardboard baffles at the ends of joists to prevent insulation from falling into vents.

Do-It-Yourself 7-2 (continued)
HOW TO INSTALL POURED LOOSE FILL INSULATION

---

metal baffles 3 inches from electrical fixtures to protect insulation from excessive heat (Fig. 7-12).

3. Pour loose fill insulation between the joists (Fig. 7-25). Be sure to fill around bracing, wiring, and so on.

4. Use a rake to fluff and level the insulation to the depth desired (Fig. 7-26). If you are filling to the top of the joists, you can use a board to level the insulation even with the top of the joists.

Walls of Existing Homes

1. No vapor barrier can be installed in existing finished walls. Consult an insulation contractor in your area regarding the possible damage that could occur from water vapor if loose fill insulation is installed. In some areas, you can apply two coats of a nonpermeable paint to the ceilings and walls to act as a vapor barrier.

2. Slowly and carefully pour loose insulation into the walls from the attic. If necessary, break up large matted lumps of insulation so that it falls down in small pieces between the studs. Fill to the top. Inspect the level of insulation after six months; if the insulation level has decreased from settling, add more to bring the level to the top again.

FIGURE 7-25. Pour loose fill insulation between the joists. Fill all nooks and crevices but do not cover light fixtures, exhaust fans, or vents (*courtesy National Mineral Wool Insulation Association, Inc.*).

FIGURE 7-26. Spread the loose fill insulation with a rake to fluff and level it.

Do-It-Yourself 7-3
HOW TO INSTALL BLOWN-IN LOOSE FILL INSULATION

Materials: loose fill insulation

Tools: pneumatic machine (contractor supplied), power drill and bits, hammer, nails

NOTE

This procedure is usually performed by a contractor; you can sometimes rent a pneumatic machine for doing the work yourself. In this case, follow the specific

---

Do-It-Yourself 7-3 (continued)
HOW TO INSTALL BLOWN-IN LOOSE FILL INSULATION

---

instructions provided with the machine and loose fill material. If a contractor does the job, you can sometimes do part of the work and save some money.

Procedure:

Loose fill insulation is applied under an unfinished (unheated) floored attic as follows (check for and repair roof leaks before insulating [Section 3-4]):

1. Insulation is placed under floored attics by blowing insulation in on both sides of floor supports. Bags of loose fill insulation are poured into a pneumatic machine that mixes the insulation nodules with air to fluff them and force them through a long flexible tube into the cavities between the joists.

2. The insulation is blown in at the truss locations between the joists. You can locate these and remove the floor boards above the trusses; then replace the boards after the insulation job is complete. By opening and closing the floor yourself, you may be able to save some contractor fees.

3. With the floor boards above the trusses open, the contractor inserts the hose on either side of each truss and blows the insulation in to completely fill the voids between the trusses (and at the trusses).

4. The contractor should use a predetermined number of bags of loose fill insulation plus or minus five bags. The number of bags should be estimated by you and the contractor to give you the insulating value for the level between the joists before an agreement is signed (and sign an agreement only with a reputable contractor).

5. After the job is finished, drill ¼-inch-diameter holes in the floor about 1 foot apart to help prevent condensation from collecting under the floor in winter.

Do-It-Yourself 7-4
HOW TO INSTALL RIGID FOAM INSULATION

Materials: rigid foam (determine the R value needed and purchase the thickness required), mastic, framing materials
Tools: utility knife

Procedure:

NOTE

Polystyrene and urethane rigid-board insulation must be covered with ½-inch plasterboard to assure fire safety. Extruded polystyrene and urethane are their own vapor barriers; bead board and glass fiber are not and therefore require the attachment of a vapor barrier.

Do-It-Yourself 7-4 (continued)
HOW TO INSTALL RIGID FOAM INSULATION

CAUTION
Residents of Alaska, Minnesota, northern Maine and other areas exposed to extreme frost penetration should contact local insulation contractors or local housing authorities regarding the procedure for insulating basement walls. Use of these methods could cause heaving of the foundation in extreme frost penetration conditions.

1. Because rigid foam is not flexible, it is best used for new construction. It is being utilized in place of sheathing, on the outside of foundations, partially under slabs, and on interior basement walls. For the homeowner, the most frequent use is in the insulation of basement walls.
2. Frame the basement, as required, for the eventual finishing off of the basement. The size of the studding (1 by 2 inches, 2 by 3 inches, 2 by 4 inches, etc.) is determined by the thickness of the foam to be placed between the studs. (Rigid foam can also be applied to the wall with a mastic prior to studding; the studding is then put in place over the foam.)
3. Insert the rigid foam and adhere it to the wall with a mastic compatible to both the foam and wall material (Fig. 7-27). Install a vapor barrier, if required.
4. Face the insulation and studding with at least ½-inch plasterboard to assure fire safety. Finish the wall.

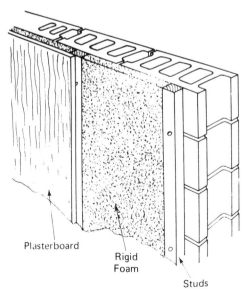

Plasterboard

Rigid Foam

Studs

FIGURE 7-27. Rigid foam is used to insulate basement walls. It is applied with a mastic between studs. If necessary, a vapor barrier is added. Plasterboard (1/2 inch or greater) covers the foam and the studs.

7-13. INSTALLATION OF VAPOR BARRIERS

When a vapor barrier is required (Section 7-5), but is not a part of the insulation, it must be installed separately. Vapor barriers are installed along studded walls, between joists, and on the ground in crawl spaces (DIY 7-5).

Do-It-Yourself 7-5
HOW TO INSTALL VAPOR BARRIERS

Materials: plastic large enough to cover the area. Use 2 mil (0.002 inch) or heavier for walls and joists; use 4 mil (0.004 inch) or heavier for ground cover of crawl spaces

Tools: stapling gun

Procedure:

Vapor barriers, if not attached to the insulation, are installed separately, as required, on walls, in joists, and on the ground in crawl spaces.

Installing Vapor Barriers to a Wall

1. After the studding and blanket insulation are in place (Fig. 7-28), cut a plastic vapor barrier of sufficient size to cover the entire wall including doors and windows (cover them over initially). Where walls join, overlap the plastic to the first stud on the adjacent wall.

FIGURE 7-28. If the insulation blankets do not have a vapor barrier, staple a sheet of 2-mil plastic over the insulation and studding. Cover with plasterboard or paneling (*courtesy National Mineral Wool Insulation Association, Inc.*).

Do-It-Yourself 7-5 (continued)
HOW TO INSTALL VAPOR BARRIERS

---

2. Fold the top, bottom, and end edges over one or two times for added strength (essentially, you are making a "hem"). Staple the barrier to the top and bottom plates, the end studs, and along some of the other studs.
3. Cut out for the doors and windows, but cut the plastic in such a manner that the edges can be folded under for added strength.
4. Staple around the doors and windows.

Installing Vapor Barrier in Floor Joists

1. Prior to placing loose fill insulation or unfaced blanket insulation between joists, install a vapor barrier, if required.
2. Lay plastic vapor barrier strips between the joists; let it wrap up the sides of the joists slightly.
3. Staple or tack the vapor barrier into place (Fig. 7-29).
4. Overlap seams 1 inch and tape them; no tape is required if the seams are overlapped 6 inches.
5. Be sure to keep the vapor barrier within 3 inches of electrical lighting fixtures and do not block vents.

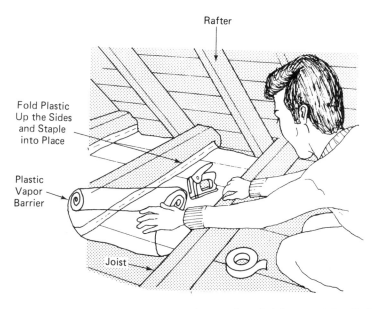

Rafter

Fold Plastic
Up the Sides
and Staple
into Place

Plastic
Vapor
Barrier

Joist

FIGURE 7-29. Install a separate plastic vapor barrier, if needed, between joists. Seal seams with tape or overlap 6 inches (*courtesy Department of Housing and Urban Development*).

Do-It-Yourself 7-5 (continued)
HOW TO INSTALL VAPOR BARRIERS

Installing a Vapor Barrier in Crawl Spaces

1. In crawl spaces, lay a polyethylene vapor barrier (at least 4-mil thickness) on the earth (Fig. 7-22). Cut the barrier to fit completely under the insulation all the way to the foundation and let it roll up the foundation a couple of inches. Do not cover over any vents.
2. Use a one-piece barrier if possible; if not possible, overlap the seams 6 inches and seal with masking tape.
3. With the insulation carefully placed on the barrier, place some old boards, rocks, or bricks along the edges of the insulation to keep the insulation and the barrier in place.

7-14. INSULATING HEATING AND AIR-CONDITIONING DUCTS

Heating ducts in unheated areas and air-conditioning ducts in uncooled areas such as cellars, attics, crawl spaces, and garages must be insulated to save you money spent on lost energy (DIY 7-6). If at all possible, relocate the ducts into the living area; at any rate, if you install additional ducts, be sure to locate heating ducts in heated areas and air-conditioning ducts in cooled areas.

Heating and air conditioning ducts should be wrapped with at least 2 inches of insulation. Duct insulation is usually sold in 1 or 2 inch blankets, but as an alternative, use the same blankets as used for wall, floor, or ceiling insulation. No vapor barrier is needed for heat-duct insulation; insulation with a vapor barrier is required on air-conditioning ducts to prevent condensation. The vapor barrier is located on the *outside* of the insulation (away from the duct). If the same ducts are used for *both* heating and air conditioning, use a vapor barrier insulation with duct tape to prevent condensation.

Do-It-Yourself 7-6
INSTALLATION OF INSULATION ON HEATING AND AIR-CONDITIONING DUCTS IN UNHEATED OR UNCOOLED AREAS

Materials: duct tape (to seal leaks in ducts at joints and to seal insulation joints), duct insulation at least 2 inches thick (vapor barrier needed for air-conditioning duct insulation)

Tools: scissors, ruler

Do-It-Yourself 7-6 (continued)
INSTALLATION OF INSULATION ON HEATING AND AIR-CONDITIONING DUCTS IN UNHEATED OR UNCOOLED AREAS

Procedure:

1. With air circulating through the ducts, check each duct joint for leaks by slowly moving your hands around the joints feeling for drafts. Where drafts are located, seal the joints with duct tape.
2. Wrap insulation blankets around the ducts but do not crush the insulation fibers as this lowers its resistance to the flow of heat (Fig. 7-30). On air-conditioning ducts or combined heating and air-conditioning ducts, the vapor barrier is on the outside (away from the ducts).
3. Butt adjacent blankets of insulation together and tape joints together with duct tape.

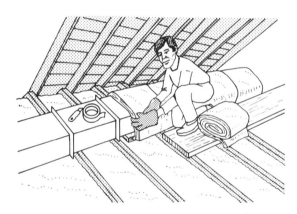

FIGURE 7-30. Air ducts in unheated or uncooled areas are insulated with blanket insulation. No vapor barrier is needed unless the ducts are used for air conditioning. Seal joints between blankets with tape (*courtesy Department of Housing and Urban Development*).

7-15. INSULATING HOT WATER HEATERS

Your water heater should be wrapped with insulation if it is in an unheated area or if its exterior feels hot to your hand. Installation of the vinyl-faced fiberglass insulation manufactured for such a job is an easy do-it-yourself task. You may also need to insulate the hot water pipes from the heater (Section 7-16) and if you haven't already done so, turn the heater thermostat down to 120 degrees F (or 140 degrees F if you have a dishwasher).

When you apply the insulation to the water heater, be sure not to block air vents. The top of the tank should be left uncovered. Butt the insulation together and secure it with masking tape. Do not allow the insulation to get near the lower part of a gas hot water heater near the flame.

One other procedure that you should do monthly is to drain out a bucket full of hot water from the base of the heater. This removes sediment from the bottom which acts to insulate the water from the heat during the heating cycle.

7-16. INSULATING WATER PIPES

Pipes for steam heat, hot-water heat, and hot-water service located in unheated areas need to be insulated to prevent heat losses and to prevent condensation or sweating of cold water pipes. Insulation also helps protect pipes from freezing if the unheated area is exposed to the elements. Pipe insulation also reduces or eliminates the noise of expanding pipes.

There are three types of pipe insulation: asbestos sleeves, vinyl-covered polyurethane sleeves, and tape (Fig. 7-31). The asbestos insulation takes the form of honeycombed sleeves that are secured with metal band clamps at the joints (DIY 7-7).

Vinyl-covered polyurethane pipe insulation consists of ¼-inch polyurethane with a 6-mil vinyl jacket. The flame-retardant insulation comes in 4-foot lengths having various inside diameters from ⅜ inch to 2⅛ inch. The insulation is preslit for rapid do-it-yourself installation consisting of snapping the insulation into place followed by a simple twist. No fittings are needed, although bands of masking tape may be used to keep the covers firmly in place (DIY 7-7).

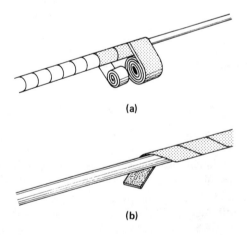

(a)

(b)

FIGURE 7-31. Pipe insulation tape may consist of (a) a roll of fiberglass with a roll of plastic that secures the fiberglass or (b) a combined vinyl foam with aluminum foil and a self-adhesive backing.

Do-It-Yourself 7-7
INSTALLATION OF PIPE INSULATION

Materials: insulation, metal straps (for asbestos insulation), masking tape (for vinyl-covered polyurethane insulation)
Tools: scissors, utility knife (for asbestos insulation)

Procedure:

Vinyl-Covered Polyurethane Pipe Insulation

1. For straight sections, open along the preslit, slip the insulation over the pipe, snap the insulation back together, and give a gentle twist (Fig. 7-32). Cut to the required length with a pair of scissors.
2. Where necessary, secure the insulation in place by a band of masking tape.
3. On wide turns or bends, apply slightly more twist.
4. At an elbow (a 90-degree turn), cut the insulation at a 45-degree angle (called a bias cut).
5. At a T- or a Y-shaped pipe joint, make a V-shaped cut (called a saddle cut).
6. At a valve or connector, first roughly fit the insulation into place with the slit around the valve or connector. Mark the location of a hole for the valve or connector and then cut the hole with scissors. Reinstall the insulation.

Asbestos Insulation

1. Place the insulation around the pipe. If necessary, cut to length and cut notches to fit around valves, connectors, T joints, and Y joints with a knife.
2. Secure adjacent sections of insulation together with metal bands.

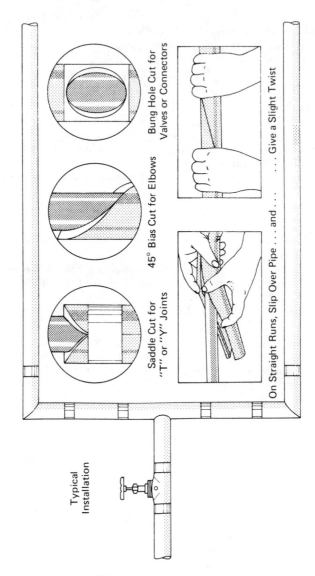

FIGURE 7-32. Vinyl-covered polyurethane pipe insulation is easy to install. It prevents hot water pipes from losing heat to unheated areas and prevents condensation on cold water pipes (*courtesy Teledyne Mono-Thane*).

chapter 8

WEATHERIZING WINDOWS AND DOORS

If you've performed the weatherizing procedures as sequentially presented in this book, you've already made significant progress toward weatherizing your windows and doors. First, you replaced broken and cracked glass and re-caulked where necessary around the glass panes to block off excessive heat transfer through openings. Second, you caulked around the window and door frames, and third, you weather-stripped along the sides, tops, and bottoms of your windows and doors. Thus, you've blocked all major paths of heat escapement in the winter and heat infiltration in the summer.

This chapter deals with the techniques of weatherizing your windows and doors. Some of the techniques are simple and the cost of materials is inexpensive. Other techniques take some work effort on your part and the results are more permanent; some techniques and materials will add to the decor of the room; others will not be so decorative, but you'll save even more energy and you'll feel warmer. Many of the techniques of weatherizing your windows and doors as discussed in this chapter can be utilized whether or not you have storm equipment in place; if storm equipment is in place, you'll block the transfer of heat even more effectively.

A lot of energy can be saved both in winter and summer (if your house is air conditioned) by using common sense. Keep all windows and doors (including storm windows and doors, if so equipped) tightly closed and latched; this is especially important at night and during periods of the day when no one is coming in and out of the doors. If a window or door does not latch tightly, move the latch plate. To do this, you'll need to remove the screws and fill the old screw holes with a wood filler. In the case of doors, you may also need to remove a little wood from the frame; use a sharp wood chisel. After the filler is hard, drill new holes and install the latch plate at the proper location.

Use the doors to the outside as infrequently as possible. This is especially important when it is windy. Winter winds which are predominately from the northeast cause a pressure difference on opposite sides of the house because of the air turbulence. For example, with a northeast wind blowing against the side of a house facing northeast, the northeast side pressure is slightly higher than the pressure on the opposite southwest side. The wind hits the northeast side and "funnels" around the sides and then beyond and behind the southwest side causing a slight decrease in pressure on the southwest side (this is the same effect as the slipstream formed behind a moving automobile). When a house door on the northeast side is opened, the wind pressure quickly blows into the house trying to equalize the pressure at the other side of the house. Opening the door on the southwest side of the house also causes the pressure to try to equalize; however, the effect is somewhat less. You can prove this to yourself by standing in the doorway of the windward side of the house and then on the leeward side of the house (not too long though, since you're letting heat escape very rapidly).

When using doors to the outside, open and close the doors quickly. If you have children, it may be difficult to get this message across to them; but make them understand that they are saving energy. If possible, close off doors on the north side followed by those on the east and west; use only southern or southeastern exposed doors to most effectively reduce heat losses. Do *not*, however, nail doors shut because this would prevent escapement through the door in case of fire; just make the use of the doors "off limits."

This chapter deals with the weatherizing of your windows and doors, inside as well as outside. Since the thermal resistance of glass is R-0.88 as compared to the thermal resistance of an *uninsulated* wall, which is about R-4.0, and since the thermal resistance of a two-inch wood door is only about R-2.0, you can see that you need to find ways to reduce thermal losses at each window and door. Information on the following areas is covered in this chapter: transfer of heat through windows and doors; indoor window accessories (decorative as well as weatherizing—shades, draperies, venetian blinds, cornices and lambricans, insulating panels, and window film), removing, replacing, and installing new windows; outdoor window treatments (overhangs, awnings, shutters, storm windows, screens, and trees); skylights, doors, storm doors, and vestibules.

8-1. TRANSFER OF HEAT THROUGH WINDOWS AND DOORS

In the winter, heat losses through windows are three to four times as great as heat losses through walls. In the summer, the total heat entering through a sunlit window may be more than 10 times as great as through an adjacent wall segment of the same area. Energy used to heat and cool homes is lost in three ways via windows and doors: *radiation, direct conduction,* and *air leaks/drafts* (infiltration). Radiation is the direct heat of the sun coming through the window. Solar collectors collect this radiant heat and use it to heat homes. In the winter, radiant heat is desirable and we encourage its entrance into our homes where we try to absorb and retain it. In the summer, however, we don't want the radiant heat and we find methods to block and reflect it away from the inside of our homes.

Indoor heated air in the winter and air-conditioned air in the summer are lost through windows and doors by *direct conduction*. All materials are conductors of heat to some degree. Your silver teaspoon conducts heat very rapidly (k=244) as is attested by leaving it momentarily in a hot cup of tea and then touching the end which has become hot. Wood (most woods have a k value of between 0.065 and 0.125) and plastic are poor conductors of heat and are therefore used as pot handles. Aluminum is a fairly good conductor having a k value of 122. You can conclude from this discussion that window frames and doors made of wood transfer less heat than ones made of aluminum. Some modern aluminum window sashes now contain *thermal breaks* to reduce heat transfer and condensation. As long as there is no metal-to-metal contact from the outside of the sash to the inside, the thermal break designs are effective.

Unnecessary energy is also expended to heat air (winter) and cool air (summer) that escapes via air leaks and drafts around window and door frames, between the glass and the frame, and through cracked or broken panes. These leaks and drafts are prevented by caulking, weather stripping, and replacement of glass (Chapters 5 and 6 and Section 3-1, respectively). Air also leaks around windows and doors simply because the latches are not secured or do not close the window or door tightly when secured. Adjust the latches as required. This adjustment may be as simple as loosening two screws and sliding the latch or it may necessitate removing screws, filling the old screw holes with wood filler, and relocating the latch with new screw holes. For an effective weather seal, latches must hold windows and doors securely shut. You can also add a hook and eye to a storm door and frame to secure it tightly.

The insulation value of a window is determined by the infiltrated heat loss and by the conducted heat loss. As mentioned previously, the infiltrated air is the air that can leak in or out between the sash and frame or between the frame and the wall. Conducted heat losses are losses through the glass, sash, and frame.

Glass is one of the best transparent materials for use in accepting solar radiation. Glass allows the short wavelengths of solar radiation to pass through it, but it prohibits nearly all of the long waves of reflected radiation from passing back through it. This is the principle upon which insulating glass and storm windows are based as a means of preventing heat loss by radiation. The sun's rays pass through the outer and inner pieces of glass; some of the rays that strike the glass are broken up into long waves of radiation. The rays broken up by the second piece of glass (the inner glass) into long waves reflect back to the first piece of glass and try to pass to the outside, but since glass prohibits the passing of most long waves of radiation, the long waves are trapped.

Before further discussion on windows and doors, let's digress a moment and define *insulating glass, glazing,* and *single, double,* and *triple glazing.* Insulating glass is two sheets of glass that are bonded together in such a manner that they enclose a captive air space. The glass edges are melted together leaving an air space between the two pieces of glass. Modern insulating glass has the air space filled with an inert gas which increases the effectiveness of the insulating quality of the window. Organic units have a metal spacer around the perimeter and the edges are sealed with an organic substance; the spaces must be filled with a desiccant (a chemical that absorbs and holds moisture trapped in the air space).

Glazing consists of the panes or *lights* (a medium such as a window or windowpane through which light is admitted) in the *sash* (the framework holding the glass in a window unit) of a window. *Glazing* is also the act of installing panes of glass in a window sash. *Single glazing* is a single sheet of glass installed in a window sash.

Double glazing consists of a single-glazed sash with an additional glass panel installed on the sash to provide an air space between the two panels; the dead air space between the two glass panels does the insulating. The second panel can either be movable or fixed and it can be on the inside or the outside of the sash. Note that *insulating glass* has a positive seal around the edges of two panes of glass to provide a true air space; double glazing does not have a positive air seal. Also, there is no desiccant to absorb and hold moisture.

Triple glazing is a sash having three panes of glass that enclose two air spaces. This can be accomplished by applying a storm panel to a sash that is glazed with insulating glass or on some units by applying inside and outside storm panels to a single-glazed sash.

Effective window glass insulation is thus provided using *insulating glass* with its two sealed panes and dead air space or inert gas between the panes. A regular storm sash placed over a single-glazed window is almost as effective. In severe weather, you can use a storm sash over an inner sash of insulating glass producing triple glazing. Some windows with insulating glass are being made with a third track for either a screen or a third piece of glass for triple glazing. Insulating glass is more convenient than single glazing with a storm sash in locations where the window is difficult to clean, such as second story windows.

Storm windows are generally more economical than double-pane insulating glass in existing homes because storm windows usually cost less to install and they reduce infiltration of air around the window sash as well.

Tinted glass may be used in large glass areas on the south and west sides of the house. These walls receive the longest and the most direct sun exposure and tinted glass will help reduce the glare. To help reduce the heat buildup from the sun in rooms with southern and western exposure, a reflective glass is necessary.

A new type of bronze mirror-like glass which you've probably seen in office towers in many cities is now available to the homeowner in thicknesses of $1/8$ and $3/16$ inch. This is PPG Industries' energy-saving reflective glass which transmits only about one-third as much visible light as clear glass; it also screens out most ultraviolet light which causes the colors of furnishings to fade. Thus, it reduces glare and increases privacy, since you can see out but others can't see in. The glass can be glazed in an existing wood or aluminum sash for remodeling or add-ons and it can be used as a removable glass panel used to reduce the radiant heat from the summer sun. In the winter, the panels are removed so that the windows allow in the sun. The $1/8$ inch glass is used for prime windows, solar shields, and storm sash; the $3/16$ inch is used for tempered patio doors.

The reflective glass is particularly effective in sunny southern climates; it saves about 15 to 18 percent on energy costs during the air-conditioning season. With the exception of exterior overhangs (Section 8-4), awnings, shades (Section 8-2), and other similar devices, the bronze reflective glass is the first product to offer the same potential savings in the summer as insulating glass provides in the winter.

Now that glazing has been discussed, something needs to be said about the window sash *rails* and *stiles* (the frame, Fig. 8-l). Since wood has a thermal conductivity k of about 0.06 to 0.13, whereas aluminum has a k of 122 and steel of 28, the transfer of heat is most effectively blocked by wood rails and stiles. Thus, wood is the most effective energy conservation frame material. It is a recognized fact that the use of insulated wood windows with storm sashes for year-round use produce the greatest energy savings.

If you like aluminum-framed windows, however, because of their need for less maintenance, an acceptable alternate to wood is being manufactured for energy-conscious homeowners. These are double-glazed insulated windows with an air space between the panes and with a thermal barrier (break) in the frame extrusion to help minimize heat loss and sweating (Fig. 8-2). The thermal barrier is injected into the channel so that the barrier is the only link between the inside and the outside.

A vinyl-covered, wood-core window is also on the market. It is designed with a vinyl outside cover to prevent rusting, pitting, corroding, chipping, cracking, and peeling.

Many homeowners cover windows and doors with screens in the summer to keep bugs out while allowing fresh breezes to blow cool air into the house and warm air out of the house. Screens also diffuse (break up or scatter) the sunlight coming through the glass of the windows and doors; this diffusion is desirable in the summer but undesirable in the winter if you expect the sun to provide you with radiant heat. Therefore, remove the screens from the windows in the winter.

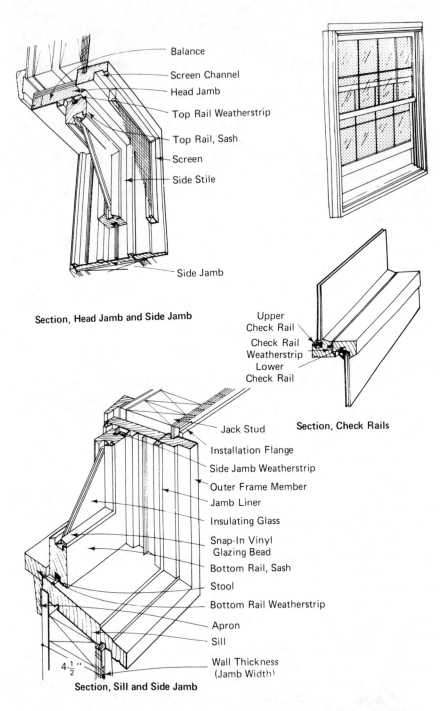

Balance
Screen Channel
Head Jamb
Top Rail Weatherstrip
Top Rail, Sash
Screen
Side Stile
Side Jamb

Section, Head Jamb and Side Jamb

Upper Check Rail
Check Rail Weatherstrip
Lower Check Rail

Section, Check Rails

Jack Stud
Installation Flange
Side Jamb Weatherstrip
Outer Frame Member
Jamb Liner
Insulating Glass
Snap-In Vinyl Glazing Bead
Bottom Rail, Sash
Stool
Bottom Rail Weatherstrip
Apron
Sill
Wall Thickness (Jamb Width)

$4\frac{1}{2}''$

Section, Sill and Side Jamb

FIGURE 8-1. Cross-sectional views of a double-hung window (a) show window construction and nomenclature (*courtesy Andersen Corp.*).

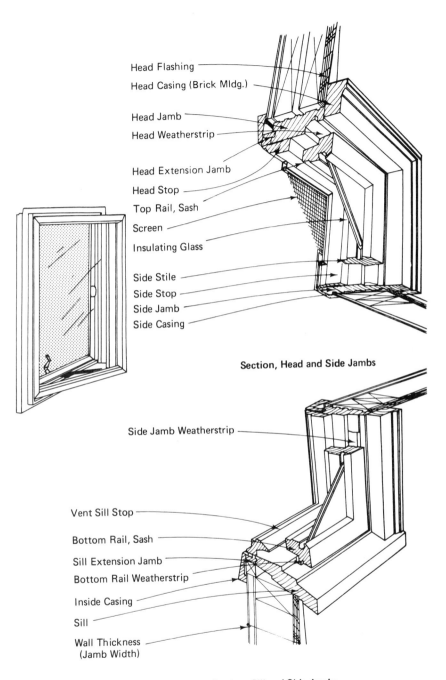

Head Flashing

Head Casing (Brick Mldg.)

Head Jamb

Head Weatherstrip

Head Extension Jamb

Head Stop

Top Rail, Sash

Screen

Insulating Glass

Side Stile

Side Stop

Side Jamb

Side Casing

Section, Head and Side Jambs

Side Jamb Weatherstrip

Vent Sill Stop

Bottom Rail, Sash

Sill Extension Jamb

Bottom Rail Weatherstrip

Inside Casing

Sill

Wall Thickness
(Jamb Width)

Section, Sill and Side Jambs

FIGURE 8-1. (continued) Cross-sectional views of a casement window (b) show window construction and nomenclature (*courtesy Andersen Corp.*).

Frame   Thermal Barrier
Extrusion

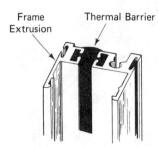

FIGURE 8-2. Aluminum-framed insulating windows must have a thermal barrier in the frame extrusion to help minimize heat loss by conduction through the frame.

8-2. WEATHERIZING WINDOWS FROM INDOORS

Radiant heat from the sun passes through glass windows into the home; this is desirable for brightening rooms. In the winter, we want as much of the radiant heat of the sun to pass into our homes as possible. Therefore, we *unblock* the windows as much as possible while the sun is shining and, in fact, we may make special efforts to absorb the radiant heat in dark rough surfaces such as floors, walls, and so on (Chapter 11). In the heat of the summer we'd like to *block* the radiant heat of the sun, but perhaps we still want to let some light into our rooms. We can block out or admit radiant heat from the sun by the use of indoor window accessories including shades, draperies, venetian blinds, cornices and lambricans, insulating panels, and window film.

Natural convection air currents occur in front of a window (Fig. 8-3). As warm air is cooled by the window, it becomes heavier and starts falling to the floor. As it falls, it cools more and falls faster until it reaches the bottom of the window. Many times, the heating device (radiant pipes, radiators, and so on) is located under the window. The cold heavy falling air reaches the heating mechanism where it is heated, becomes lighter, and rises to the ceiling. These naturally occurring convection currents can be minimized by the use of shades, draperies, venetian blinds, and cornices and lambricans.

A study by the Illinois Institute of Technology concluded that the use of window shades can save up to 15 percent of the cost of heating and cooling a house. In the winter, a properly-installed drawn window shade can prevent from 24 to 31 percent of the heat loss through the glass. In the summer, a sunlit window with a drawn, light-colored opaque shade admits 47 to 54 percent less total heat than an unshaded window; a translucent shade achieved a 44 percent heat reduction. These reductions include both solar radiation and outdoor temperature components. Thus, even though a home is not air conditioned, the inside temperature will be appreciably lowered. The study concluded that 8 percent of the fuel bill in cold weather and 21 percent of the bill in hot weather could be saved; this is based on a well-insulated house with the average amount of window glass and with no prior shading devices utilized.

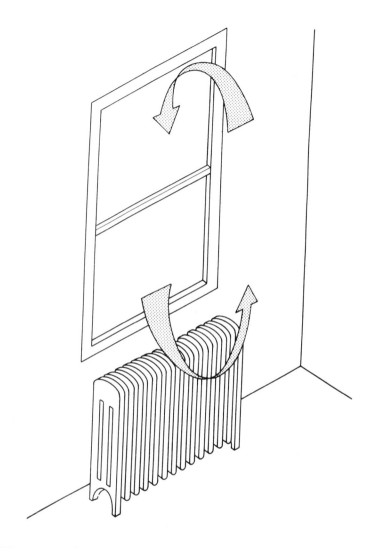

FIGURE 8-3. Natural air currents occur in front of a window even if a radiator or other heating device is not below the window.

In comparison to the ability of a shade to reduce heat losses in winter, a typical drapery and a typical venetian blind were found to reduce the heat loss only 6 to 7 percent. In the summer, a light-colored drapery with a white surface backing reduced heat gain by 33 percent, a fully-closed venetian blind by 29 percent and a half-open (45 degree angle) blind by 18 percent. Part of this difference in the amount of energy saved is attributed to the method of installation; effective shades are attached *inside* the window casing. Blinds are attached outside the casing and pass air between horizontal slats that cannot be effectively closed. Draperies allow air to pass within the pleats (they do not fit tightly against the window frame).

Shades

Window shades are available in an enormous range of colors, textures, and designs to match any decor. They can be used alone or with draperies or shutters (Fig. 8-4) for added effect, both from decorative and weatherization standpoints. Shades can be pulled from the top down or from the bottom up and are effective for regular windows, peaked, cathedral, or other unusual windows as well as for overhead and studio skylights. The cords of bottom-up shades slip through the hem slat about 3 inches from either side of the side; the cords rise to pulleys mounted at the top of the window, then drop to a convenient level where the cords secure to a cleat (Fig. 8-5).

Shades can be mounted within the window casing or on the outside of the casing. To be effective in reducing heat losses, however, the shades must be

FIGURE 8-4. Window shades help prevent heat losses through glass in the winter and block solar radiation from entering a window in the summer. White opaque shades mounted within the window casing are the most effective (*courtesy Window Shade Manufacturers Association*).

attached to the inside of the window casing as close to the top as possible (shades mounted to the outside of the casement are about 57 percent less effective than inside the casement). There should not be more than a ¼-inch gap between the shade and the casing when the shade is drawn. The shade is drawn to the sill. Even more efficiency can be attained if the shade edges can be drawn through channel strips. Perfectly sealed shades provide the best protection against heat transfer by convection and radiation.

As far as color is concerned, a white, opaque shade is the most effective shading device; this is followed by a white, translucent shade, white-lined

FIGURE 8-5. Shades can be mounted so they are pulled from the bottom up. These are particularly effective on windows or skylights of high–peaked or cathedral walls and ceilings (*courtesy Window Shade Manufacturers Association*).

draperies, and venetian blinds. An aluminum reflective surface on the outside is the most effective of all.

Shades are also more effective than draperies or venetian blinds in reducing heat transfer by air movement in the summer; this is again due to a tighter air seal around the window. Shades are more opaque than draperies or blinds and therefore more effective at blocking incoming solar radiation. Shades that have a light side and a dark side can be reversed; the light side out to reflect radiant heat in the summer and the dark side outward in the winter to absorb the sun's radiant heat.

Shades are measured and installed as described in DIY 8-1.

Do-It-Yourself 8-1
MEASURING AND INSTALLING SHADES

Materials: shades of proper width and length (see below), mounting brackets, ½-inch round head screws (4 for each shade)
Tools: rule or yardstick, awl or ice pick, small drill (a little smaller than the screw diameter), ³/₁₆ inch tip conventional screwdriver

Procedure:

NOTE
For the most effective weatherizing, mount the shades inside the window frame. Take the measurements with you when you buy the shades; the salesperson will cut them to the proper length.

Inside Window Frame Mounting

1. Measure the width of the window from jamb to jamb (Fig. 8-6).
2. For the length, measure the full height of the window from the top frame to the sill and add 12 inches.
3. Specify "inside bracket mounting" to the salesperson so the shades will be cut to the proper size.

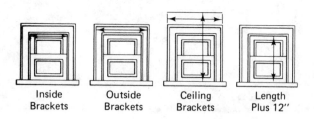

| Inside | Outside | Ceiling | Length |
| Brackets | Brackets | Brackets | Plus 12" |

FIGURE 8-6. Measure as shown depending on where you will mount the window shade brackets; inside mounting is the most effective in preventing heat losses (*courtesy Window Shade Manufacturers Association*).

Do-It-Yourself 8-1 (continued)
MEASURING AND INSTALLING SHADES

Window Frame Mounting

1. Measure and mark the position where the brackets are to be placed. The shade should overlap the frame 1½ to 2 inches on each side.
2. For the length, measure the full height of the window from the top frame to the lower sill and add 12 inches.
3. Specify "outside bracket mounting" to the salesperson.

Installation

1. Locate the mounting brackets as near the top of the window as possible. Mark the location of the holes (the mounting flange faces inward on both brackets).

NOTE
Nails are usually provided to hold the brackets in place. Screws are recommended, however, to provide a long-lasting installation.

2. Start the holes with an awl and then drill holes.
3. Mount the *slotted* bracket on the left side and secure it with screws. Mount and secure the other bracket on the right side.
4. Place the shade roller in the bracket on the right and then into the slotted bracket on the left.

Draperies

Draperies provide a method of preventing heat losses while simultaneously accentuating windows to add a decorative touch to your living space (Fig. 8-7). Floor-length draperies that fit tightly at the top and sides are the most effective (tack the sides of the draperies to the window casing and the bottoms to the molding to close off drafts).

The heaviness and color of the drapery fabric, as well as the correct opening and closing, influence the effectiveness of draperies as weatherizing devices. A heavy fabric that is close woven most effectively blocks the transfer of heat. A lining on the drapes provides an additional insulation effect that reduces heat losses and gains.

Light-colored, opaque draperies are effective in both winter and summer. In the winter, the light color reflects the heat back into the house; in the summer, the light color reflects the sun, and hence the heat, away from the window.

In the winter, keep the draperies closed as much as possible, opening them only when there is bright sunshine coming through the window or when

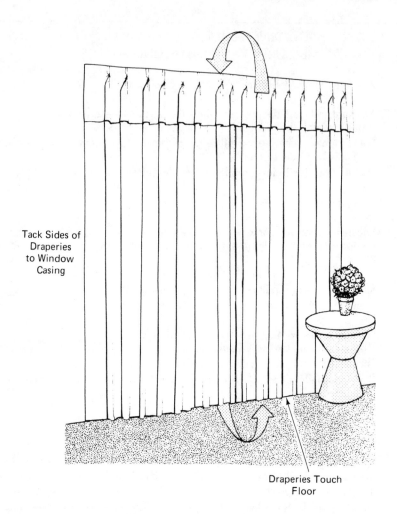

Tack Sides of
Draperies
to Window
Casing

Draperies Touch
Floor

FIGURE 8-7. Draperies should touch the floor and the sides can be tacked to the window
casing to reduce air currents.

you need the natural light; keep them closed at night. On cloudy days and
especially cloudy, windy days, keep them tightly closed. In the summer, close
the draperies in the daytime to reflect the sun coming through the windows.
Open the draperies wide at night, as well as the windows (unless you are
air conditioning), to let the cooler outside air replace the warmer, inside air.

Draperies are mounted on drapery rods as described in DIY 8-2. The
rods are expandable in length and therefore you need only rough measurements when you buy the rods and mounting hardware.

Do-It-Yourself 8-2
DRAPERY ROD INSTALLATION

Materials: drapery rods, mounting brackets, screws (supplied with the brackets for mounting onto the casing), toggle bolts, screw anchors, or molly bolts for wall mounting

Tools: awl or ice pick, drill, masonry drill (or an old drill bit for drilling into plaster, if brackets are to be wall mounted)

Procedure:

NOTE

Drapery mounting brackets can be installed to the window casing or to the wall. Mounting the brackets higher on the wall or further to the sides gives an appearance of a bigger window. Specific bracket mounting directions are provided with drapery rods and hardware.

Attachment to Window Casing

1. Locate the mounting brackets to the top outsides of the casing and mark the locations of the hole centers. If the rods are longer than 48 inches, you'll have a center bracket(s) as well as end brackets.
2. Start holes with an awl or an ice pick. Then drill holes with diameters a little smaller than the diameters of the screws provided.
3. Mount the brackets securely with the screws.
4. Put the rod in place, adjusting it to the correct length.

Attachment to Walls

1. Locate the mounting brackets, usually about 4 inches above the window and from 6 to 24 inches from each side (usually located so that draw draperies on traverse rods can be opened to completely clear the window).
2. Drill holes for the toggle bolts or screw anchors (the package will have instructions telling you the diameter of the holes).
3. Perform steps 3 and 4 as described for window casing mountings.

Venetian Blinds

Venetian blinds control the amount of light and radiant heat coming into the room. By opening the blinds in the winter, the sunlight with its radiant heat can come streaming in. Closing the blinds at night provides a shield against cold air infiltration; however, since the blind slats do not close tightly, the blocking of infiltrating air is not as effective as a more solid barrier such as shades, lined draperies, or insulating panels.

In the summer, white blinds can effectively be used to reduce the amount of radiant heat gain from the sun. The slats can be angled such that the direct sunlight is reflected, but pleasant indirect light is admitted through the slots between the slats. Dark-colored slats are tilted toward the sun in the winter in such a manner that the radiant sunlight passes into the interior and the heat of the sun is also absorbed by the dark material. Ideally, blinds should have slats with one white side and one dark side. Slats are made of aluminum or steel.

Blinds are most effective when mounted inside the window casing. However, a minimum mounting distance (usually from 1 to 2½ inches) is required for the brackets (be sure that the window casing mounting surface will be large enough to accept the attaching brackets before the blinds are measured for custom fitting by the manufacturer). For inside casing mounting, measure the width and length of the window with a wood or metal rule as shown in Fig. 8-8A. The width (W) is measured where the mounting brackets will be placed. The length (L) is measured from the top inside casing to the window sill.

If the window casing is not wide enough for inside mounting, the blinds can be mounted outside (though they will not be as effective in reducing the amount of heat transfer). Measure the overall width of the area to be covered (Fig. 8-8B); the measurement should overlap the window at least 1½ inches on each side to allow for bracket mounting. The length should be at least 3 inches longer than the length to be covered to allow for the bracket mounting. Brackets are mounted to the casing with screws. Predrill the screw holes at a diameter slightly smaller than the diameter of the screw shank.

(a)   (b)

FIGURE 8-8. Inside window casing measurements for blinds are made as shown in (a). Be sure there is sufficient surface on the casing for inside bracket mounting. Make outside casement measurements as shown in (b).

Cornices and Lambricans

A cornice or a lambrican is used to decorate windows and add to the mood of the room. It also aids in weatherizing windows by cutting down on the natural convection air currents that fall between a window and draperies (Fig. 8-9). The cornice or lambrican blocks the flow of air from the ceiling down the window to the floor by blocking off the top space between the window and the

FIGURE 8-9. A cornice (shown) or lambrican blocks the flow of air currents from the ceiling down between the window and the draperies.

drapery; in essence, the cornice or lambrican provides a "cap" around the top of the window to stop drafts. The drapery blocks air between the window and the room.

Cornices and lambricans differ only in their physical shape and size. The basic structure is made with hand tools from plywood. Simpler units are painted with several coats of a good quality paint or are covered with wallpaper or contact paper. More decorative units are covered with cotton or dacron fill which is held in place with muslin. The muslin is covered with a piece of fabric to match or accentuate furniture. Complete plans for making cornices, valances, and lambricans are found in home decorating magazines.

In the absence of a cornice or lambrican, material can be tacked into the frame and drooped over the curtain. Cornices, lambricans, and material aid in reducing air currents around a window. In the summer, they help to keep the room cooler too; just open the top of the window a little to let the hot air out.

Insulating Panels

You can reduce heat transfer through windows by using foam insulating panels (Section 7-1); this is a temporary method. Place the panels over or in your windows during extreme cold, hot, and windy weather. You can also use a Homosote or a cork board mounted on closet door runners; these are particularly effective in front of large glass areas such as sliding glass doors. Decorations can be placed on the panels so that they match the room decor whether in front of the glass or along the wall when slid away from the door.

Window Film

Window film (Scotchtint® by 3-M Company) is used to block the ultraviolet light, heat, and glare from the sun coming through your windows. Window film is a 0.001-inch polyester film with a transparent layer of aluminum that is vapor deposited on the film's surface. The film is applied to the inner surface of either regular glass, insulating glass, or storm windows and produces a mirror-like finish which makes the glass one-way for privacy.

The film is useful in southern climates where air conditioning is used often. In winter in colder climates, most of the heating benefits of the sun could be lost; however, the film does reflect the interior heat back into the house. The film is not for use on exterior glass surfaces, glass with burglar alarm tapes, car windows, textured, patterned, or frosted glass, or plastic sheet.

The film is applied by cleaning the glass, drying, then wetting the window and film. The film is then put in place and smoothed and the backing is peeled off. The backside is then wet and squeezed smooth.

8-3. REMOVING, REPLACING, AND INSTALLING NEW WINDOWS

If you don't expect to move from your present home, knowing the best orientation of your home and the best location for windows doesn't help you much. However, you may consider room additions, installation of larger windows, or the closing up of needless windows.

Ideally, the largest amount of glass in your home should be on the south side. The rooms that are used most often during daylight and in which there is not a lot of activity are best situated along the south side. In the winter, the sun is at a low angle and streams into the south, glassed area. In the summer, the angle of the sun is higher and house overhangs block the direct rays of the sun from entering the southside windows. In contrast, the north side of the house should have the least amount of glass; if you don't need the glass for light or summer ventilation, you might consider removing some of the north-exposed windows. The east and west sides of the home should have more glass than the north side, but less than the south side. The morning sun shines in the eastern windows and the evening sun in the western windows.

If you decide to put in new or replacement windows, you must consider whether to buy insulating glass or single-pane glass with a tight-fitting storm window. In most cases where the winter temperatures are chilly, the additional investment in either one can be recovered in the form of heat-cost savings for most homes in a few years' time depending upon the severity of the winter weather and fuel costs. Insulating glass provides two panes of glass melted together along the edges with an air space or gas-filled space between the glass; thus, there are only two surfaces to clean whereas with storm windows, there are four glass surfaces to clean.

New, insulating windows (Fig. 8-10) or replacement, insulating windows are available in several styles including aluminum sash with a vinyl thermal break, wood sash with a white aluminum exterior covering, wood unfinished on the interior and exterior, wood with a baked enamel exterior, wood sash with a vinyl surface on the exterior and the interior, and steel.

In colder climates, the costs of triple glazing can also be recovered over a reasonable number of years. Triple glazing can be accomplished by placing a storm window over an insulating glass window or two storm windows on either side of a single glass window.

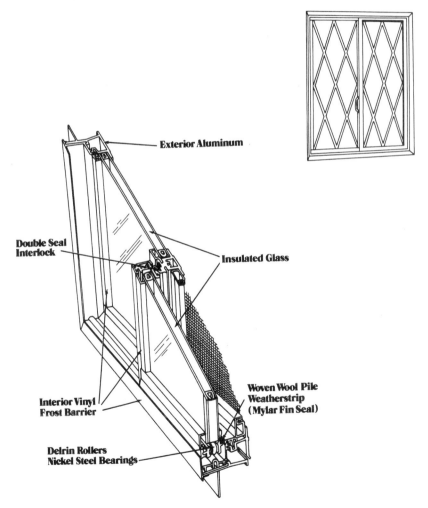

FIGURE 8-10. This sliding window has insulated glass, a screen over one-half of the window, weather stripping, and a frost barrier (*courtesy Winter Seal of Flint, Inc.*).

Before deciding to reduce the size or remove any windows, you must consider cross-ventilation for summer cooling, safety, and local and federal regulations. For adequate, natural, summer cooling by cross-ventilation, there should be at least two operating windows in each room on opposite walls (preferred) or on adjacent walls. This permits the entry of a breeze through one window and an exit through the other.

From the standpoint of safety, every room should have at least one window large enough and close enough to the floor to allow a person to easily exit through it in case of an emergency such as a fire. The amount of glass area, the amount of ventilation area, the dimensions of emergency egress openings and their proximity to the floor are all controlled by local and FHA regulations. Be sure to check your local code before remodeling or building.

Fixed windows may be used anywhere that operating windows are not necessary. The fixed windows give a larger, unrestricted viewing area and cut down on possible air infiltration.

Replacing Existing Windows

Existing double-hung wooden sashes that are damaged from rot can easily be replaced with new double-hung windows. The old sashes and cords are removed and the new sashes in frames are placed inside of the existing wood frames. First measure the *full* width and height of the existing frame (the parting strips and stops will be removed). Take these dimensions to your window dealer who will make the windows to fit your dimensions if premade sizes are not already available. The width measurement is critical as there is no compensation for error, but the top of the new window has an expansion strip that will take up a little extra in the length dimension.

When you receive the new windows, remove the stops and parting strips first from the inside of the old window to gain access to the sashes; use care in removing these strips as they are to be reused. Then remove the sashes and the sash cords and old weights. Install the new window into place. Check it with a level and shim as necessary on the bottom to level it. Screw the window into place in the side frames, making adjustment to the height, as required for a snug fit. Then replace the stops and parting strips.

Installing a Window in a Wall

Instructions for installing windows in a wall are provided by window manufacturers and do-it-yourself books. You need to prepare a rough opening in a wall (Fig. 8-11), install the window, and close up any remaining area around the window. Roughing out includes laying out the window opening width between regular wall studs, removing internal plaster or plasterboard, removing external siding and sheathing, cutting and installing a header, installing jack studs, and cutting and installing cripples. Be sure to use only

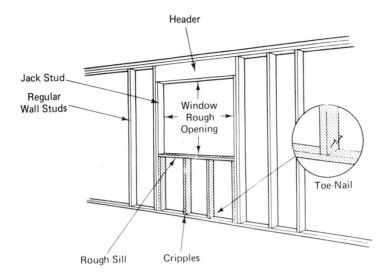

FIGURE 8-11. When studs are partially removed in a wall for a window, a header, jack studs, and cripples are added to support the window and the structure above the window (*courtesy Andersen Corporation*).

kiln-dried lumber; green lumber warps as it dries out. The window is installed next by setting the window into place from the exterior, leveling and plumbing the window, and nailing the window into place. Flashing is then installed followed by caulking. The frames of wooden windows are then painted or stained (Fig. 8-12).

Closing Up a Window Opening

If after considering all factors you decide to remove a window (probably from the north side), start by removing the casing from the interior and the exterior. Then remove the sashes and the frame of the window. On the inside, install plasterboard, plaster, or paneling. Finish the wall to match the decor. On the exterior, add on the necessary siding or shingles.

As an alternative to removing the window, you might consider simply blocking the window off from the inside. Remove the inside casing. Paint the back of plasterboard or paneling a black or gray color and close up the window space. From the exterior, it will still resemble a window.

8-4. WEATHERIZING WINDOWS FROM OUTDOORS

Exterior window accessories include overhangs, awnings, shutters, and storm windows. The accessories are used to provide protection to your windows.

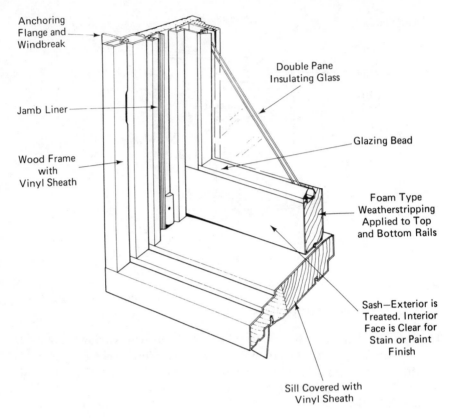

Anchoring Flange and Windbreak

Double Pane Insulating Glass

Jamb Liner

Glazing Bead

Wood Frame with Vinyl Sheath

Foam Type Weatherstripping Applied to Top and Bottom Rails

Sash—Exterior is Treated. Interior Face is Clear for Stain or Paint Finish

Sill Covered with Vinyl Sheath

FIGURE 8-12. The wood frame of this window is treated with a water repellent preservative and covered with a rigid vinyl sheath (*courtesy Andersen Corporation*).

In the winter, you'll want shutters or storm windows to protect the windows against rain, snow, and infiltrating cold winds. In the summer, you'll want overhangs, awnings, or shutters to block out the radiated heat from the sun. Any method that stops the sun before it gets in through the glass is seven times as good at keeping you cool as having shades, blinds, or curtains inside. Awnings and shutters also protect the windows against summer storms. Don't overlook trees either; deciduous trees provide shade in the summer and allow sunlight to reach your windows in the winter (refer to Chapter 11).

Overhangs

An overhang is a horizontal projection from your home over your windows. Overhangs are most advantageous on the south, southeast, and southwest sides of the house because they block summer sun from entering your windows; but, because the angle of the sun is lower in the sky during the late fall, winter, and early spring, the overhang permits the sun to shine through the window to warm the interior (Fig. 8-13). An overhang should be designed to completely shade a south window from May until September; the overhang reduces the heat gain by about 50 percent.

If your home is of modern design (or the architect took advantage of the sun's angle), you may have a "natural" overhang from the eaves or from an extended second floor. If you're not that lucky, you can build your own overhang of wood, metal, or other suitable material. The overhang should be between 40 inches for a latitude of about 30 degrees (Fig. 2-1) to 54 inches for a latitude of 50 degrees. It should slope 5 degrees downward from the house to allow rain runoff and must be sufficiently strong in northern climates to support a heavy snow.

An extension panel that either slides into the main overhang frame or folds from it on a hinge (Fig. 8-14) is effective on southeastern and southwestern exposures in the fall and spring (when the sun is at a lower angle than in the

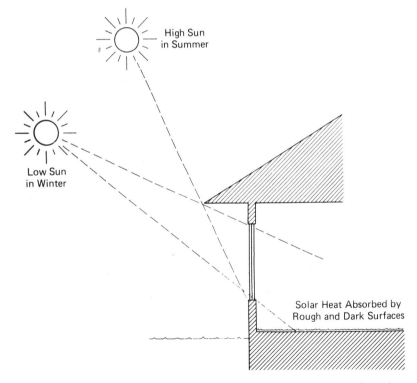

High Sun
in Summer

Low Sun
in Winter

Solar Heat Absorbed by
Rough and Dark Surfaces

FIGURE 8-13. A natural overhang from the eaves or a projected second story allows the sun to provide solar heating through southern exposed windows in the winter, but blocks the sun in the summer because of the higher angle of the sun.

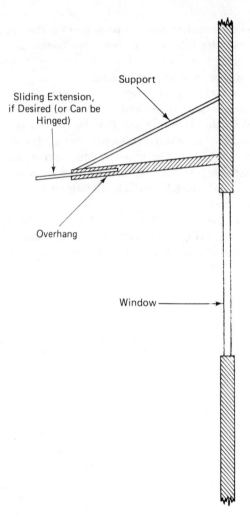

FIGURE 8-14. An overhang will be built from wood to shade southern exposed windows in summer.

summer). By periodically adjusting the length of the overhang, the amount of direct sun into the house can be controlled.

Awnings

In the summer, the rising sun streams through windows on the east and southeast sides and the setting sun streams through the southwest and west sides of the house. Overhangs or awnings are used on the south side to prevent the summer sun from shining into the house (the angle of the sun in the sky is high during the summer). Awnings can be used on the east, southeast, south, southwest, and west sides to block the sun when it is at a low angle at sunrise and sunset.

Canvas or vinyl awnings can be raised or lowered, as required, to admit as much sunlight as desired or to protect the awnings during high winds. Canvas or vinyl awnings are easily removed and stored during the winter. Aluminum awnings provide shading and at the same time have slots through them to let cool breezes circulate into your windows. If aluminum awnings remain on all winter, they block some radiant heat from the sun, but they also partially block cold wind, rain, and snow, and thereby afford some protection to the windows from the elements.

In purchasing aluminum awnings, consider their length and overhang. A compromise should be found depending upon the desired exposure so that the summer sun is blocked but the winter sun can at least partially come through the window. Aluminum (as well as canvas or vinyl) awnings are screwed securely into the house frame. Aluminum awnings are more or less permanent installations and are difficult to remove, store, and replace each year. Canvas and vinyl awnings have detachable rods and the material can be rolled for storage.

Secure awnings with wood screws or lag bolts of sufficient length (about three inches or longer) to anchor the awnings to the house.

Shutters

If the outside windows have shutters, hopefully the shutters are functional as well as decorative. (Lucky is the person who has an older house with working shutters!) To be functional, the shutters need to have hinges and rods to move them and hook them closed and open. The shutters should fit inside the window casing and should meet and overlap at the center when closed.

Close the shutters during adverse weather conditions. In the winter, they help block wind trying to infiltrate the windows. In the summer, they block the sun, but let the breezes in; in any season, they'll keep the rain out.

You can buy plastic, vinyl-covered wood, aluminum, and wooden shutters. The plastic, vinyl-covered wood, and aluminum shutters are decorative rather than functional; wood shutters are decorative, functional, and the most difficult to maintain since they'll occasionally need a coat of paint. To determine the size of functional shutters needed, measure the inside of the casing with a wooden rule; subtract $1/8$ inch to allow for an opening and closing clearance. If necessary, buy shutters that are larger than the opening. The bottoms of the shutters are easily cut off with a hand saw and the sides are easily planed with a hand plane for proper fitting.

Purchase mounting hardware (hinges, rods, and brackets) to match your home decor. Install the hinges according to the directions given with the shutters and hardware. Locate the hinges approximately one-sixth of the length of the shutters from the top and from the bottom. Predrill screw holes at a diameter slightly less than the diameter of the screws. Screw all hardware into place (use galvanized or brass screws).

Some energy-conscious do-it-yourselfers with older homes which have shutters are retaining the original look of the house without adding storm windows. One technique is to remove the slats from the shutters, install retaining strips, and glaze glass into place. A second technique is to keep the slats in place and screw clear plexiglass over the shutter. Both methods retain the shutter appearance, but provide double glazing. One advantage of the plexiglass is that it can be removed in the summer so that the shutters can block the sun but still let in the cool breezes.

Storm Windows

This section discusses the addition of storm windows to aid in the reduction of heat losses through window glass in the winter (Fig. 8-15). The addition of a storm window over the normal window is *double glazing*; one air space exists between the two panels. In *triple glazing*, there are three panels and two air spaces (Section 8-2). The advantage of using double glazing over single glazing is an increase in the resistance value, from about R-0.88 for single glass to R-1.79 for a storm window placed 1 to 4 inches from the window (QRC 7-2). The storm window also reduces infiltration of cold air around

FIGURE 8-15. A storm window reduces the infiltration of cold air and doubles the resistance value of single glazing (*courtesy National Mineral Wool Insulation Association*).

window rails, stiles, jambs, casing, sills, and glazing. Storm windows of any type do the same job; the more expensive windows provide more convenience and are more attractive, but this doesn't mean expensive windows are more effective. Removing, replacing, and installing *new* windows having insulating glass are discussed in Section 8-3.

Storm windows may be made of clear sheet vinyl, plastic sheet, acrylic, or glass. They range from single panels that must be put in place each fall and removed each spring to triple-track assemblies that include upper and lower windows and a screen. Triple-track windows (called combination windows) are used during heating and cooling periods and can be opened for natural ventilation at other times.

Clear sheet vinyl storm windows are the cheapest; however, they will probably last only one year and therefore their best application is indoors over attic or cellar windows in older or second homes and for doubling up on existing storm windows in northern exposures. You can purchase sheet vinyl in rolls or in kits with extruded vinyl frames. To install sheet vinyl, refer to DIY 8-3.

The sheet vinyl storm window kits with extruded vinyl frames snap together. The frame pieces are first measured to fit the window; miter joints

---

Do-It-Yourself 8-3
HOW TO MAKE AND INSTALL VINYL STORM WINDOWS

Materials: roll of sheet vinyl; use at least 6-mil-thick material; wood or cardboard strips ½ inch wide, ⅛ inch thick, by width and length of windows; brads, staples, or small screws

Tools: hammer, stapling gun, or screwdriver, as applicable; ruler; scissors

Procedure:

1. Measure (Fig. 8-16) the width of the window to the outsides of the casing (molding). Measure the length of the window from the top of the casing to the sill (or to the bottom of the casing if there is no sill). Add 4 inches to each of these dimensions.
2. Using the dimensions, cut the sheet vinyl with scissors.
3. Cut two strips equal to the length dimension and two strips equal to the width dimension minus 1 inch.
4. Wrap the bottom edge of the sheet vinyl two turns around one of the width strips. Fasten this strip to the bottom of the window at the sill.
5. Stretch the sheet vinyl to the top casing of the window, wrap the vinyl around the other strip and fasten it in place.
6. In similar manner, attach the side (length) strips. The wrapping of the vinyl around the strips makes an airtight fit against the window.

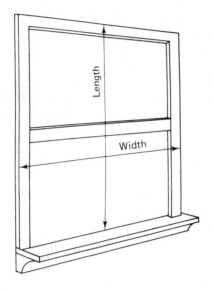

FIGURE 8-16. Obtain the width and length measurements as shown. For sheet vinyl, add 4 inches to each dimension for wrapping around wood or cardboard strips.

are not needed because the extruded pieces simply snap together. The sheet vinyl is then cut to the size of the frame. The sheet vinyl fits between a two-piece tongue and groove pressure-fit channel strip on each side of the window. The completed window is then attached to the window with brads, staples, or small screws.

Sheet plastic (0.04-inch thick) mounted in vinyl mounting trim is also applied indoors (Fig. 8-17). It is more permanent than sheet vinyl and is handy for covering windows on second or third stories or other places where access from the outside is difficult. One manufacturer provides sheet plastic in standard sizes to fit most windows (24 by 36 inch, 38 by 56 inch, 44 by 64 inch, and 40 by 80 inch); the manufacturer also sells self-adhering "snap-open" trim molding and sill molding. The sheet plastic is "cut" by scribing it with an awl or ice pick and then snapping it along the scribed line. Trim strips are cut with a small hand saw. The sheet plastic is inserted in the trim and the storm window placed against the existing window frame for a size check. The backing is then peeled from the self-adhering tape and the storm window is set into place. To remove it in the summer, simply snap open the trim and remove the sheet. The manufacturer provides complete installation instructions.

Acrylic plastic in 1/4 inch widths can be cut to fit a window for storm protection; the acrylic is easily installed on the inside making it useful for windows that are high or hard to reach from the outside. It is also excellent for storm doors, cellar windows, and garage door windows. Since the acrylic is not fragile like glass, it is safe to handle. The acrylic is somewhat flexible and therefore seals effectively against the window frame. Make storm windows from acrylic as described in DIY 8-4.

FIGURE 8-17. This storm window consists of a rigid, clear plastic sheet and vinyl mounting trim. The trim adheres to the window frame permanently and snaps open for removal of the plastic sheet when not needed (*courtesy Plaskolite, Inc.*).

Do-It-Yourself 8-4
HOW TO MAKE AND INSTALL SOLID ACRYLIC STORM WINDOWS (NO FRAME)

Materials: sheet of ¼-inch acrylic plastic; brass or galvanized 1-inch round-headed screws

Tools: screwdriver; one of the following cutting tools (1. steel crosscut, circular saw blade recommended for finished cuts on plywood, veneers, laminates, etc., six teeth per inch, 2. saber saw blade with 14 to 24 teeth per inch, 3. special scribing tool manufactured by Plastic Plus™ or Red Devil® available at hardware stores); drill and bits

Procedure:

1. Locate the indoor mounting surface along the window on which the storm window is to be mounted. You can use the casing or jambs (but be sure that an effective seal can be made) for the mounting surface.

2. Measure the width and length of the area to be covered (Fig. 8-16 for casing). Mark the cutting guidelines on the masking paper covering the acrylic. Do not remove the masking paper until after cutting; it protects the surface from scratches and reduces friction and gumming behind the blade.

3. Cut the acrylic. If a circular saw is used, set the blade just higher than the thickness of the acrylic to prevent excessive chipping; do not force-feed the acrylic into the blade. If a saber saw is used, clamp a straightedge to the material to guide the material along a straight line. If the material is to be scribed and

---

Do-It-Yourself 8-4 (continued)
HOW TO MAKE AND INSTALL SOLID ACRYLIC STORM WINDOWS (NO FRAME)

---

broken along the cutting line, scribe each line with moderate pressure about 10 times. Place a ¾-inch dowel under the scribed line (keep the scribed line face up). Wearing protective gloves, hold the material firmly with one hand and apply downward pressure on the short side of the dowel with the other hand until the acrylic breaks along the scribed line. Sand edges to remove saw marks; this assures maximum breakage resistance.

4. Drill holes slightly larger than the screw diameters at each corner and about 12 inches apart, equally spaced about ¾ inch from the edges along all sides (locate the holes so that the screws will go into a flatsurface of the window). Use a standard metal twist drill in a hand drill; back the acrylic with a piece of wood where the drill is to come through. Use a sharp drill, very slow speed, and minimum pressure.

5. Place the acrylic storm window in place at the window. Mark the location of the screw holes and remove the acrylic.

6. Use a drill bit of diameter slightly smaller than the screw to drill pilot holes in the window casing or jamb.

7. Place the acrylic in position and screw it into place. If the acrylic doesn't fit snugly, place a piece of plastic foam tape weather strip (Section 6-1) against the inside edges of the acrylic. In the summer, you can easily remove it by removing the screws. If more than one storm window is made, identify each one so they can be quickly returned to the proper location the following winter.

You can make exterior storm windows for your cellar window wells that will let in lots of sunlight, but will keep heat in and leaves and snow out (Fig. 8-18). Make a wooden frame from exterior plywood. Paint the wood with an exterior paint and add mounting brackets so that the frame can be removed in the spring. Cover the frame with sheet vinyl or, for a more permanent job, use acrylic plastic. Do *not* use glass because there is a chance someone could fall onto the frame over the well, breaking the glass, causing serious injury to themselves.

You can make aluminum framed storm windows at a fraction of the cost of buying them; also consider buying salvaged storm windows available for a couple of dollars. (Take your window dimensions with you though because you must have a correct fit.) You'll need to visit your hardware store and purchase aluminum sash that comes fitted with a glazing channel that protects and seals the glass, or acrylic placed in the sash and friction-fit corner locks that secure mitered corners. Determine the exact location of the storm window on the window; storm windows for some casement windows, awning windows, and difficult to reach locations will be located on the inside rather than on the outside. Build the aluminum storm windows as described in DIY 8-5.

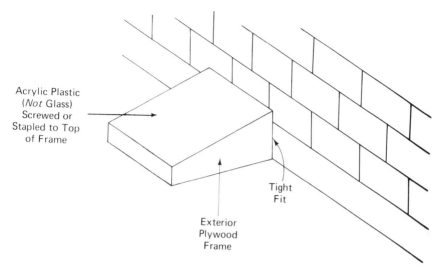

Acrylic Plastic
(*Not* Glass)
Screwed or
Stapled to Top
of Frame

Tight
Fit

Exterior
Plywood
Frame

FIGURE 8-18. A cellar window well storm window keeps heat in and water, snow, and leaves out of the well. Weather-strip where necessary.

Do-It-Yourself 8-5
HOW TO BUILD ALUMINUM FRAMED STORM WINDOWS

Materials: aluminum sash (with glazing channel), double-strength glass or acrylic plastic, friction-fit corner locks, hanging brackets

Tools: hacksaw, glass cutter (or have it cut for you), acrylic cutting tool (DIY 8-4), utility knife, screwdriver (to install brackets)

Procedure:

1. Make window measurements for length and width of the storm window.
2. Remove and save the glazing channel from the sash section.
3. Cut the sash to length making 45-degree miter cuts at each corner with a hacksaw.
4. Cut the glass or acrylic to the required dimensions to fit inside the sash.
5. Place the glazing channel around the glass and tape it temporarily into place. Cut 90-degree notches out of the glazing channel at the corners so that the channel fits around the glass corners.
6. Attach two friction-fit corner locks to two sash sections of the same size.
7. Place two opposite sash sections over the glass. Then install the other two sides joining at the friction-fit corner locks.
8. Attach the mounting brackets to the frame and to the window.
9. Place plastic foam tape weather stripping around the perimeter of the window to make a tight seal between the storm window and the mounting frame.

You can also build storm windows using wood furring strips, angle irons, and sheet plastic or acrylic (DIY 8-6).

If you are not especially handy or you'd rather not build storm windows yourself, perhaps your best alternative is to have a storm window company measure for and make your windows. Then you can do the installation yourself, thereby saving installation costs. If the window is too big when it arrives, scribe along one of the V-grooves on the side or bottom fins with a sharp tool and break off whatever width is needed with pliers. File off any burrs. Before placing the storm windows into position, place a bead of acrylic latex caulking (Section 5-1) around the top, sides, and bottom of the house window where the storm window will mate the window (if any painting needs to be done, do it before caulking and let the paint dry). Drill screw pilot holes with a diameter slightly less than the screw shank and screw the storm window into place with aluminum or galvanized screws.

If you don't want to tackle this project yourself, have a contractor install the storm windows. Watch for sale prices, especially in the spring and summer when storm window contractors are not as busy. You'll make up the cost of the

---

Do-It-Yourself 8-6
BUILDING WOOD FRAME STORM WINDOWS

Materials: 1- by 3-inch straight wood furring strips; four galvanized angle irons per window; sheet vinyl, sheet plastic, or acrylic; weather stripping; attaching brackets, paint (good exterior grade)

Tools: saw, rule, screwdriver, drill and bits, paintbrush

Procedure:

1. Locate the mounting position of the storm window. Measure the length and width; subtract ¼ inch from each dimension.
2. Cut the furring strips to the length (two required) and width (two required) needed; cut the ends at 45 degrees so that a width and length fit together to form a 90-degree angle.
3. Secure each corner with angle irons. Locate the irons, drill holes slightly smaller in diameter than the screws, and screw the angle irons in place.
4. Attach the sheet vinyl, sheet plastic, or acrylic. The sheet vinyl should be rolled over several times along each edge so that the vinyl makes an airtight fit. You can use ⅛-inch-thick wood strips ½-inch wide inside the rolled plastic to make a better fit. Tack, staple or nail the plastic strips into place. If sheet plastic or acrylic is used, screw it tightly into place against the frame.
5. Paint the wood.
6. Screw the storm window into place with long galvanized or brass screws. As an alternative, you can attach hanging brackets.
7. Place weather stripping against the frame.

---

storm windows in energy savings in a few short years. Remember, storm windows are practically permanent and maintenance free.

Most combination storm windows come with two or three $1/4$ inch holes (or other types of vents) drilled through the frame where the frame meets the window sill. The holes keep winter condensation from collecting on the sill and causing rot, and also let rainwater that may come through the screen in warm weather escape. Keep these holes clear, and drill the holes yourself if your windows don't already have them.

If you install combination storm windows, remove the screens in the winter. The screens disperse (scatter) the sunlight with its radiant heat that could come through the window to provide warmth. You should install the screens in spring and leave them in during early fall so that you can let in fresh air to either warm or cool your house as necessary. If you have air conditioning, you will leave the storm windows in all summer to decrease the infiltration of warm air; if screens remain in place, you can open the windows on cool days rather than using the air conditioner.

You can, incidentally, decrease the effort needed to raise and lower combination storm windows (and interior windows as well) if you spray the tracks and grooves with a silicon spray. You can also rub the tracks and grooves with a dry bar of soap. Never paint any surface that acts as a guide for double-hung or sliding windows. Instead, use linseed oil for wood preservation and wax for easy operation.

Screens

Although not necessary to aid in weatherizing your home, screens are a welcome outdoor window treatment in areas where flying insects are a problem. Installed properly, screens allow the opening of windows for the admittance of cool air and the exiting of hot air without letting in mosquitoes, flies, gnats, and other bothersome insects. Screens are used in the spring, summer, and fall, but are removed during the winter because they diffuse the sunlight that could come through the window to heat the home interior by radiant heat.

Screens are made of plastic, aluminum, bronze, copper, or galvanized steel. Small holes can be patched by cutting squares of screening from old screening. Make the squares oversize; then pull strands of wire from along the edges so that strands of wire project from each edge like fingers (Fig. 8-19). Bend these fingers at right angles to the patch. Place over the hole and pass the fingers through the screen. Then bend the fingers back to secure the patch. If needed, you can use a long strand of screening and sew it in and out along the patch to secure the patch in position.

Damaged screening is removed from wooden framed screens by removing trim pieces, removing tacks or staples, and then removing the screening. The new screening is cut to size, placed in position, tacked or stapled taut, and the trim strips are replaced.

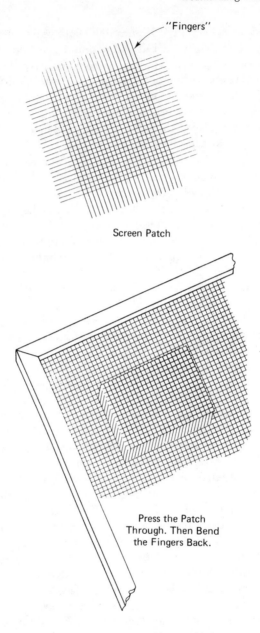

"Fingers"

Screen Patch

Press the Patch
Through. Then Bend
the Fingers Back.

FIGURE 8-19. Screen patches can be made from old screening.

Damaged screening in aluminum frames is removed by carefully prying out the spline material and removing the screen. The new screening is fit in place (slightly larger in dimension) and the spline is placed back into the frame with a roller installation tool to hold the screen in place; one side is put in place followed by the opposite side. Then another side and its opposite is fastened into place. Excess material is trimmed off.

Trees

This section cannot be completed without mentioning the fact that trees can be used effectively to control the amount of light at a window. Deciduous trees are trees such as oak, maple, and fruit that shed leaves and grow new leaves with the change in seasons. The leaves of deciduous trees growing in the vicinity of windows on southern exposures can effectively block sunlight during the late spring, summer, and early fall. In the winter, the leaves are gone and the sun's rays pass through to provide radiant heat to the window.

Evergreen trees, on the other hand, keep their leaves (needles) all year-round. Planted judiciously on the northern exposure and the windward side of the home, the trees provide a natural windbreak which reduces the pressure against the windows and doors and thereby reduces infiltrated air. Nature can help weatherize your home in other ways too, as discussed in greater detail in Chapter 11.

8-5. SKYLIGHTS

One way to put light and radiant heat into your home is by placing skylights in the roof (Fig. 8-20). Thus, any room which has a roof directly over it can be flooded with light: studies, porches, bedrooms, bathrooms, kitchens, workshop, or any other room. To be the most effective for heat gain, the roof where the skylight is installed should face the southern direction.

Make sure that the skylight you purchase has double glazing or insulating glass. Also, a skylight sold with flashing included helps ensure against water leaks around the skylight and through the roof (Fig. 8-21).

Some skylights can be tilted open to let in fresh air and for easy cleaning. If the skylight opens, then you'll need some type of screening to keep flying bugs and insects out. One manufacturer has an insect screen available that attaches in place with "hook and loop" material; one corner of the screening has to be unfastened for opening and closing the skylight. If the skylight is installed out of reach, a manual pull string or pole, or a motorized opening and closing mechanism is available for some skylights.

The insulated or double-glazed glass of the skylight, weather stripping, and caulking take care of protection from the elements during the winter. In the summer, though, the radiant heat may become too strong. In this case, outside awnings, roller shades, or venetian blinds can be installed, as preferred, to regulate the influx of the sun.

FIGURE 8-20. A skylight provides plenty of light and radiant heat to a room (*courtesy Velux-American, Inc.*).

You need to have some carpentry talent or be an experienced do-it-yourselfer to install a skylight. Directions are enclosed with the skylights. Installation procedures include laying out the location of the skylight from the inside to fit between rafters, cutting the roof open, framing the opening, installing the skylight, waterproofing the installation, and trimming the inside ceiling. The proper installation of the skylight is important to the total heat loss of the house. It is essential that the roof insulation is continued right up to the skylight frames and that the moisture barrier and windproof layer, if any, are connected to the skylight frame.

FIGURE 8-21. Skylights can be installed in a day. Be sure flashing is included to ensure against water leaks (*courtesy Velux-American, Inc.*).

8-6. DOORS

If your exterior doors are warped (so they don't close correctly), chipped, or otherwise damaged, you might consider replacing them with new weatherized doors (Fig. 8-22). Steel-faced doors with polystyrene or urethane foam cores are available in many attractive and distinctive colors and patterns. Costing about the same as a standard solid wood door plus a storm door, the metal-covered foam doors provide thermal and infiltration resistance. Magnetic weather stripping, weather-stripped thresholds with base flashing and caulking, interior and exterior steel skins separated by a thermal break, and preglazed insulating glass features contribute to the total effect of the weatherized doors with thermal resistance values of about R-8 (Fig. 8-23). Insulated steel doors provide better security against forced entry and fire than a wooden door, and there is generally less maintenance.

The doors are installed by one of two methods. One method is an easy do-it-yourself project in which the new door is simply hung on the existing hinges. The second method is to purchase the door as a prehung unit complete with its own frame and hinges. You install the prehung unit by nailing it into the rough opening; the prehung unit is usually installed by a carpenter or an

FIGURE 8-22. Decorative weatherized doors of steel-faced polystyrene or urethane cores are used to replace warped and weather-beaten doors (*courtesy Pease Company, Ever-Strait Division*).

TOP VIEW

Fire Protective
Return Wings

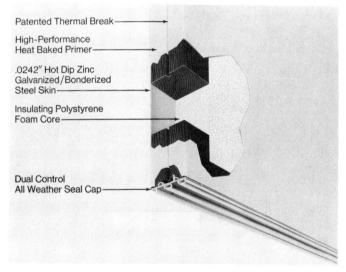

Patented Thermal Break

High-Performance
Heat Baked Primer

.0242" Hot Dip Zinc
Galvanized/Bonderized
Steel Skin

Insulating Polystyrene
Foam Core

Dual Control
All Weather Seal Cap

BOTTOM VIEW

Moisture Barrier

Fire Protective
Return Wings
and Bottom Seal
Housing Retainer

FIGURE 8-23. In addition to being weatherized, metal-sheathed doors provide some fire protection (*courtesy Pease Company, Ever-Strait Division*).

experienced do-it-yourselfer when the entryway is entirely changed. Manufacturers sell not only the doors and frames, but also door trim and matching sidelights.

You don't need to winterize interior doors that close off unused areas, but if you air-condition an individual room, you can weather-strip the door; also add a strip to the door bottom (Chapter 6). If you don't have a door on a room that you want closed off, you can add a wood door or a folding door (Fig. 8-24). Wood doors are available already prehung in a framework.

Sliding doors (Fig. 8-25) out onto the balcony or patio are, of course, beautiful; they also have large heat losses through the glass and have about three times the heat loss around the sliding doors as does a hinged unit. These doors also have metal frames that transmit several times as much heat as

FIGURE 8-24. Interior heated areas can be closed off from unheated areas with folding doors. Fit the sliding doors as close to the ceiling and the floor as possible.

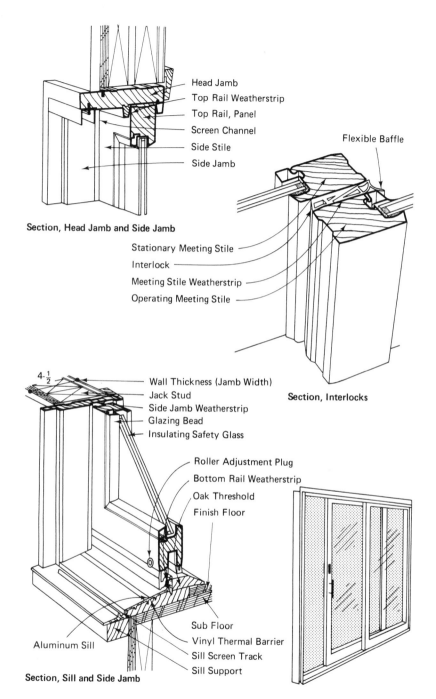

Head Jamb
Top Rail Weatherstrip
Top Rail, Panel
Screen Channel
Side Stile
Side Jamb

Section, Head Jamb and Side Jamb

Flexible Baffle

Stationary Meeting Stile
Interlock
Meeting Stile Weatherstrip
Operating Meeting Stile

Section, Interlocks

$4\frac{1}{2}$

Wall Thickness (Jamb Width)
Jack Stud
Side Jamb Weatherstrip
Glazing Bead
Insulating Safety Glass

Roller Adjustment Plug
Bottom Rail Weatherstrip
Oak Threshold
Finish Floor

Sub Floor
Vinyl Thermal Barrier
Sill Screen Track
Sill Support

Aluminum Sill

Section, Sill and Side Jamb

FIGURE 8-25. Insulating safety glass and vinyl thermal barriers provide protection against heat losses through sliding glass doors. Close curtains or insulating board in front of the doors when the sun is not shining through (*courtesy Andersen Corporation*).

wooden frames unless the metal frames have built-in thermal breaks between the inner and outer frames. Be sure to weather-strip sliding glass doors and consider placing draw drapes over them in the winter or install sliding insulating panels that can be moved into place (Section 8-2).

8-7. STORM DOORS

Provided they are properly weather stripped, storm doors can decrease the loss of heat and cut down on drafts through and around the doors of your home. Storm doors also let you open the inside door without the home receiving the blast of a blizzard. If a *combination* (storm windows and/or screens) door is installed, the main home door can be opened and the screens in the storm door allow the breezes to blow in while keeping the bugs out.

Most storm doors are made of aluminum but wood and steel are also available. The aluminum doors may be bright, anodized, or covered with a baked-on paint. Bright aluminum indicates a plain mill finish that will tarnish and pit quickly. An anodized finish is silver-gray in appearance; it protects the aluminum from corrosion. The baked-on finish allows you to decorate with color and to rid your home of the aluminum look. Wood doors must be painted occasionally as they wear from the weather. Wood doors are easily fit into nonstandard-size openings by sawing excess material from the top or bottom and by planing the sides.

Combination storm doors are the most useful type; they contain glass panels and screen panels. There are two designs in combination storm doors. With one design, one of the two glass panels slides up behind the other for storage; this has the disadvantage that only one panel is open and screened for the summer season. With the second type of storm door, the glass panel(s) are removed and stored and screened panel(s) are inserted in their place (Fig. 8-26). This provides an opening nearly double the size of the stored panel door to allow cool air to circulate during the summer.

Storm doors are also available with decorative, colored, acrylic, side-light panels. Some have jalousie glass that allows full or partial ventilation, but the jalousies are not very effective weather seals in the winter.

The easiest type of storm door for the do-it-yourselfer to install is the prehung type. All holes are predrilled for hardware and the door comes hinged on an aluminum Z-shaped extrusion that fits over the existing casing.

You can't get a quality storm door without paying for it and to be effective, you need quality. Look for these features: piano hinge or three strong hinges, removable storm panels of tempered safety glass or plexiglass held in a vinyl U-shaped channel, removable screen panels with aluminum or fiberglass screen, aluminum wall thickness greater than 0.05 inch, strong rigid kickplate (now available in double insulated fiberglass), reinforced corners

FIGURE 8-26. This prehung, aluminum door contains acrylic side-light panels. The glass panel is replaced with a fiberglass screened panel in the spring.

secured with screws, and an expander strip at the bottom for a weathertight fit against the threshold. The door should include a drip cap to mount over the door to keep water out, a pneumatic closer at the center of the door to close the door slowly and securely, and a safety chain with a snubber (spring) mounted at the top of the door to keep the door from slamming wide open if caught by a high wind.

To install a storm door, carefully measure the height and width of the opening. In regard to height, order the door a little less than the height needed; the bottom expander is adjustable to fit snugly over the threshold. If you can't buy the exact width of door needed, buy one a little oversized and use a wood chisel to remove a little of the wood on the casing. This is easier than shimming the door. Install the storm door as described in DIY 8-7.

Do-It-Yourself 8-7
PREHUNG STORM DOOR INSTALLATION

Materials: door kit

Tools: drill and bits, screwdriver, hacksaw, wood chisel (if necessary to remove wood from casing)

Procedure:

1. Measure and cut the Z bar to size.
2. Screw the predrilled frame to the existing door frame. If necessary, use a wood chisel to remove wood from the casing. Also use a level to check that the door is level. If it is not level, shim it with thin wood strips.
3. Install the latch set and striker. Position, set, and adjust the striker so that the door latches tight.
4. Install the door closer and adjust it so that it closes the door quickly, but does not cause the door to slam.
5. Install the wind chain and snubber.
6. Close the door. Adjust the bottom expander for a snug fit against the threshold.
7. If necessary, add weather stripping around the door (Chapter 6).

8-8. *VESTIBULES*

Another excellent method (and incidentally, one used in older homes) of reducing the influx of cold air through an opened exterior door is to add a vestibule (a hall or room between the outer door and the interior of a home). The vestibule (Fig. 8-27) acts as an "air lock," may be permanent or portable, and may be installed on the exterior or the interior. When the outer door is opened, cold air enters the vestibule but not the house; the pressure of the wind enters only into the vestibule. If so designed and provided, you can use the vestibule as a place to remove snowy, wet boots and clothes. If the vestibule contains sufficient glass and faces the sun, it will be somewhat warmer than the out-of-doors and you may even be able to grow some plants there. It doesn't need insulation, but it must be caulked and weather-stripped to keep drafts out.

With the exterior door of the vestibule closed, the interior door is then opened for admittance into the house. The wind is blocked by the air-locking vestibule. Placing the vestibule on a southern facing side of the house and making the other doors off limits will be the most effective.

Ideas and plans for vestibules/air locks are contained in Popular Science, September 1975 and September 1974, and in Popular Mechanics, April 1973. A person with a bit of construction experience can make one. Basically, the vestibule is framed with 2- by 3-inch studs, 2- by 4-inch studs, or posts.

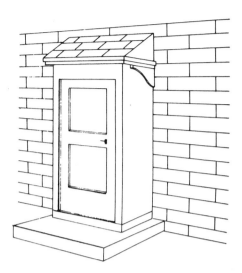

FIGURE 8-27. A vestibule acts as an air lock to reduce the influx of cold air into the house when the door is opened. A vestibule can be located outdoors or indoors.

Exterior walls are constructed of exterior plywood while interior walls are made of plasterboard or paneling. If your exterior already has a canopy or a porch, you'll have less work. You can buy doors completely prehung in a frame to make your job of hanging a door easier.

FOR FURTHER READING

How to Replace a Window, Family Handyman, March 1976.
How to Install a New Wood Window, Popular Science, Sept. 1975.

chapter 9

HUMIDIFICATION AND DEHUMIDIFICATION

Have you ever noticed that outdoors on two successive days at about 70 degrees F you feel too hot on one of the days, yet cool enough on the other to slip on a light jacket? Assuming that the wind conditions on both days were about the same, what caused you to feel hot one day and cool the next was the *humidity*–the degree of wetness or dampness of the air.

Relative humidity is the ratio of the amount of water vapor in the air at any one time to the greatest amount of water vapor possible in the air at the same temperature. When the relative humidity is low (up to 50 percent), water evaporates into the air rather quickly; when the relative humidity is high (50 to 100 percent), the rate of water evaporation decreases until, at 100 percent, no evaporation takes place. When the surrounding temperature is high and our bodies are active, we perspire; our bodies secrete and emit perspiration, a saline (salty) fluid. If the relative humidity is high, the perspiration does not evaporate into the air and our bodies feel excessively hot. If the relative humidity is low, the perspiration evaporates rapidly, lowering our skin temperature; this makes us feel cooler. Putting more moisture into the air slows the evaporation process from the skin so that we feel warmer. We are most comfortable with a relative humidity of about 35 percent.

In the winter, our homes are sealed from the outside weather. The better the caulking, weather stripping, storm doors and windows, insulation, and vapor barriers, the better is the seal from the outside temperature and humidity. We further dry the inside air by heating it to keep us warm. As the moisture from the air is evaporated, so is the moisture from woodwork, furniture, plants, books, and so on, until the total moisture within our homes drops below a comfortable level (below about 30 percent). Similarly, moisture in our skin dries rapidly, decreasing our skin temperature, making us feel cooler.

A low relative humidity in the home causes dry skin; breathing dryness (waking up in the morning with a dry throat and nostrils); loss of moisture from hygroscopic materials (materials that absorb and retain moisture) such as natural wood fibers in woodwork, furniture and most foods; static electricity, and the drying of living plants. Low relative humidity also causes books, musical instruments, and furniture glue to become brittle. Without putting additional moisture into a heated home in winter in some manner other than everyday methods such as cooking and showering, the relative humidity in most homes is about 15 percent. This is too dry and in fact is drier than Death Valley or the Gobi Desert.

In contrast, more moisture in the air makes you feel better, is better for your health, helps eliminate static electricity, keeps your pets and plants healthy, and keeps your furniture, veneers, musical instruments, and wallpaper from drying out and becoming brittle. With a higher humidity in the home, there is less evaporation of moisture from your skin and you can lower the house temperature without feeling cooler. Experts agree that if the humidity is increased from 20 to 40 percent and the temperature is decreased by 3 degrees F, the comfort level remains the same; you also realize about an 8 percent energy saving. At 68 degrees F, a comfortable, healthy, relative humidity is between 30 and 35 percent.

You have to guard against excessive moisture in the home during the heating season too. Excessive moisture causes cold windows and other surfaces to fog or perspire; wood, paint, and other finishes to peel; plaster to crack; moisture in insulation to freeze, thus becoming ineffective; metal to rust; paneling, carpets, and draperies to mildew; and causes possible damage to structural members of the house. Thus, although adding moisture into the house is recommended to increase the relative humidity to about 35 percent, humidity above this level is not recommended. Experts recommend the maximum safe indoor relative humidities listed in QRC 9-1.

In the summer you may find that the humidity in your home is excessive, causing painted surfaces to flake, hygroscopic materials (wood fiber) to swell, metal to rust, and mildew to form on books, walls, and similar places. Mildewing and rotting begin to occur when the relative humidity exceeds 70 percent. These conditions usually occur in areas where air does not circulate and where you do not air-condition; the basement is the most likely place for excessive moisture in the summer.

The remainder of this chapter deals with measuring relative humidity and how to humidify and dehumidify, as required. Specifically, measuring relative humidity, natural ways to humidify, humidifiers, natural ways to dehumidify, and dehumidifiers are discussed.

		Maximum Safe
		Recommended
Outside	*Outside*	
Temperature °F	Temperature °C	*Indoor Relative Humidity*
−10	−23	20%
0	−18	25%
10	−12	30%
20	− 7	35%
30 & above	− 1	35%

Quick Reference Chart 9-1
RECOMMENDED INDOOR RELATIVE HUMIDITY

9-1. MEASURING RELATIVE HUMIDITY

You have already learned that relative humidity is the ratio of the amount of water vapor in the air to the greatest amount possible at the same temperature. Relative humidity is measured as a percentage and we feel the most comfortable indoors when the temperature is 68 to 70 degrees F with a relative humidity of 30 to 35 percent. You can measure relative humidity easily with a relative humidity gauge or by means of a wet and dry bulb thermometer and a conversion chart.

A relative humidity gauge has a needle that points to the relative humidity on a calibrated scale between zero and 100 percent. These gauges are available in combination with a thermometer and a barometer and are mounted for desk or wall placement. Prices range from $15 to $150 depending upon accuracy and decor.

Relative humidity can also be measured using wet and dry bulb thermometers and a psychrometric chart. Obtain two mercury thermometers reading in degrees Fahrenheit. Make a wet bulb thermometer by using a piece of thread to attach a piece of white cotton cloth around one thermometer bulb. Securely attach a string to the other end of the thermometer so that the thermometer can be slung in a circle.

To use the wet and dry bulb thermometers to determine relative humidity, read the dry thermometer termperature. Wet the cloth on the wet bulb thermometer with water and rotate (sling) the thermometer rapidly in a circle to partially evaporate water from the cloth. Read the wet bulb thermometer every minute or so until the temperature does not decrease any further. Then make the final wet bulb thermometer reading.

Figure 9-1 is a basic psychrometric chart from which relative humidity can be obtained if dry and wet bulb temperatures in Fahrenheit degrees are known. Dry bulb temperatures are plotted horizontally (20 to 105 degrees

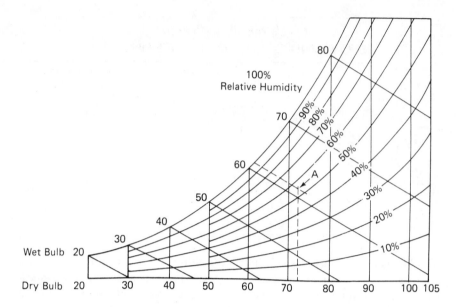

FIGURE 9-1. A psychrometric chart is used to determine relative humidity when the dry and wet bulb temperatures are known.

F), wet bulb temperatures are plotted diagonally (20 to 83 degrees F), and relative humidity is shown by the curved lines. To determine the relative humidity, locate the dry and wet bulb temperatures on the respective scales. Move along the scale lines until they intersect; the intersection indicates the relative humidity (you have to interpolate the approximate percentage). For example, if the dry bulb temperature is 72 degrees F and the wet bulb temperature is 61 degrees F, the relative humidity is 55 percent as shown at point A, Fig. 9-1.

9-2. NATURAL WAYS TO HUMIDIFY

There are two ways to humidify: one is by natural ways which cost you nothing and the other is by adding a humidifier which will cost you some money. Natural methods of adding humidity are both unintentional or intentional. Unintentional natural methods of putting moisture into the home include showers and bathing, plumbing devices, cooking, perspiration, laundry, and pets. Sometimes these sources are desirable and other times they are not. Intentional natural methods of humidifying are discussed in the following paragraphs.

When air is warm, it holds more moisture. Therefore, if water is placed near a source of heat, the water will evaporate more readily, placing moisture into the air. One of the easiest natural methods of humidifying is to place metal pans of water on radiators or on top of the heater; but do not place water over or on an electric heater. If possible, you could also place water pans in hot air ducts.

WARNING

Do not place water pans over or on electric heaters. Placing your hand in water and touching an electric circuit could cause dangerous shock and possible death.

You can also put moisture in the air by keeping some water in your sinks and tubs. Warm or hot water is best, but keep in mind that you are using energy to heat the water. One suggestion is to let shower water accumulate in the tub and then let it (or bath water) remain there after your shower (or bath) is completed. You can also let the last rinse from your dishwasher or your clothes washer remain in the sink or laundry tub so that it partially evaporates into the air as the water cools. Be careful though that you don't overflow the sink or laundry tub onto the floor.

Two other natural ways of humidifying are via hobbies–the raising of indoor plants and flowers and tropical fish. If you already raise plants and flowers, recall how often you water them; frequently, perhaps daily, in the winter months when the heating system is drying out the air, and much less frequently in the other months. Thus, by growing more plants, you are naturally adding moisture to the air of your home. One effective method of growing some plants is to place the planters in a sunny window in a metal tray containing small stones. Place water in the tray; this allows for water evaporation into the air and also the plants can absorb some of the water. If the tray is placed on a radiator, so much the better.

Tropical fish tanks have heaters that keep the water at about 72 degrees F which helps the water evaporate. Tanks also have filtering and air injection systems to place oxygen into the water for the fish. The rising air bubbles disturb the water surface causing water to be atomized into the air.

If you have room for clothes lines in your basement or other parts of the home, hang washed clothes up to dry. To increase the drying rate, hang them near a radiator, heater, or hot air duct.

If you have an electric clothes drier, you can disconnect the normal exhaust hose to the outside and let the drier vent inside. Be careful though that this doesn't cause excessive moisture. Do *not* vent a gas drier to the inside because of the fumes of combustion exhausted by a gas drier.

WARNING

Do not vent the exhaust of a gas drier to the indoors. The fumes of combustion from a gas drier could be hazardous to your health and life.

You can also add water to the air by using a vaporizer normally intended to add moisture and medication to the air during bronchial illness. Don't add the medication to the vaporizer. If a hot water vaporizer is used rather than a cold water vaporizer, insure that the vaporizer is located so that the hot water cannot fall on anyone.

9-3. HUMIDIFIERS

If you've tried the natural methods of humidifying and you cannot reach the recommended safe humidity recommended in QRC 9-1, consider adding a humidifier. Humidifiers are manufactured in a variety of types and models all having the same purposes–to add moisture to the air to make you feel more comfortable and to protect your home and furniture from excessive drying. This section describes the various types of humidifiers and how to install a typical humidifier. Quick Reference Chart 9-2 is a guide to the number of gallons of water that needs to be added each 24 hours to a home having the approximate areas shown.

There are two major types of humidifiers. One type is quite easily installed permanently into a hot air heating system in such a way that warm, moist air is distributed to all areas via the duct distribution system. The second type of humidifier is the portable unit which is centrally located between a number of rooms needing humidification. Hydronic heating, steam, and electric home heating systems do not lend themselves to the addition of a humidifier; a portable humidifier must be used. Humidifiers consist of a water tank (reservoir), a *medium* for carrying water into a stream of warm air, a fan (or natural air current) for blowing warm air through the medium to pick up moisture, a water inlet (manual or from a cold water pipe), and a humidistat that automatically turns the humidifier on when the ambient air becomes too dry.

Hot Air Heating System Humidifiers

If you have a hot air heating system, you're not so lucky because the moving hot air tends to dry out the air more than other heating systems (not to mention blowing dust); but you are also lucky in the sense that it is relatively easy to add a humidifier into the hot air ducting system (Fig. 9-2). Thus, moisture is carried in the warm air to every room serviced by the ducts. The hot air duct humidifiers all operate fundamentally on the same principle: placing water into the stream of hot air from the furnace. Because of the fact that the warmer the air the more moisture can be held in it, it makes good sense to locate the hot air humidifier as near the hottest part of the heater as possible; therefore most hot air humidifiers are located at the plenum which is the large duct area immediately above the furnace (or they draw air from the plenum).

Quick Reference Chart 9-2
HUMIDIFICATION SELECTION GUIDE

Find your humidification requirements by first determining the size of your home and its type of construction. Then read across to the necessary humidification requirements. For example, if you have a 1500-square-foot home with average construction, you have a humidification requirement of *at least* 7.7 gallons/24 hours.

Construction of Residence	Size of Residence in Sq. Ft.					
	500	*1000*	*1500*	*2000*	*2500*	*3000*
Tight (Well Insulated, Vapor Barrier, Tight Storm Doors and Windows with Weather Stripping, Dampered Fireplace)	0.1	2.1	4.2	6.3	8.4	10.4
Average (Insulated, Vapor Barrier, Loose Storm Doors and Windows, Dampered Fireplace)	1.2	4.4	7.7	10.9	14.1	17.4
Loose (Little Insulation, No Storm Doors or Windows, No Vapor Barrier, Undampered Fireplace)	2.5	6.9	11.4	15.9	20.4	24.8

NOTE

If there is uncertainty as to the type of home construction, the values shown in the average category may be used.

An amount of approximately 2 gallons per 24 hours provided by internal sources of humidity (based on a family of four) has already been deducted from the above values.

To prevent overhumidification, make certain you properly adjust the humidistat, preferably when there is a major change in outdoor temperature. Condensed moisture or frost on inside windows is a good sign that your controls are set too high. You can place your humidifier almost anywhere in the living area as long as the moist air can circulate freely throughout the house. Avoid placing the unit near cold, outside walls or in the bathroom, kitchen, or laundry area. When your humidifier first begins to operate, it may run constantly for a day or so in order for dry furniture, drapes, etc. to absorb moisture. Normal changes in indoor and outdoor conditions will cause the unit to cycle on and off to maintain necessary humidity levels. When determining the square foot area to be humidified, include only those areas where normal, daily living occurs. (Data from Association of Home Appliance Manufacturers)

Most hot air furnace humidifiers consist of a number of similar parts including a water-transfer medium, water-inlet valve, water-holding tank, water shutoff valve, and a motor. Water-transfer media are made of absorbent material such as fiberglass, plastic, or other material. The base of the medium sets in the water holding-tank and soaks up water that evaporates into warm air

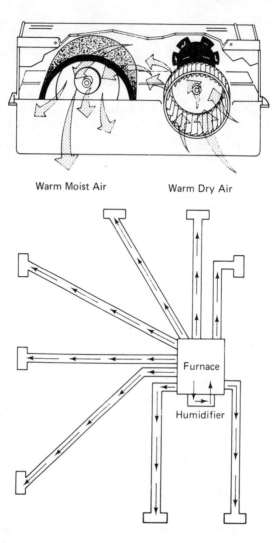

FIGURE 9-2. This humidifier is installed on the outside of a hot air furnace plenum. The fan draws warm air from the plenum, directs the air over the rotating medium where it absorbs moisture, and then returns it to the plenum. The furnace blower circulates the warm moist air through the ducts to individual rooms.

as it passes by or through the upper part of the medium. The water-inlet valve admits water from a cold water pipe valve into the water-holding tank. When the water in the tank is at the full level, a float valve cuts off the inlet valve. On some models, a motor is used to drive the medium; on others the flow of warm air rotates the medium. Several typical hot air furnace humidifiers are discussed in the following paragraphs.

The major difference in the design of hot air furnace humidifiers is in the method of getting the water into the hot airflow. Some humidifier designs use fixed absorbent plates, movable discs, rotating drums, rotating discs, fixed filters, a motor-driven slinger, or an atomizer/pulser. Each has its own advantage and selling point and the selection is up to the buyer. However, prior to purchasing, you must determine the amount of moisture that your home needs (QRC 9-2) so that you purchase a humidifier of adequate capacity. You must also know the physical limitations of the area in the plenum or duct where the humidifier is to be located.

A typical *fixed-plate* hot air furnace humidifier uses highly absorbent fiberglass or plastic plates as the media. The plates sit side-by-side (with air space between the plates for warm airflow) in a water tank inside the heater plenum. A water line is connected to a cold water pipe and a float valve cuts the water off when the water-storage tray is full. No electricity is needed. As water evaporates because of warm air circulating through the plates, replenishing water is drawn into the plates like water is absorbed into a sponge. This type of humidifier has a limited capacity of about 3 to 4 gallons per day. If you have central air conditioning with coils in the plenum, you cannot use this type of humidifier.

A second type of hot air furnace humidifier is a hollow, rotating cylindrical frame that is covered with a plastic or fiberglass medium. When air passes through the plenum, it makes the cylinder rotate through the water; water evaporates from the top of the media. This type of humidifier evaporates 6 to 7 gallons of water into the air per day making it adequate for small homes. If air-conditioning coils are located in the heater plenum, this type of humidifier cannot be used.

If your home needs 10 to 30 gallons of water added per day, then you are in need of an electrically powered humidifier in your hot air furnace. There are two popular types of electrically powered humidifiers: motor driven and spray. Each of the powered-type humidifiers may be turned on by one of four methods: a furnace blower relay which turns on the humidifier whenever the furnace blower motor is on, by a *sail* switch which is an airflow-sensitive switch that is installed in the return air duct, by a humidistat which is a humidity-sensitive switch placed in the return air duct, or by a temperature-sensitive switch located in a hot air duct.

The most popular power humidifier uses a motor to drive an evaporator cylinder or disc (Fig. 9-3). It is simple to install, is reliable, and mounts external to the plenum. A self-contained fan draws warm air out of the plenum across the medium and returns the moist air to the plenum where it is blown by the heater blower through the ducts to all areas. The water-holding tray is easily removed for cleaning and for access to the medium for replacement. Most units contain a lime control which allows you to periodically flush the humidifier water tank of impurities collected. If the lime control is used, an overflow tubing connection to a drain (or sink) must be installed.

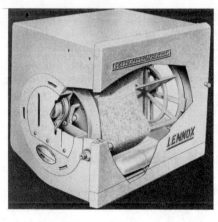

FIGURE 9-3. This humidifier installs on the return air plenum or duct and is connected to the supply air plenum. The bypassed warm air flows through the cylindrical foam filter media, is moistened as it rotates through the water reservoir, and then re-enters the air stream on the return air side (*courtesy Lennox Industries Inc.*).

The spray-type, hot air furnace humidifier has a spray nozzle mounted in the plenum (Fig. 9-4). A humidistat plus a sail switch or blower relay causes a solenoid to energize which opens the water valve on the spray nozzle. The fine spray is evaporated by the warm air from the plenum and is carried through the ducts. Minerals in the water may be carried by the hot air, causing white powder to be deposited on everything.

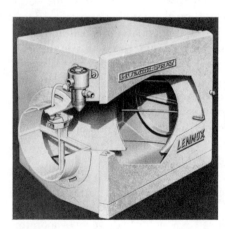

FIGURE 9-4. This humidifier is also attached to the return air plenum or chamber. The bypassed warm air flows through a polyester foam media filter, is moistened by a fine spray of water from the nozzle assembly, and then re-enters the air stream on the return air side (*courtesy Lennox Industries Inc.*).

Installation of a humidifier into a hot air furnace is not difficult. Installation consists of three steps: water connection, plenum cutout and mounting (Fig. 9-5), and electrical connection. Follow the complete installation instructions provided by the manufacturer for your humidifier. The work is basically described in DIY 9-1.

Typical Applications

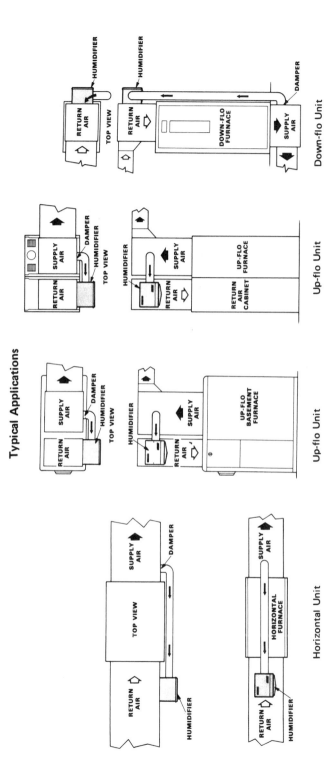

FIGURE 9-5. The humidifiers in Figs. 9-3 and 9-4 are installed in such typical applications as shown here (courtesy Lennox Industries Inc.).

Do-It-Yourself 9-1
HOT AIR FURNACE HUMIDIFIER INSTALLATION

Materials: saddle valve, length of copper tubing, brass compression nut(s) and sleeve(s), nylon compression nut, electrical wire, humidifier (most or all of these parts may be included with the humidifier)

Tools: drill and bits, tin snips or saber saw with metal cutting blade, wire strippers, electrical tape, adjustable open-end wrench

Procedure:

WARNING

Before beginning installation of the humidifier, remove all electrical power to the heater. Use an electrical drill that has a third wire for ground and be sure the drill is properly connected to a grounded outlet. If you are not qualified in electrical work, have an electrician make the electrical connections.

1. *Saddle Valve Connection to Cold Water Pipe*
   a. Turn off the supply of cold water to the cold water pipe where the saddle valve is to be located. Open a cold water spigot downstream of the shutoff valve so that the water pressure is relieved in the pipe.
   b. Using a properly grounded drill and a drill bit slightly larger than the saddle valve stem (usually a $3/16$ inch drill bit), drill a hole in the cold water pipe at a location as close to the humidifier as possible.
   c. Place the valve stem in the drilled hole so that the rubber gasket is flat against the pipe. Install the two bolts through the clamps that hold the saddle valve to the pipe. Tighten the bolts sufficiently so that the rubber gasket seals around the pipe.
   d. Slide the brass compression nut and brass compression sleeve onto one end of the copper tubing. Push the tubing fully into the fitting on the saddle valve. Tighten the compression nut, but do not over-tighten. If water leaks, it can be tightened until the leak stops.

2. *Humidifier Installation to the Plenum*
   a. The humidifier comes with a full-size metal plate that is attached to the plenum to strengthen it. Hold the plate to the plenum and mark its location with a pencil. Mark the location of required plenum metal cutouts.
   b. Drill pilot holes for the attaching machine screws. Using a drill and tin snips or saber saw, remove the metal cutouts required in the plenum metal.
   c. Mount the plate to the plenum with sheet metal screws.
   d. Mount the humidifier to the plate with sheet metal screws.
   e. Mount the humidistat template. Drill the required mounting holes and make the cutouts in the ducting. Attach the humidistat with sheet metal screws.

3. *Electrical Connections*
   a. Insure that all power is removed from the heater.
   b. Make the electrical connections as instructed by the manufacturer. There are normally two sets of connections: from the humidistat to the humidifier and

---

Do-It-Yourself 9-1 (continued)
HOT AIR FURNACE HUMIDIFIER INSTALLATION

---

from the humidistat to the blower motor. All wiring is to be done according to local and national electrical codes and ordinances.

4. *Final Connections*
   a. Secure all parts of the humidifier together.
   b. Connect the water copper tubing to the inlet connection on the humidifier. Secure with a brass compression nut and sleeve or with a nylon compression nut.

The hot air furnace humidifier needs to be cleaned halfway through the heating season (about February 1) and at the end of the heating season to remove lime and other impurities that have accumulated in the water tank and on the media. Turn the water saddle valve off and place the humidifier power switch to off. Remove the water tank and the media. Rinse the tank and media under a full force stream of warm water. Pour out all water and then pour white vinegar into the tank to the normal level of the water. Place the media in vinegar. After 24 hours, the acidic properties of the vinegar will have softened and dissolved the lime deposits and other impurities. Use an old toothbrush to brush the sides and corners to loosen all impurities. Then rinse the tank and media in warm water and replace them in the humidifier. Turn on the electrical power switch and the water saddle valve. If the media cannot be adequately cleaned, order a replacement (it's a good idea to start off each heating season with new media).

Console Humidifiers

If you don't have a hot air furnace, you can humidify your rooms with a console humidifier (Fig. 9-6). The console humidifier is plugged into any convenient electrical outlet and the water reservoir is filled daily. The console humidifier can be moved from one room to another or can be centrally located between two rooms.

Console humidifiers are of three types: one type has a motor-driven cylinder that contains an evaporating pad (medium); another type has a pump that wets a fixed evaporating pad, and the third type of console humidifier has a motor that drives a pad like a treadmill. Each type has a fan that draws air into the humidifier, circulates it through or over the evaporator pad, and exits the moist air into the room; a three-speed fan is recommended. Console humidifiers are capable of adding from 3 to 12 gallons of water into the air per day.

Methods of increasing humidity in the home are summarized in QRC 9-3.

FIGURE 9-6. This portable humidifier utilizes a rotating cylinder with a foam pad to place water in the stream of air provided by a fan. Water is added to the humidifier daily, or as required (*courtesy The West Bend Company*).

Quick Reference Chart 9-3
METHODS OF INCREASING HUMIDITY IN THE HOME

1. Place metal pans of water on radiators or heater (but not on an electric heater) or in air ducts.
2. Leave warm standing water in sinks and tubs after use.
3. Raise indoor plants.
4. Raise tropical fish.
5. Hang wet, washed clothes near a radiator, hot air duct, or heater to dry.
6. Vent electric clothes dryer indoors instead of outdoors.

CAUTION
Do *not* vent a gas drier indoors; always vent it outdoors.

7. Add moisture to the air with a vaporizer.
8. Install a humidifier in a hot air furnace.
9. Place a portable room humidifier between adjacent rooms.

9-4. *NATURAL WAYS TO DEHUMIDIFY*

While insufficient humidification is usually the problem in the winter, rooms such as the bathroom, laundry, and the kitchen may actually have too much humidification from vaporized water, causing local damage. In the warmer, muggy months of the year, you may find over-humidification in your basement as well as other rooms in your home. Remember that over-humidification causes flaking paint, wood swelling, mildewing, mold, corrosion, and rotting. The solution to over-humidification is to find natural ways to rid the area of excess moisture; if natural methods cannot solve the problem, then a dehumidifier (Section 9-5) is necessary.

The simplest solution to excessive moisture is to provide air circulation that will carry the moisture to less moist parts of the house or to the outdoors. If the kitchen or/and bathroom collect excessive moisture when meals are cooked and showers are taken, allow the doors of these rooms to remain partially open so the moist warm air can circulate to other parts of the house. If this doesn't solve the problem, crack a window to let in dry air from the outside. As a last resort, you can use an exhaust fan to draw the warm, moist air out of the room and vent it outdoors (exhaust fans in kitchens get rid of smells as well as the heat and moisture from cooking). As a suggestion, you might consider venting a humid bathroom into a dry living area located above the bathroom (you'd be moving excessive moisture to an area of insufficient moisture).

Other areas in the house that have excessive humidity are treated naturally in the same way—by providing a flow of air. You can use a fan or you can open an interior door and partially open a window. If you are using a humidifier, turn it off until the areas dry out; then turn it back on but at a reduced level. Be sure that attic vents are open and unobstructed so that moisture caused by the warm air inside meeting the cold air on the roof does not cause condensation (more on this in Chapter 10).

Another natural way to get rid of excessive moisture is to partially open two windows on opposite sides of a room or rooms for a short period each day. Be sure to vent your clothes drier to the outside because it can put a lot of moisture into the air (gas driers are *always* vented to the outside). Put all of your plants into one room and keep the door closed so that the moisture does not move to the other rooms. Don't overheat this room. Damp basements can sometimes be dried out naturally by opening the basement windows; you can increase airflow by placing an exhaust fan in the wall or in the window (you can remove a pane of glass and install a fan with a wood frame to fit the pane area). If moisture still cannot be removed from the basement, you'll need a dehumidifier.

A note about condensation and storm windows before we proceed to a discussion on dehumidifiers. Condensation is formed when two sides of an object are subjected to greatly different temperatures: for example, a window at 68 degrees F inside and 10 degrees F outside. This difference in temperature causes condensation on the glass in the form of frost or ice. A properly installed storm window or insulating glass prevents condensation because the cold and warm surfaces are separated by an air space. Metal frames of storm windows have a similar problem and need a thermal barrier to act as a temperature insulator to prevent condensation and conductive heat transfer.

If a storm window is in place and condensation accumulates on the inner window, the storm window needs caulking and weather stripping; cold air is leaking past the storm window to the inner window. If, on the other hand, the storm window has condensation on it, the inner window needs to be weather-stripped; warm air is leaking past the inner window to the storm window. (Refer to Chapter 5 for caulking and to Chapter 6 for weather stripping.)

9-5. DEHUMIDIFIERS

A dehumidifier is a device used to remove excessive moisture from a closed area such as a room or basement. A dehumidifier is needed whenever natural means of dehumidification (Section 9-4) are insufficient to lower the humidity to a level where occupants are comfortable, moisture is minimal, mildew does not grow, and the moisture does not effect paint, wood, etc. Dehumidifiers are seldom needed in air-conditioned rooms because the air-conditioner dehumidifies the area while it cools the area and the extracted water flows to the outdoors. (Remember that less moisture can be held in cool air than in warm air.) A dehumidification selection guide is shown in QRC 9-4.

Quick Reference Chart 9-4
DEHUMIDIFICATION SELECTION GUIDE

Values in table indicate dehumidification required in pints per 24 hours, based on the area of the space to be dehumidified and the conditions that would exist in that space when a dehumidifier is not in use.

* Condition Without Dehumidification	Area–Sq. Ft.			
	500	1000	1500	2000
Moderately damp—Space feels damp and has musty odor only in humid weather.	10	14	18	22
Very damp—Space always feels damp and has musty odor. Damp spots show on walls or floor.	12	17	22	27
Wet—Space feels and smells wet. Walls or floor sweat, or seepage is present.	14	20	26	32

* During warm and humid outdoor conditions.

Dehumidification variables also include such other factors as climate, laundry equipment, number of family members, number of doors and windows, and degree and intensity of area activity.
(Data from Association of Home Appliance Manufacturers)

Dehumidification is dependent upon warm air circulating over a cold coil (Fig. 9-7). Moisture from the warm air is condensed when it comes in contact with the cold coil surface. The coils are kept at a temperature below the dew point of the air so that moisture condenses out of the air; but, the temperature of the coils must be high enough that frost or ice formation does not occur, causing a blockage to airflow.

A dehumidifier is usually a small hermetic refrigerating system having both a *condenser* and an *evaporator* in a cabinet. Room air is drawn by a fan over an evaporator which is cold. As the air touches the cold surface of the evaporator copper tubing, the air cools below its dew point, causing water to condense and collect on the evaporator. The cooled air is then moved over the condenser to cool the condenser and rewarm the air. The condensed water from the evaporator drips into a direct drain, runs through a tube to a drain, or collects in a condensate moisture container which is emptied when full.

A humidistat should be a part of the dehumidifier. The humidistat automatically turns the dehumidifier off when the desired humidity level is reached. As the humidity level again increases, the humidistat turns the dehumidifier on. Full range control of 20 to 80 percent humidity is available on most units. A shutoff switch is included in dehumidifiers having a condensate drain pan. When the pan becomes heavy because of accumulated condensate, the dehumidifier is automatically turned off so that overflow of the

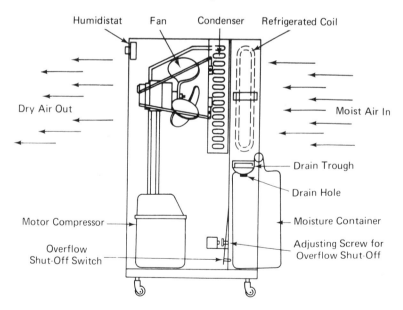

FIGURE 9-7. When moist air contacts the refrigerated evaporator coil of a dehumidifier, the moisture is condensed. Accumulated moisture from this dehumidifier is collected in a container.

moisture container does not occur. A light indicating a full moisture container also turns on. Compressor motors are $1/6$ to $1/5$ horsepower.

Methods used to remove excessive moisture from a home are summarized in QRC 9-5.

Quick Reference Chart 9-5
METHODS OF REDUCING HUMIDITY IN THE HOME

1. Provide air circulation through the humid area.
   a. Crack windows or doors open.
   b. Locate a fan to circulate air.
   c. Install an exhaust fan.
2. Lower the setting of any humidistat on a portable humidifier or a humidifier installed in a hot air furnace.
3. Be sure attic vents are open and unobstructed.
4. Vent clothes driers to the outside.
5. Do not hang wet clothes indoors.
6. Move all plants to one room and close the room door.
7. Put covers over fish tanks.
8. Install storm windows and be sure they are caulked and weather stripped.
9. Install a dehumidifier.

chapter 10

AIR CIRCULATION

Humidity is one of the factors that makes us comfortable. Temperature and air movement are two other factors. At about 68 to 70 degrees F, with a humidity of 30 to 35 percent, our body temperatures are quite comfortable. If the temperature decreases by a degree or so, while the humidity increases a few percent, our comfort level remains the same. If the temperature rises or our body activity increases, our bodies provide for natural cooling by making us perspire. If the humidity is low enough, the perspiration rapidly evaporates from our skin lowering our skin temperature so that we feel somewhat more comfortable. If the humidity is high, the perspiration does not evaporate so readily and we feel hot.

Our movement causes moisture to evaporate more quickly. Think of outdoor swimming experiences. If the temperature is 75 degrees F in the morning and there is no wind when you come out of the water, you feel quite comfortable. Perhaps after lunch the temperature has risen to 85 degrees F and the wind has started to blow steadily at 10 to 15 knots. When you come out of the water now, you feel chilly and uncomfortable. You quickly grab a towel and dry the water off. Although the air temperature has increased, your body feels uncomfortable because the wind causes the water on your body to evaporate quickly, carrying away body heat.

In addition to air movement increasing evaporation, air movement also equalizes temperatures within a confinement and exchanges stale air for fresh air. Many of us have hot air heating systems in which cooler air from rooms is drawn into the furnace plenum where it is heated and recirculated to the rooms. The constant circulation of the air nearly equalizes the temperatures in the confined area of the home. True equalization is not possible in a heated home, however, because of natural air currents in which warm air rises and cold air falls, the infiltration of cold air and the exiting of warm air at windows and exterior doors, and the natural cooling of air along cold exterior walls. Stale air and moisture-laden air are exchanged for fresh air by exhaust fans in the kitchen, bathroom, and laundry. In the warmer months of the year, we are more comfortable and satisfied when we can open the windows somewhat so that we don't feel "all cooped up" and so that a little fresh air can circulate into our homes.

Proper air circulation does more than just make us feel comfortable. Proper air circulation reduces humidity, reduces sweating and frosting of roofs in the winter, and removes unwanted cooking odors, fumes, and hot air. Cooking, laundering, and taking hot showers and baths put excessive moisture into the air which causes sweating of walls and fixtures, mildew, and corrosion of metal parts. Proper air circulation removes the moisturized air from the room and replaces it with drier air by the opening of windows and/or doors, or by cracking a window open and using an exhaust fan.

If your home is built with an adequate ventilation system having inlets in the soffits and outlets in the gables, roof vents, or turbines or power ventilators, the ventilation system is probably adequate and capable of expelling excessive hot air in the summer and moisture that might enter the attic from the living area in the winter. If your home was built a number of years ago, however, it was not built knowing that a lot of moisture would someday be put into the living area from modern dishwashers, clothes washers, and driers. Similarly, the homes were not insulated as well as they are today and since you have taken steps within your home to weatherize it, you may find that you have sweating or frost caused by excessive moisture on the inside of your roof on cold days. You've "buttoned up" your home to keep the warm air inside and to keep a relative humidity at the recommended level (Chapter 9) through insulation, vapor barriers, weather stripping, caulking, double glazing, storm doors, additional humidity, etc. The moisture in the air is in the form of vapor (gas). The vapor pressure is sufficient to force its way through wood, plaster, brick, concrete, and other building materials. The vapor that bypasses barriers creates condensation problems; if there is no significant change in temperature between two areas, then there is no condensation. However, in the winter, when the warm, moist air from the house passes through the ceiling, building materials, insufficient vapor barriers, and insulation, it meets the cold roof, which should be vented to let out the moisture-laden air. However, if the ventilation is inadequate to carry off the warm moisture-laden air, it condenses on the cold interior roof and rafters forming moisture or, if cold enough, frost (Fig. 10-1). This has very detrimental effects that can include damaged plaster, ineffective insulation, rotted wood, and cracked roofing shingles. The cure is additional air circulation sufficient to keep both sides of the roof at the same temperature and to carry away the vapor before it

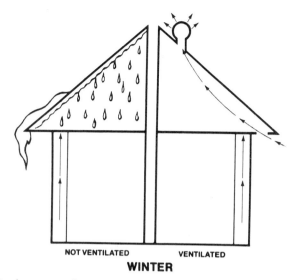

NOT VENTILATED   **VENTILATED**
WINTER

FIGURE 10-1. Inadequate ventilation in the attic in the winter causes moisture-laden warm air from the living area to condense on the roof. The resulting condensate causes many problems (*courtesy Triangle Engineering Co.*).

can condense. Keeping the underside of the roof at the same temperature as the outside may seem like it would cause heat losses from the house. But experts agree that with proper ceiling/attic insulation, the cold, circulating air above the insulation does not significantly cause heat losses. In other words, it is better and more cost-effective in the long run to lose a slight amount of heat by proper ventilation than it is to ruin insulation, crack plaster, rot wood, or crack roofing shingles from condensation or frost.

In the summer, without adequate attic air circulation, the hot sun on the house roof causes a thermal blanket of hot air to heat the interior of your house (Fig. 10-2). Think about it. The hot sun beats down its solar energy on the roof. If you have a light colored roof, a lot of the energy is reflected away; if it's a dark or black roof, more solar radiation is absorbed into the roofing materials. By radiation and conduction of heat from the roofing material, the roof wood, rafters, and supports become hot. By radiation and convection, the air in the attic becomes hot and transfers the heat to the insulation on the attic floor (the ceiling of the floor below). With the intense trapped heat in the attic (about 160 degrees F with an outside air temperature of 95 degrees F) the heat penetrates the insulation and heats the ceiling below it by radiation and conduction. The hot ceiling heats the air in the rooms by radiation and convection, raising the temperature (to about 85 degrees F in the example). In a well insulated home, with windows and doors closed and the shades drawn, the temperature rises at about one degree F per hour. If exterior doors are opened, warm air rushes in and the interior temperature rises more rapidly.

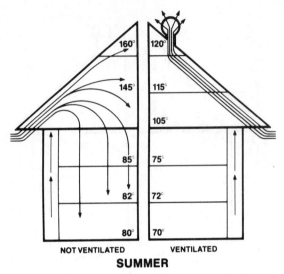

FIGURE 10-2. Inadequate ventilation in summer causes a thermal blanket of hot air to heat the house interior (*courtesy Triangle Engineering Co.*).

Air circulation, both natural and powered, reduces air-conditioning costs. Heat trapped in the attic reaches temperatures of 160 degrees F because of the radiated heat from the sun. This heat blanket works its way past the ceiling insulation to the living space below. The outside air is at a much lower temperature, and our objective is to replace the hot air in the attic with outside air via air circulation. When the heat-blanketing effect in the attic above is removed, the temperature of the living space below is lowered. Therefore, the air conditioner does not have to come on as often nor stay on as long. This saves energy costs. Secondly, since there is not as high a temperature to be lowered, the size of the air conditioner can be smaller; this saves on cost too.

The temperature of 85 degrees F is of relative importance in maintaining a comfort level during the warm seasons of the year. Without air conditioning, the temperature of the inside of the home can become greater than the outside temperature due to the radiant heat of the sun and to the heat blanket placed over the living area by the attic. Relatively few days across the United States are over 85 degrees F (for example, 8 percent of the days in Boston are over 85 degrees F, 10 percent in Chicago, 22 percent in Denver, 5 percent in Los Angeles, 15 percent in St. Louis). Thus, with the radiant heat causing the temperature in our homes to rise, we can reduce the inside air temperature by replacing the inside air with cooler outside air. When the ambient temperature is less than 85 degrees F, the body loses heat and we feel comfortable. If the air is too cool, we lose too much heat and we feel chilly. We can thus cool the whole house on all those days except those above 85 degrees F without the use of air conditioning. Instead of the air conditioner, we bring in the cooler air from the outside, cool off the house and then keep the cool air in the house while the outdoor temperature rises above 85 degrees F. If the temperature in the house is 70 degrees F at 6 A.M., it will be approximately 82 degrees F by 6 P.M. (as previously stated, the temperature rises about one degree F per hour in a well insulated house).

When the inside of the home is heated (by heat from the attic, or the furnace), the walls, ceiling, floors, and other structural materials hold or store the heat. Similarly, the furniture, draperies, etc., reach the temperature of the room and store the heat. Thus, all of the objects in your home store and retain heat. When the sun sets and the house cools, this stored heat is dissipated into the living space within your home, keeping the living space at the hottest temperature that the living space reached. This is the reason that an air conditioner continues to run even after the heat of the day is gone. The objective of air circulation is to pull the cooler evening air from the outside into your home and carry the hot air from the living area, the stored heat from the structural materials, and the stored heat in the furniture to the outdoors. As the heat is extracted and the temperature decreases throughout the night, our objective is to retain or store the cooler air in the living space and to keep the structural materials and the furniture at a lower temperature level the next morning. Thus, we continue to circulate the air throughout the night and the early hours of the morning. Then the living area is sealed off by closing the storm windows, and the doors to hold in the reservoir of cool, outdoor air that we brought in during the night.

Air circulation then makes us feel comfortable, reduces humidity, reduces sweating of roofs by removing excess moist air, removes unwanted hot air, cooking odors, and fumes, and reduces the costs of air conditioning. Circulating air increases the well being and health of people, animals, and plants. When used properly it saves energy and thus, money. This chapter deals with the proper and effective use of circulating air, both naturally and by powered force.

10-1. NATURAL AIR CIRCULATION

You can increase your comfort, decrease attic and living space temperatures, reduce sweating roofs, remove unwanted hot air, cooking odors, and fumes, and decrease costs of air conditioning through natural air circulation at little or no cost. All you have to do is provide the correct ventilation for the flow of air at the proper time. Let's look at the attic first: if we can create enough natural air circulation in the attic in the summer to prevent a thermal heat-blanketing effect from heating the living area and if we can prevent condensate from forming on the inside of the roof in the winter, we've met our objective for the small cost of some ventilators. With good airflow through the attic, we'll increase natural air circulation through the living area for additional cooling.

Natural air circulation is dependent upon *wind* pressure and direction, *thermal effect, stack effect,* and ventilators. Wind pressure from the prevailing wind causes a positive (high) pressure area to be present on the side of the house facing the wind and a negative (low) pressure to be present on the leeward side of the house (Fig. 10-3). If windows are properly opened on the windward and leeward sides of the living area, natural air circulation takes place. Similarly, if proper ventilators are placed in the soffits and a gable or ridge vent is located at the peak of the house, natural air circulation takes place.

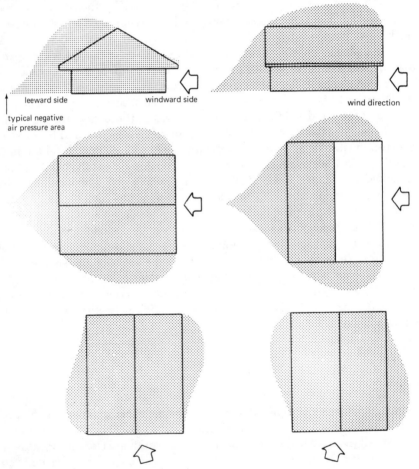

FIGURE 10-3. Air patterns surround a residence and cause high and low pressure areas. Predominate surface wind directions are shown in Figs. 2-10 and 2-11.

Thermal effect is the airflow caused by the difference in the temperature of the air outside the home with respect to the temperature of the air inside the home. When there is a temperature differential, the cold air will move, pushing the warm air upwards. Since the outside air temperature is usually lower than the inside attic temperature, air will circulate naturally because of the thermal effect; the hotter inside air flows from one ventilator while cooler outside air flows inward through the ventilator.

The stack effect (also called chimney effect) is the airflow caused by the difference in elevation between the lowest and highest ventilator opening (Figs. 10-4 and 10-5). Since warmer air rises, it rises to the top of the stack and

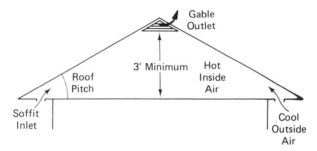

FIGURE 10-4. Cool air enters the soffit ventilators as warmer air flows out of the higher gable outlet ventilators.

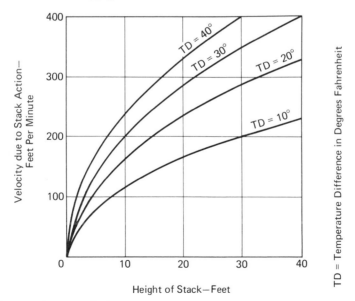

FIGURE 10-5. The greater the height distance between the inlet and outlet ventilators and the greater the temperature difference, the greater is the velocity of the natural air circulation between the ventilators.

is expelled to the exterior while cooler outside air enters at the bottom of the stack. The stack effect works together with the thermal effect causing *gravity ventilation*. Gravity ventilation increases in effectiveness as the air intake supply increases, the roof pitch becomes higher, and the inlet and outlet ventilators are placed further apart. The temperature of the outside air is cooler than the inside air temperature which builds up as stored heat. The cooler air coming in at the lower vent pushes the lighter, warmer air as it rises out of the upper ventilator. This stream of air up the stack causes the attic air to become cooler.

Ventilators

The amount of air circulated naturally is also dependent on the free area (opening) and location of the ventilators. Ideally, inlet ventilators are placed in the direction of the prevailing winds and are the lower ventilators of a ventilator system. The outlet ventilators are either roof ventilators (preferred) or are ventilators placed opposite the intake on the low pressure (leeward) side of the home. If the ventilators are located on opposite sides of each other and there is a wind shift, an intake ventilator could become an outlet ventilator and vice versa. Ventilator openings are to be equal in area at both the inlets and the outlets (high and low pressure areas).

Ventilators must keep out rain, snow, and bugs. They usually contain slotted openings, called louvers, which may be adjustable as to the amount of opening, but the louvers usually are in a fixed position. The louvers keep out the rain and snow, but allow air to circulate under the openings; screening is placed inside the louvers to keep out bugs. The mesh of the screening should be a minimum size of ⅛-inch openings; slightly larger mesh up to ¼ inch allows for better air circulation. The free area in square inches is usually marked on the ventilator and it is usually about 50 percent of the gross size or maximum area of the ventilator; the area blocked by louvers and wire mesh is *not* a part of the free area. The airflow through the free area sometimes becomes blocked by fibers of insulation and dirt in the screening. Fibers and dirt should be removed periodically by brushing the screen with a soft bristle brush such as a paint brush.

Ventilators, also often referred to as louvers (Fig. 10-6), may be triangular in shape for mounting in the gable at the peak of the house or may be rectangular for mounting on exterior walls. Ventilators mounted in the soffits under the eaves are rectangular, usually 8 by 16 inches, or they may be a continuous strip that is 2 to 4 inches wide by any length.

The free area required for use of attic ventilators (louvers) is relative to the attic floor area in square feet. Measure the attic floor width and length and multiply these together to get the total attic area (in square feet). The FHA recommends that there be one square foot of free ventilated area for each 150 square feet of attic floor. If a vapor barrier is in place, and there is low moisture

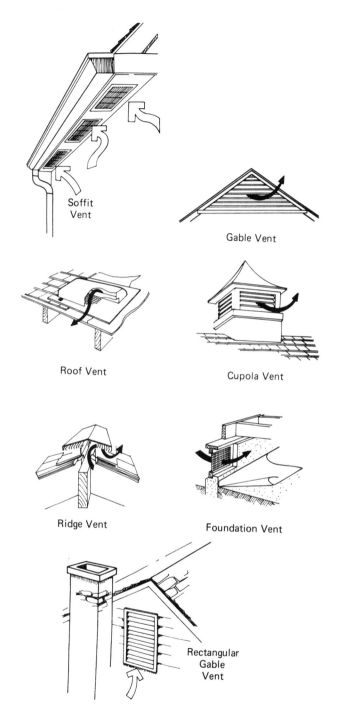

Soffit Vent

Gable Vent

Roof Vent

Cupola Vent

Ridge Vent

Foundation Vent

Rectangular Gable Vent

FIGURE 10-6. Ventilators come in a variety of types, shapes, and sizes for any ventilation application.

transmission through the vapor barrier, then the FHA recommends one square foot of free ventilated area for each 300 square feet of attic floor area. When the ventilated area is calculated, 50 percent of that area is required as the upper part of the ventilation system for exhaust and the other 50 percent is required as inlet ventilation. The upper exhaust ventilator must be located at least 3 feet above the under eaves or cornice ventilators. This distance is required to make effective use of the stack and thermal effects.

The American Ventilation Association recommends (as a minimum) two times the amount of free area recommended by the FHA. The AVA recommendation is one square inch of free area for every square foot of attic area. For example, if your attic measures 600 square feet, you must provide 600 square inches of ventilation; 300 square inches are to be provided at the inlet vents and 300 square inches at the exhaust vents.

Attics are usually ventilated by placing ventilators in the soffits around the perimeter of the home. Additional ventilators of equal area to the soffit ventilators are placed in the gables at the ends of the home (Fig. 10-7) or through the roof; this allows the stack and thermal effects to take place. Soffit and roof ventilators should be as continuous as possible for maximum effectiveness.

Rectangular, Gable End, Vent with Undereaves Vents

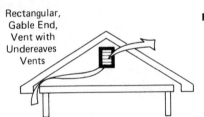

FIGURE 10-7. Many older homes have air circulating from soffit ventilators through gable ventilators. With all of the modern appliances presently in homes, the natural ventilation may be inadequate for removing heat and moisture (courtesy Leigh Products Inc.).

Sufficient ventilators must be placed in the soffits so that a natural air stream from the soffit ventilators to the gable or roof ventilators passes by and contacts all of the surfaces of the roof where moisture tends to collect in the winter and which become hot in the summer. The underside of the roof is the most important surface for air to circulate past during the summer to get rid of heat and during winter to get rid of excessive moisture.

Continuous soffit vents and a ridge vent (Fig. 10-8) are especially effective as a means of natural air circulation. This combination provides a more rapid and complete change of attic air than any other combination of nonpowered ventilators. Ridge vents are easy to install when the home is built; however, builders often do not put ridge vents in because of the added expense. Once a home is built, ridge vents are difficult to put into place. Soffit vents can be installed quite easily.

In natural air circulation, the best means of ventilating is to have a combination of soffit ventilators, gable ventilators (louvers), and roof ven-

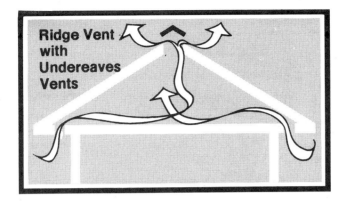

Ridge Vent
with
Undereaves
Vents

FIGURE 10-8. Continuous soffit and ridge ventilators provide the best natural attic ventilation. Ridge vents are installed during home building (*courtesy Leigh Products Inc.*).

tilators all working together to create a ventilating system. In the event the natural air circulation does not provide sufficient ventilation to keep the living area as cool as you would like, does not satisfactorily vent the attic space to rid it of excessive heat blanketing in the summer, or if excessive moisture-causing condensation and frost forms in the winter, then you must consider the addition of power ventilation to your home.

Roof Ventilators

Instead of using gable ventilators in the attic, roof ventilators (Fig. 10-9) may be used. Roof ventilators operate on an *ejection action* principle in which the outside air passes over the surface of the ventilator and then speeds up (Fig. 10-10). As the air passes the surface, it creates a low pressure and air is drawn from the attic by induction. Roof ventilators work even without the wind; they start working as soon as there is a temperature differential; the rising air through the heat stack causes the ventilator to work (Fig. 10-11).

The roof ventilator is weatherproof so that rain and snow cannot be blown into the attic space. The ventilator that you install must have sufficient capacity to let the hot air of summer and the moist air of winter exit through it. It is located on the top rear slope of the roof (this is for aesthetic reasons only) in the center of the house. The higher the roof ventilator is placed, the greater is the efficiency of ventilation because of the stack effect. Also, if the ventilator is at the top of the roof, it is in a direct line to receive and make maximum utilization of the wind which further induces the flow of air by the ejector action.

A roof ventilator is easily installed by cutting a hole through the roof and then nailing the ventilator and roof flashing into place. The nails are caulked to prevent water leakage. The manufacturer provides full installation instruc-

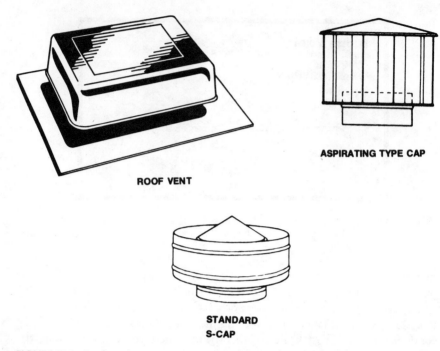

ROOF VENT

ASPIRATING TYPE CAP

**STANDARD
S-CAP**

FIGURE 10-9. Roof ventilators are easily added to increase natural airflow from soffit ventilators to the roof ventilators.

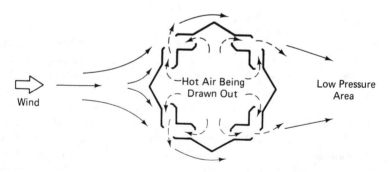

Wind

Hot Air Being
Drawn Out

Low Pressure
Area

FIGURE 10-10. The wind creates an aspirating action in roof ventilators.

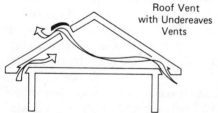

Roof Vent
with Undereaves
Vents

FIGURE 10-11. Roof ventilators work even without wind (*courtesy Leigh Products Inc.*).

tions (DIY 10-1 describes the installation of a *turbine* ventilator which is the same as the roof ventilator, but the turbine must be aligned).

Do-It-Yourself 10-1
TURBINE VENTILATOR INSTALLATION

Materials: turbine ventilator, galvanized roofing nails, roof cement or caulking to cover nailheads

Tools: drill and bits, keyhole or saber saw, ruler, utility knife for cutting roof shingles, hammer and screwdriver

Procedure:

NOTE

Follow the manufacturer's specific installation instructions. The following procedures cover most installations and provide you with a knowledge of the tasks that must be performed.

1. The turbine ventilator is installed near the peak of the roof on the rear slope with the turbine top exposed to the wind from all directions (Fig. 10-13). When installing two ventilators, place each one-fourth of the distance from the end of the house; when installing three, place one in the center and the other two one-sixth of the distance from each end of the house. The ventilator must be located in the center between two rafters; make the measurements inside the attic. Drill a quarter-inch hole from the inside through the sheathing boards and shingles. It is not necessary to cut rafters or add bracing for a good installation.

2. Lay the base of the turbine ventilator over the attic air exit location and mark a hole for the base through the roof. Use the outside of the base for the template (Fig. 10-14A).

3. Cut shingles away with a sharp utility knife. Cut ½ inch outside of the marked circle.

4. Use a drill to start a hole for the saber or keyhole saw. Cut through the roofboards following the circle (Fig. 10-14B).

5. Loosen the locking screw in the adjustable pitch base. Place the base in position and turn the top section sufficiently to make the top of the base level (Fig. 10-14C).

6. Coat the flashing with roofing cement and slide the top half of the flashing up under the shingles; remove any nails that are in the way. Secure the base to the roof with nails driven through the flange (Fig. 10-14D). Caulk the nailheads.

7. Place the ventilator over the base. Check that the ventilator top is level; if not level, slightly rotate the adjustable pitch base (Fig. 10-14E).

8. Use the base ring holes as a guide in drilling matching holes around and through the base flange. Screw the ventilator to the base with sheet metal screws.

9. Caulk all flashing and shingle edges with roofing cement. Make sure all raised shingles have been thoroughly caulked with roofing cement and pressed back into position.

Turbine Ventilator

A turbine ventilator (Fig. 10-12) operates on a centrifugal action in which spinning blades produce an area of low pressure which draws air from the attic through the roof and vent. The turbine blades begin to spin with the slightest breeze and a flywheel effect keeps the turbine spinning even after the wind stops. The turbine blades may spin even without apparent outside air movement because of the thermal effect of cooler outside air entering the soffit vents and pushing the warmer air above the vents through the top of the turbine.

One 12-inch turbine ventilator is sufficient to ventilate 600 square feet of attic floor area, provided that at least a two-foot-square inlet opening is located a minimum of 3 vertical feet below the turbine ventilator. Thus, if your attic floor measures 580 square feet, you must provide two square feet (288 square inches) of inlet opening around the perimeter of the attic at the soffits. DIY 10-1 describes the installation of the turbine ventilator.

FIGURE 10-12. A slight breeze starts the turbine spinning; the spinning blades cause a low pressure that draws air from the attic (*courtesy Triangle Engineering Co.*).

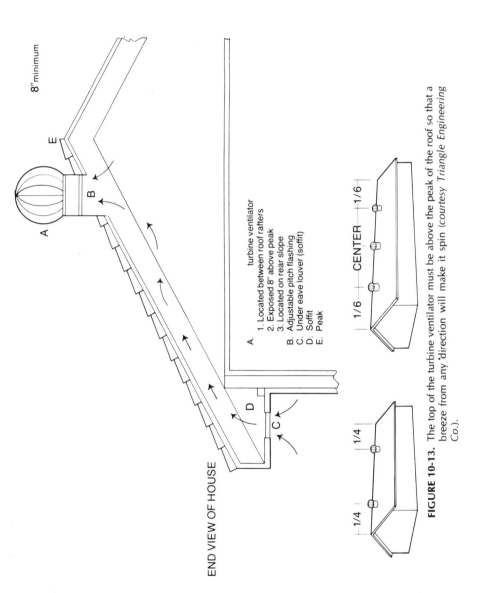

8" minimum

END VIEW OF HOUSE

A. turbine ventilator
  1. Located between roof rafters
  2. Exposed 8" above peak
  3. Located on rear slope
B. Adjustable pitch flashing
C. Under eave louver (soffit)
D. Soffit
E. Peak

1/4 — 1/4

1/6 — CENTER — 1/6

FIGURE 10-13. The top of the turbine ventilator must be above the peak of the roof so that a breeze from any direction will make it spin (*courtesy Triangle Engineering Co.*).

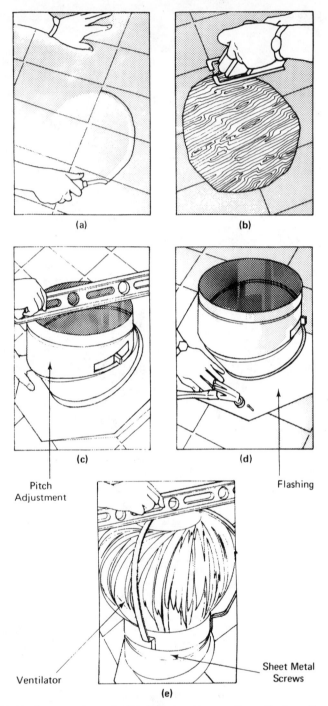

(a)

(b)

(c)

(d)

Pitch
Adjustment

Flashing

Ventilator

Sheet Metal
Screws

(e)

FIGURE 10-14. Steps are shown for installing a turbine ventilator. Roof ventilators are similarly installed.

Living Area Natural Ventilation

Natural air circulation is provided through the living area by opening windows on the windward and the leeward sides of the home (Fig. 10-15). Ideally, windows on the windward or high pressure side should be opened from the bottom about 6 inches. The same number of windows giving you approximately the same amount of area opening on the leeward side of the home should be opened from the tops about 6 inches. The pressure difference between the windward and leeward sides of the house will cause air to flow. You will have to experiment with the amount of openings of the windows and doors to establish the best method of circulating the outside air through your

Placement of Windows for Direct Cross Ventilation is Best. Where this Cannot be Done, Windows Should be Placed in the Adjacent Walls and Away from Corners to get Best Possible Cross Ventilation.

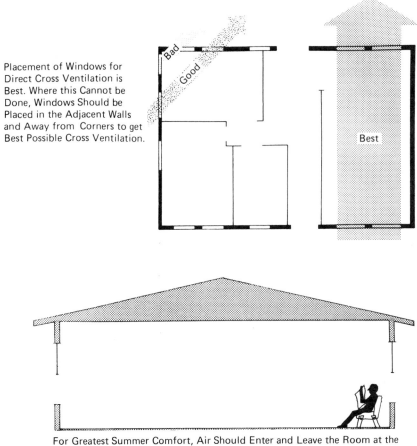

For Greatest Summer Comfort, Air Should Enter and Leave the Room at the Level of Occupancy to Maintain the Air Flow at that Level.

FIGURE 10-15. Natural air circulation should not be blocked. Equal openings should be made at inlets and outlets *(courtesy Andersen Corporation)*.

home. Opening windows on one side of the house only is not effective because there is no pressure difference. Remember that rooms not frequently used should be closed off so that air circulates only through the living area of your home. Be sure to keep closet doors closed too.

When opening windows in a room or house, if the total size of the inlets equals the total size of the outlets, a breeze provides a full cooling effect. If the inlet area is larger than the outlet area, cooling is diminished; if the inlet size is smaller than the outlet, there is a high velocity of air that passes through the house; this is ideal for high humidity conditions. Opening windows from the bottom on the inlet side and from the top on the outlet side lets more air circulate due to the stack effect.

One inexpensive way to add additional air circulation to your home is to add vents through the exterior walls of your home to the outside. Vents are placed near the floor on the prevailing windward side of the home (if the prevailing evening wind is different than during the day, place the vents on the side of the evening prevailing wind). A like number of vents totaling the same amount of area as the inlet vents is placed on the leeward exterior walls of the home near the ceiling. These vents should be constructed by cutting out an area of the inner and the outer wall between studs. Use pine wood to make a frame for the vent. You can put an adjustable louver on the inside and a hinged door on the outside with control lines that pass through the wall so that the outer door can be opened and closed as desired. The inside of the vent and frame should be insulated with Styrofoam™ so that the cold air of winter cannot infiltrate through the opening.

10-2. POWERED AIR CIRCULATION

To be effective, natural air circulation systems depend on wind velocity and direction, inlet and outlet ventilators, open windows, thermal effect, and the stack effect. Powered air circulation systems also depend on these components and factors plus an electric motor powered fan that provides a *positive* means of forcing air through the living space and through the attic space to remove hot air in the summer and moisture in the winter. The fan motor is usually controlled by an automatic thermostat that causes the fan to turn on and off at preselected temperatures. An optional humidistat is used to control the fan in the winter; it switches the fan on when the humidity in the attic space reaches a predetermined level and switches it off again when the humidity level drops.

Typical curves of indoor and outdoor temperatures, one with natural ventilation only and the other with forced ventilation, are shown in Fig. 10-16. Note that the outdoor air temperature and the indoor air temperature are the same at about 6 P.M. Thus, after 6 P.M., we want to move the outdoor air inside to replace the hotter indoor air and store up a reservoir of cool air inside for the next day.

Powered air circulation systems include an air furnace heating system, a portable fan, paddle fan, window fan, and powered roof ventilators which are attached to a window or wall in the attic or through the roof. Another type of fan used very successfully to cool a house by powered air circulation is known as a *whole-house fan*.

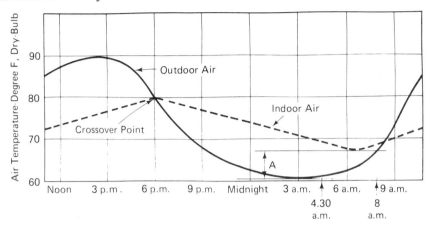

Typical Curves of Indoor and Outdoor
Air Temperature, with Natural Ventilation Only.

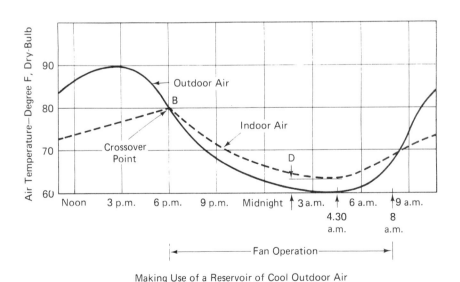

Making Use of a Reservoir of Cool Outdoor Air
During the Night, with Forced Ventilation

FIGURE 10-16. Forced ventilation during the night provides a reservoir of cool air in the home for the next day.

10-3. HOT AIR FURNACE BLOWER

With no additional expense, assuming you have a hot air duct furnace system, you can circulate air through your home to provide a degree of cooling. The circulating air causes the moisture in your skin to evaporate, making you feel cooler. The ideal method of using your heating duct system is to have a ventilator in the cold air return at the base of your heater. You then close off the other cold air return ducts in the rest of the house and open the basement windows or the window nearest the furnace so that it draws in the cooler outside air. Fully open the wall heating vents in the house and lower the windows on the walls opposite your heating vents a few inches from the top. To cool your home, turn the heating system blower to "On" with the manual switch located on your thermostat (Fig. 4-6) or furnace. (The heating system should not turn on. Be sure you are not heating the air with your furnace.) With the basement windows open, the furnace draws in the cool outside air and blows it through your ducting system forcing the hotter room air through the open windows to the outdoors.

When circulating air through the house via the heat ducts of the hot air system of your furnace, keep all vents clear of curtains, chairs, furniture, etc., to promote the free circulation of air. Close off vents in rooms that are not used. Vacuum the ducts occasionally to remove accumulated dust.

10-4. PORTABLE FAN

A portable fan (Fig. 10-17) provides a cooling effect by stirring up the air around you. The fan actually does no cooling, and if anything, adds a slight amount of heat to the air produced by the fan motor. Thus, the portable fan is inefficient and ineffective as it does not produce cooling. The portable fan can be moved from one location to another and directed so that it can blow air directly on you causing evaporation of perspiration from the skin; this lowers the skin temperature, making you feel cooler.

Portable fans are available in different diameters, multiple speeds, and with tilting and oscillating features. Assuming that two fans are operating at the same speed, the fan with the larger diameter circulates a larger volume of air per minute. A multiple speed feature allows you to slow the fan to a quieter speed for sleeping. A tilting feature permits the fan to be directed to any desired level and an oscillating feature permits the fan to slowly move back and forth within an arc of approximately 90 degrees.

10-5. PADDLE FAN

Reminiscent of the ice cream parlor days, large slow moving paddle fans suspended from the ceiling are making a comeback in homes (Fig. 10-18). Not only is the propeller fan decorative because of its large blades, but it also

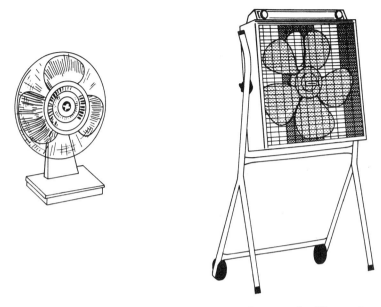

FIGURE 10-17. Portable fans are available in one to three speeds, different sizes, tilting mechanisms, and with thermostats. Oscillating fans cycle back and forth through 90 degrees.

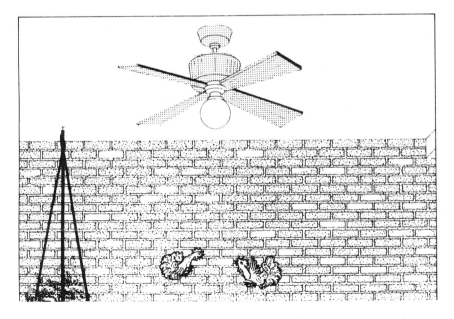

FIGURE 10-18. Paddle fans provide air circulation as well as decoration. This model incorporates a light below the motor (*courtesy Wind-Wonder, Inc.*).

causes a circulation of a large volume of air. Like the portable fan, it provides air movement, but no air cooling; the larger the blades, the greater the amount of air circulation. Models are available with two speeds.

10-6. WINDOW FAN

A fixed window fan (Fig. 10-19) is easily installed and provides air circulation and cooling as it draws in cool air from the outside through open windows in *other* rooms, forcing the warm air out through the fan to the exterior. The best place to install a window fan is in a hallway window because it can then cool more than one room. When a window fan is installed in a room, all windows in that room need to be closed so that a short circuit airpath from the window to the fan does not bypass the rest of the rooms and hallways to be cooled. In use, the windows near the fan are closed and windows on the other side of the house are opened about 6 inches from the bottom. Choose windows in rooms in which air movement will occur; do not cool rooms that are not being used. Some window fans have two or three speeds which create different amounts of air circulation. Some window fans also have a thermostat which can be set to turn the fan on when the air temperature exceeds the preset temperature and turns the fan off again when the inside temperature drops below the preset temperature. A window fan is not as effective as a whole-house fan (Section 10-8).

A large window fan can be placed in an attic window to cool several rooms. The fan pulls the cool evening air through windows in the rooms to be cooled that are opened from the bottom about 6 inches. The air comes from the exterior through the windows and the warm air flows to the attic via an attic door which is opened part way, an attic stairway door, or a louver installed in a hallway ceiling that is opened when the fan is operating.

The fan must fit the window frame snugly to prevent air from "short circuiting" around the edges of the fan. Fill air gaps around the fan frame with pieces of Styrofoam insulation or cloth.

FIGURE 10-19. Window fans mount easily with four screws into the window frame. Window fan features may include multiple speeds, a reversing switch, and a thermostat.

10-7. POWERED ROOF VENTILATORS

Powered ventilators (Figs. 10-20 and 10-21) are used to rid the attic of excessive hot air in the summer and moist air in the winter; they are more effective than natural air circulation because they provide positive air movement. A powered ventilator consists of a motor-driven exhaust fan in a weatherproof housing which prevents rain and snow from entering the attic; the powered ventilator is mounted most practically through the roof. Flashing is provided with the ventilator to prevent water leakage through the roof after

FIGURE 10-20. Powered roof ventilators contain a motor-driven fan, weather protection, and a thermostat. Some have an optional humidistat (*courtesy Leslie-Locke*).

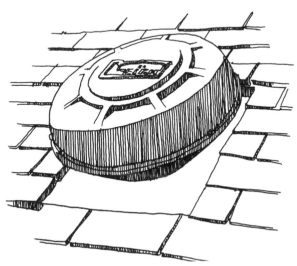

FIGURE 10-21. Powered roof ventilators move much more air than natural air circulators (*courtesy Leigh Products Inc.*).

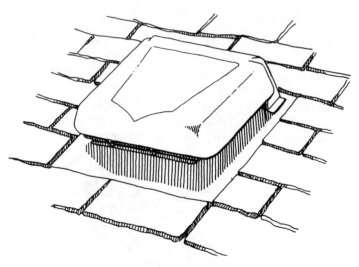

FIGURE 10-21. (Continued).

the powered ventilator has been installed. The powered ventilator is most often placed on the top of the rear sloping side of the roof (for aesthetic appearance) and centered along the roof. If two powered ventilators are used, they are spaced so that the roof is divided into three equal areas. Under-eave soffit air intakes are used in conjunction with the powered ventilators; cool air is drawn in through the soffit eave air intakes and is exhausted through the powered ventilator. If it is not practical to place a powered ventilator through the roof, then a *power gable* ventilator should be installed vertically on a gabled, side wall (Fig. 10-22).

The powered attic ventilator can be controlled by a thermostat to turn the ventilator on when the air temperature reaches about 110 degrees F and turn it off when the temperature decreases to about 90 degrees F. A humidistat can also be connected to the ventilator so that the ventilator comes on in the winter when the humidity level exceeds a preset value. When the humidity level under the roof decreases below the preset level, the humidistat turns the ventilator off.

To determine the size of the power attic ventilator needed to remove excessive hot air and moisture from the attic, determine the square foot area of the attic to be ventilated and multiply this by one CFM per square foot. This gives the total CFM required for light- and medium-colored roofs. For a dark-colored or black roof, add 15 percent more to the square foot area.

The intake ventilators for a power attic ventilator system should provide one square foot of free opening per 300 CFM of fan capacity. A second recommendation based on *square feet* of attic area is to provide one-half

FIGURE 10-22. This attic ventilator is designed for installation inside the attic behind an existing gable louver (*courtesy Wind-Wonder, Inc.*).

square inch of free area per one square foot of attic area. For example, if your attic is 500 square feet, 250 square inches of free ventilation area must be provided. Locate ventilators all around the perimeter of the attic to ensure good airflow under the roof. This will ensure that the air passes all places where moisture accumulates and where there is excessive heat (Fig. 10-23).

The physical installation of a powered roof ventilator is similar to that of a turbine ventilator, except simpler (no leveling is required). The powered roof ventilator must also be wired to 117 vac house power; be sure to follow all electrical wiring codes. Refer to DIY 10-1.

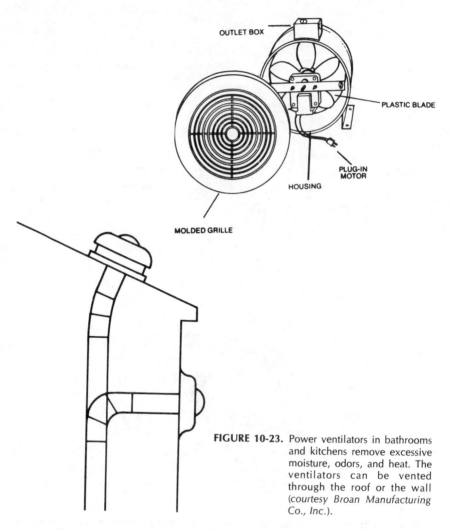

OUTLET BOX

PLASTIC BLADE

PLUG-IN
MOTOR

HOUSING

MOLDED GRILLE

FIGURE 10-23. Power ventilators in bathrooms and kitchens remove excessive moisture, odors, and heat. The ventilators can be vented through the roof or the wall (*courtesy Broan Manufacturing Co., Inc.*).

10-8. WHOLE-HOUSE FAN

On days or nights when the temperature is below 85 degrees F, the entire house and the attic can be ventilated quite adequately at the same time using a whole-house fan (Fig. 10-24). This arrangement is very effective for homes that do not need air conditioning, cannot afford air conditioning, and for homeowners interested in energy savings. The whole-house fan concept can be used all around the country, because all around the country there are relatively few days in which the temperature exceeds 85 degrees F. On days less than 85 degrees F, and all evenings and nights when the temperature is below 85 degrees F, the whole house and attic can be cooled without the need

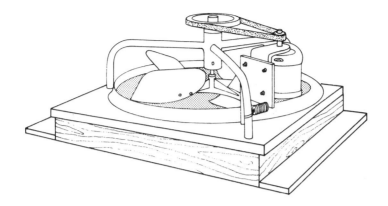

FIGURE 10-24. The whole-house fan draws cooler outside air through inlets and windows and exhausts the hotter inside air through outlets (*courtesy Triangle Engineering Co.*).

for air conditioning. When used in conjunction with air conditioning, the whole-house fan used in the spring and the fall when the temperatures are mild and in the summer when the temperature is less than 85 degrees F, provides an opportunity for comfort at a greatly reduced cost. The whole-house fan gets rid of the stored heat in the house structure and in the furniture, draperies, and similar items. The whole-house fan cools the house down so that the house temperature effectively coasts into the next day with a reservoir of stored coolness.

A whole-house fan requires sufficient openings through windows and doors, sufficiently large exhaust openings in the attic and a fan having adequate capacity to change the air in the home (Fig. 10-25). It is recommended that at least one air change is made every minute.

The size of the whole-house fan required is determined by finding the volume (in cubic feet) of the living space to be cooled. Do not include the attic, basement, garage, closets, storage rooms, or unoccupied rooms. Find the volume of each room to be cooled by multiplying the room length times the width times the height. Then add all of the individual volumes together to obtain the total volume. Obtain a fan that will move at least that many cubic feet per minute or more. For example, if the living area is 5,900 cubic feet, obtain a fan with a 5,900 or greater cubic-feet-per-minute rating. A two-speed fan is best; the fast speed should equal or exceed the required rating. The lower speed is used during the cooler evenings and cooler days of the year.

To determine the net, free area of the attic vents for a whole-house fan, add up the net, free area of all existing outlets including roof ventilators, soffit vents, end gable louvers, attic windows, etc. Remember that the free area is about 50 percent of the gross area. At least a minimum of one square foot of opening must be provided for each 750 CFM of air delivered by the fan. The

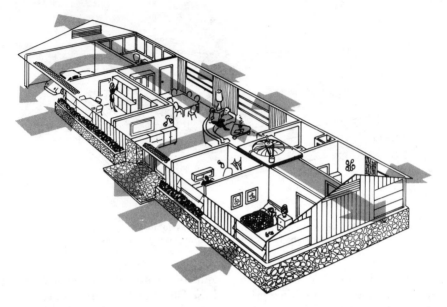

FIGURE 10-25. The whole-house fan greatly reduces the load on air conditioners and deletes the necessity for air conditioning in some regions (*courtesy Triangle Engineering Co.*).

exhaust outlets should be split up in the attic along several walls so that there is a well-balanced ventilation system.

The whole-house fan can be installed in a hallway ceiling even though the hallway may be as narrow as 36 inches (you can put a 42-inch fan above the opening) (Fig. 10-26 and DIY 10-2). Alternately, the whole-house fan can be located in the kitchen if the area above the kitchen is sufficient; in this case the whole-house fan also can get rid of kitchen odors. Other alternate locations for

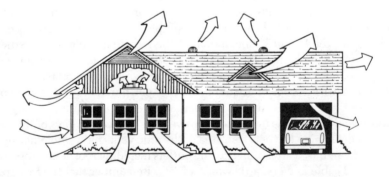

FIGURE 10-26. Whole-house fans are best located in the ceiling at the top of the living area. Alternate locations are shown (*courtesy Triangle Engineering Co.*).

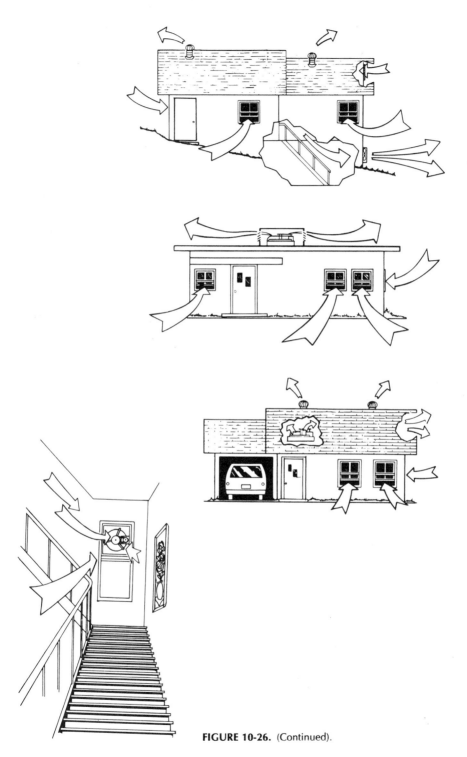

FIGURE 10-26. (Continued).

Do-It-Yourself 10-2
WHOLE-HOUSE FAN INSTALLATION

Materials: whole-house fan, electrical tape, electrical boxes and switches, wire
Tools: saw, hammer, ruler, screwdriver

Procedure:

NOTE
Follow the manufacturer's specific installation instructions. The following procedures cover most installations and provide you with a knowledge of the tasks
that must be performed.

1. Frame in the opening in the attic by cutting out joists and installing header as
   shown in Fig. 10-27A. The inside dimension of the framed opening should be
   the same as the opening size dimensions of the shutter. The shutter frame is
   wider than the opening size.
2. Make a platform of 1- by 6-inch lumber (Fig. 10-27B).
3. Push the fan through the opening and set in position (Fig. 10-27C). The wood
   frame on the fan gives the proper fan height from the platform.
4. Fasten the fan to the platform by toe-nailing through the wood frame (Fig.
   10-27D).
5. Install the automatic ceiling shutter (Fig. 10-27E) snug, but not tight. Use screws
   (usually supplied with the shutter).
6. Make the electrical circuit connections; be sure to follow the applicable electrical codes (Fig. 10-27F).

whole-house fans are the stairwell of a two-story house, the walls of a
playroom, garage/living space wall, or the roof of a flat-roofed house.

With the whole-house fan properly installed, insure that there is
sufficient ventilation in the attic for the air to exit. Then, raise the windows in
the living area approximately 6 inches from the bottom. The speed of the air
coming through the rooms to the whole-house fan varies depending upon the
size of the window opening, the size of the room door opening and the distance
of the window from the fan. You can control the speed of the air by varying the
opening of the window, by closing off parts of the house, etc. By trial and error
of the openings of different windows and doors, you will establish a pattern
and a schedule which is satisfactory for your comfort. The less rooms cooled,
the greater is the airflow through the other rooms. Later in the evening, you
may want to zone cool by closing off some of the rooms that have sufficiently
cooled down leaving more air to circulate through the bedrooms. Don't turn
on the fan until at least one window is open; if the house is completely closed
up and the whole-house fan is turned on, sufficient air could be drawn through
the chimney to bring soot and dust into the living area.

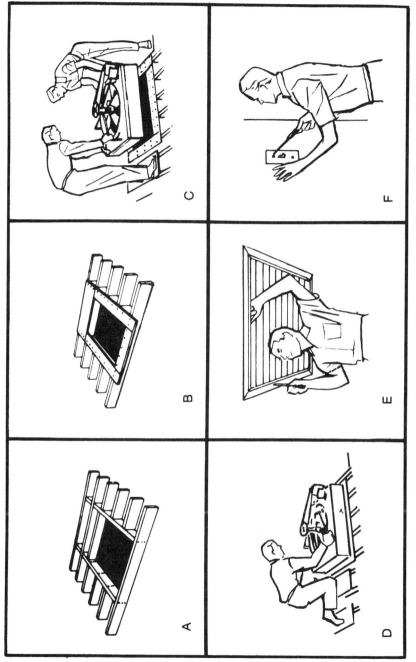

FIGURE 10-27. Whole-house fan installations are relatively easy to accomplish.

Whole-house fans are controlled by means of a thermostat control or a timer switch. For example, you may desire that the fan is turned off after the inside temperature reaches a temperature of 65 degrees F. Similarly, you may want the fan to turn off at some time, such as midnight, so that any noise produced by the fan does not interfere with sleeping.

10-9. COMPARISON OF VENTILATION TECHNIQUES

As you can see, natural air circulation by simply opening windows on the windward and leeward sides can cost you absolutely nothing and can give you some degree of comfort. Ventilators are very inexpensive items and easy for the do-it-yourselfer to install. Turbine ventilators have some cost, but they are free to operate once they are installed, and the cost of maintenance is nearly zero. A power ventilator is slightly more expensive than a turbine ventilator, but it has the added bonus of positive air displacement and will give you a lot more cooling than natural air circulation. Power ventilators do not use much electricity and are inexpensive to maintain. The whole-house fan, or a large window fan cost more than other types of air-circulating devices; however, as compared to air conditioning, they are relatively inexpensive. Similarly, the maintenance costs are low and the cost of operation is very low as compared to air-conditioning costs.

The graph in Fig. 10-28 illustrates the differences in temperature of the outside air, an unvented attic, a turbine vented attic, and a power vented attic. The temperature in degrees Fahrenheit is shown on the vertical axis and the time of day from 6 A.M. of one day to 6 A.M. of the next is shown on the horizontal axis. The power ventilator is turned on at 110 degrees F and turned off at 100 degrees F. A 1/8 horsepower motor using 150 watts of electricity is used to power the power ventilator. Note that the outside temperature is always equal to or lower than the inside attic temperature. This is because, as we previously discussed, once the heat gets into the house, it is absorbed and retained in the structure. Thus, with any venting, whether it is natural air circulation or powered circulation, it is difficult to get the temperature inside down to a temperature equal to that on the outside. The top line of the graph illustrates the unvented temperature. Notice that at approximately 3 P.M., the attic temperature is about 50 to 60 degrees F higher than the outside temperature (please note that these are attic temperatures, not living space temperatures). The turbine-ventilated attic is at times the same temperature as the outside temperature and the greatest difference is about 15 to 20 degrees F. A more stable temperature in the attic is provided by the power-ventilated attic. Remember that the power ventilator turns on at 110 degrees F and off at 100 degrees F; this keeps the attic from cooling to the actual outside temperature. With adequate insulation between the ceiling living space and the attic floor, the attic temperature can remain at about 90 degrees F without much effect on the living space below.

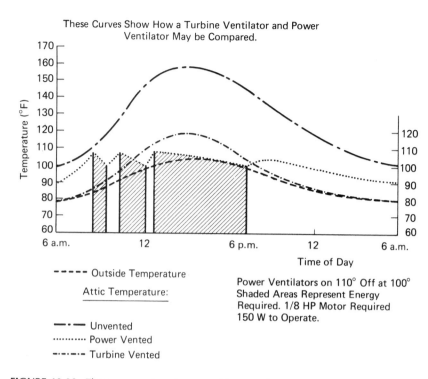

These Curves Show How a Turbine Ventilator and Power
Ventilator May be Compared.

----- Outside Temperature

Attic Temperature:

—·— Unvented

·········· Power Vented

—··—··— Turbine Vented

Power Ventilators on 110° Off at 100°
Shaded Areas Represent Energy
Required. 1/8 HP Motor Required
150 W to Operate.

FIGURE 10-28. These curves show how a turbine ventilator and a power ventilator may be
compared.

chapter 11

LET NATURE HELP WEATHERIZE YOUR HOME

So far, we've discussed the physical changes you can make to the inside and outside of your home to weatherize it. This chapter discusses the use of nature in providing a 25 to 30 percent reduction in heating and air-conditioning loads by the correct use of trees, vines, shrubs, grass, ponds, pools, hills, and man-made paving and fences to landscape your property (Fig. 11-1). Natural plant growth, water, and man-made paving and fences not only decorate your home, they also cut fuel bills by serving as insulation and windbreaks in the winter and as wind funnels and moisture evaporators to cool in the summer.

Landscaping is beneficial in reducing heat gains or losses in homes in the following ways: reduction of direct solar radiation, sky radiation, and reflected ground radiation at windows during the summer; reduction of air leakage in all

FIGURE 11-1. Retain or plant evergreen and deciduous trees, vines, shrubbery, and ground cover to reduce the energy needed for heating and cooling (*courtesy Andersen Corporation*).

seasons through cracks and joints around windows and doors, at roof eaves, roof corners, and at the foundation line by lowering the wind velocity at the building surfaces; and the reduction of the heat transmission of windows and to a lesser degree of other building materials. Correct landscaping can modify the amount of wind, sun, and rain that strikes the home's surfaces and thereby can decrease the temperature difference between the indoors and the outdoors. Preventing radiant heat from entering a house by means of external shading is seven times more effective than using shades or curtains on the inside.

Specific information as to the type of trees, vines, shrubs, and ground covers that grow best in your locality can be obtained by contacting your local Cooperative Extension Service or the Department of Agriculture, county agriculture agent.

11-1. WINDS

A windbreak is any barrier to the flow of the prevailing surface wind. A windbreak can consist of a row of evergreen trees (Fig. 11-2) spaced closely together, hedges, vines, shrubbery, fences, walls, sheds, or a garage. The windbreak is used to control the wind direction and speed. It can act as a barrier, reducing wind pressure around a house, or it can act as a funnel, increasing the wind pressure and improving the air circulation through a house. When the wind strikes the windbreak, there is a small calm area on the windward side and a large calm area on the leeward side.

The prevailing surface winds can be an asset during summer, a deficit during the winter. By making effective use of the prevailing winds in the summer, you can use nature or man-made structures to funnel the winds into the windows and doors of your home to promote natural air circulation. In the winter, you can use nature or man-made structures to block the prevailing

FIGURE 11-2. A windbreak planting of evergreens will enable the heating system to heat your home with less energy expended (*courtesy Andersen Corporation*).

surface winds to reduce the amount of infiltrated air and to reduce the wind pressure difference between the windward and leeward sides of the house (which also decreases the amount of infiltration).

11-2. WINDBREAKS

To make effective use of the winds, you need to know the directions and the speeds of the *prevailing* surface winds during the winter and summer seasons. Although the prevailing winds of winter are generally from the north, while the summer prevailing winds are generally from the south, southwest, or southeast, these winds are *not* from these directions in all places across the country. For most localities, with the exception of coastal areas, the direction of the prevailing winds during the winter differ from the prevailing wind direction during the summer. These prevailing surface winds are predictable in both direction and mean speed. Figures 2-10 and 2-11 illustrate the mean resultant surface wind direction and speed for midwinter (January) and midsummer (July).

A solid windbreak such as a wooden or masonry wall deflects the wind in such a way as to produce a wake of air which contains large, organized eddies. These eddies contain high pressures which return to the ground downstream and can often produce infiltration on the leeward side of a home. To prevent these eddies, and to extend the protected area, penetrable windbreaks of trees or shrubs are preferred to solid windbreaks (Fig. 11-3).

Since most winter winds in the United States are from the north and west, natural plant windbreaks should be planted on the north and west sides of the home with an extension on the east side where space permits. The natural plant windbreaks lower the wind velocity, cutting down on the amount of infiltration, and thus lower the heating bill. Trees, vines, and shrubs all make excellent natural windbreaks.

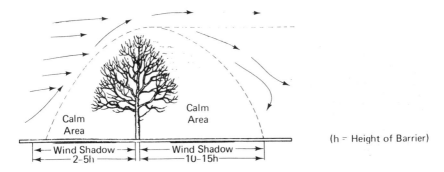

FIGURE 11-3. Research has shown that a natural windbreak is more effective than a man-made wood or masonry windbreak.

11-3. TREES

Trees provide a natural windbreak producing a drag in the wind that causes a decrease in windspeed. Trees also provide effective shading of the sides and roof of a home. Finally, the leaves of trees evaporate moisture as the wind blows, by causing local temperature decreases because of the evaporation of the moisture.

Two types of trees are used in weatherization landscaping: the *evergreen* and the *deciduous*. Evergreens have foliage which remains green and functional all year round, and therefore make excellent windbreaks. The leaves of the deciduous tree shed seasonally or at a certain point of development in the life cycle of the tree. Since deciduous trees lose their leaves during the cold season, deciduous trees are useful for providing shade during the warmer seasons while allowing the sun to shine through the barren branches during the winter months to provide radiant heat to the home (Fig. 11-4). A grove of deciduous trees provides shade, wind protection, and also decreases the outside surrounding temperature through evaporative cooling of moisture from their leaves. This evaporative action creates cool areas under the branches of the trees.

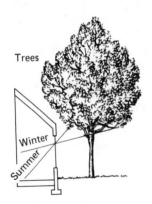

Trees

Winter

Summer

FIGURE 11-4. The leaves of deciduous trees block sunlight in the summer, but since the leaves fall off in the winter, the winter sunlight can warm the house.

Trees that are used as windbreaks (Fig. 11-5) should have a dense growth and grow to a height equal to or preferably about one and one-half times the house height. The maximum distance from the house to the windbreaking trees should not be more than five times the house height as measured from the leeward side of the house. The windbreaking trees may be placed somewhat closer to the house, but the distance should be limited by the radius of the trees and by taking into consideration the root system which can go deep into the ground spreading out from the tree at about the same size as the tree branches. Roots can cause cracks in house foundations and damage to sewer and water pipes coming from or to the house. (For example, a house 20 feet

FIGURE 11-5. Windbreak trees should be dense and grow to 1½ times the house height at maturity.

high requires a row of shade trees 30 feet high planted not more than 100 feet from the house.)

To act as a windbreak, evergreen trees should be planted along the northern exposure. Also, plant evergreen trees in the path of gusting winds and where winds whip around surrounding structures, hills, or valleys. A single row of evergreens is often adequate, but a double or triple row is more efficient. The idea of using evergreens as a windbreak works when the trees are small, and works even better as the trees mature.

In the summer in southern latitudes, the home is over-heated mostly by heat from the sun on the east and west sides as the sun rises in the morning and sets across the sky in the evening. During midday, the sun is high overhead and overhangs built onto the house provide shading on the southern exposure. Deciduous trees are planted to provide shade on the east and west sides of the home and also, if necessary, on the southern side if the overhangs are inadequate. In more northern latitude climates, where houses have their southern walls exposed to the sun during the summer, deciduous trees are planted on the western and southern sides of the house.

11-4. VINES

Deciduous vines such as Boston Ivy and Virginia Creeper are grown on the southern and western exposed walls of the home to deflect the summer sun and thus provide a cooling effect. The cooling effect is intensified when the vine is grown on a trellis attached to the wall, thus permitting air to circulate between the vine and the home. A trellis is especially recommended on framed homes as a way of preventing clinging ivy or other vines from growing in between boards and causing rot.

Evergreen vines such as English Ivy are effective for winter warmth when it is grown on a sunless north wall. On the north wall the leaves deflect the cold winds and the leaves and stems provide an insulating effect.

11-5. SHRUBBERY

Dense shrubbery four feet or higher planted on the eastern, western and northern exposures act as energy savers. Dense shrubbery such as arborvitae, hemlock, or spruce, when planted close to a home, affect the outside surface

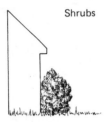

Shrubs

FIGURE 11-6. Dense shrubbery blocks wind, creates shade, and provides an insulating air space between the shrubbery and the foundation.

temperature by blocking the wind, by creating shade, and by providing an insulating dead-air space between the shrubbery and the building (Fig. 11-6). When planted close to the foundation, shrubbery keeps the foundation walls, and hence the basement, warmer. Low shrubberies are of limited benefit.

11-6. GROUND SURFACES

Ground surfaces include nature's brown dirt, grass, ground-covering ivy, and similar plants and flowers (Fig. 11-7). Ground surfaces also include man-made surfaces such as concrete paving and macadam. The color and the texture of the ground surface affect both the radiant and conductive heat losses of the surface. Surfaces that are light in color reflect the sun's radiant light directly into windows and walls. This increases the heat load in the summer by indirect radiation. Dark-colored surfaces store large amounts of solar warmth affecting local temperatures around them; by retaining large amounts of solar radiation, a dark surface delays the heat load.

If you are considering putting in driveways or sidewalks, you need to consider your climatic (geographic) location, the orientation of your house with regard to the sun, and whether you need to provide extra reflected light from the paved surface to the house to help provide radiant heat through windows or whether you need to store the heat in the surface near the house for warming. Concrete paving, for example, on the south side in a warm climate can increase the air-conditioning load significantly, whereas in a cold climate, the radiant reflection from the concrete may be unimportant in the summer, and in a temperate climate, paving may be an asset if exposed to the sun in the winter and shaded in the summer.

If you're already stuck with concrete paving or black macadam in areas where they are undesirable with regard to weatherization of your home, there are some techniques you can use to partially alleviate the problem. Plant

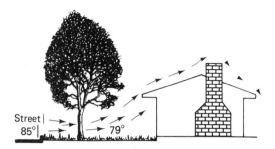

FIGURE 11-7. Ground cover reduces temperature by evaporative cooling.

shrubbery to block the reflection of light from the concrete paving; place the shrubbery between the paving and the house, but closer to the concrete than to the house. Decrease the heat stored in macadam by sprinkling water on it so that evaporative cooling takes place. You can also plant trees or shrubberies in the direction of the prevailing surface winds to provide a funneling effect of the wind across and away from the home so that the heat coming off the macadam is carried away from the house.

Ground-cover plants blanket a large surface area with their large leaves, but allow for air circulation through their leaves. The dark color of the leaves prevents reflections and the natural evaporative action of moisture from their leaves provides cooling. Plants do not reflect heat nor do they store heat for later reradiation to the home.

11-7. PONDS AND POOLS

Additional summer cooling comfort can be provided by a pond, a pool, or simply a grass lawn on the windward side of the house. As the wind blows toward the house across the pond, pool, or damp grass, it causes moisture to evaporate causing evaporative cooling; thus, the air is cooled before it reaches the house. However, keep in mind that if the prevailing wind is from the same direction in both the summer and the winter, the pond or pool could cause colder air to be blown toward the house during the winter also.

11-8. HILLS

Hills and large outcrops of rock also provide wind shelter to your home (Fig. 11-8). As the wind hits the windward side of the hill, its pressure on the leeward side is reduced. In addition, hills usually have grass and trees and other natural growth which also provide wind protection as previously discussed. Your home is ideally located if it is on the prevailing winter surface wind leeward side of a hill.

FIGURE 11-8. Building into a hillside provides natural insulation. Use the contours of the terrain (*courtesy Andersen Corporation*).

11-9. OTHER WAYS OF PUTTING NATURE TO WORK TO WEATHERIZE YOUR HOME

The outside walls of your home can be insulated with tar paper and leaves that have fallen from the trees in the fall. Place the tar paper against the exposed walls; the tar paper provides a moisture barrier between the leaves and the foundation of the house. Then rake leaves against the tar paper and cover the leaves with a tarpaulin or some other method to hold them from blowing away in the wind. You can also put large, plastic bags stuffed with fallen leaves against the foundation.

In the dead of winter, when there is a lot of snow on the ground, you can shovel snow against the leaves, or simply against the foundation of the house as a means of keeping the house warmer. The snow aids in reducing the amount of wind pressure against the house.

SUMMARY

If you are buying a piece of property for building a new home or buying a used home, select a property on which you can make the best use of nature–hills, rocks, evergreen and deciduous trees, vines and shrubbery, ponds, and streams. Also plan to make effective use of man-made pools, walls, driveways, walkways and so on. Use the information in this chapter to help plan your landscaping. Figure 11-9 and QRC 11-1 give you some ideas.

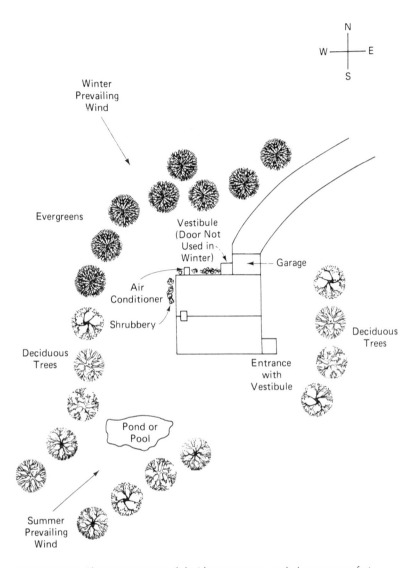

FIGURE 11-9. Plant evergreens and deciduous trees to maximize your comfort.

Quick Reference Chart 11-1
ENERGY CONSERVING LOCATIONS FOR LANDSCAPE ELEMENTS

Landscape Elements	Cold Climate	Temperate Climate	Warm Climate
Ground cover or grass	Negligible effect on all sides	On south	On east, west and south
Paving	On south	Shaded if on south	Shaded if on east, west and south
Shrubs against house wall	On east, west and north	On east, west and north	On all sides
Deciduous shade tree	Negligible effect on all sides	On south and west	On east and west
Evergreen trees	On east, west and north	On east and west	On east and west
Windbreak (trees, bushes, fences)	On sides exposed to winter winds	On sides exposed to winter winds	Undesirable effect on all sides
Windbreak used to funnel wind	Undesirable on all sides	On sides exposed to summer winds	Where cross ventilation is possible

chapter 12

MORE EFFICIENT HEATING

Now that your house is sealed tightly and you've taken the necessary steps to let nature help weatherize your home, it's time to check up on your furnace. By performing some easy maintenance tasks to your furnace, balancing your heating circulation system, and by having a qualified serviceman tune your system, you can heat more efficiently. Then, further increase your furnace heating efficiency by adding heat reclaimers to the system. Some of these heat reclaimers can be built by the do-it-yourselfer; others can be purchased and installed by the do-it-yourselfer or by a serviceman.

Home heating is accomplished by conduction, convection, radiation, and combinations of these methods. In conduction, heat flows from a warmer body to a cooler body such as heat along a silver teaspoon or from a warm, inner wall to a colder, outer wall. Heat always travels in a direction from warmer to cooler. Warm air rises because it is less dense than cooler, dense, heavier air which falls to take its place causing heat transfer by convection air currents. The transfer of heat from a hot object into the air is heating by radiation. A radiator heats a room by radiation.

Central heating systems are used in most homes. In a central heating system, the heat is generated at a central point and is then distributed to various rooms by means of a distribution system. Distribution may be by air ducts (which carry hot air by gravity or forced air from a blower to grilles or registers) or by pipes (which carry steam or hot water to radiators) to the rooms to be heated.

Most central heating systems consist of a furnace, a distribution system, and a thermostat (Chapter 4). When the home temperature drops sufficiently to close the contacts of the thermostat, an electrical signal is sent to the furnace. Heat is produced in the furnace and the heat is then distributed via the distribution systems to the rooms of the home. When the temperature rises sufficiently to open the thermostat contacts, the furnace is shut off, removing the source of heat. The distribution system then stops distributing heat to the home.

Fuels used in central heating systems include oil, gas, coal, the sun, and wood. Oil and gas are the most common fuels used in burners. Coal is mainly a fuel of the past but it is making somewhat of a comeback, particularly in power generation. Wood is being used more and more where it is available; it is being used primarily as a supplementary source of heat in fireplaces and stoves (Chapter 14). And most of us are familiar with solar heating. Though not practically developed as a complete source of heat, solar heat is finding application in all parts of the country and is being supplemented with a gas, oil, or electric system when needed (Chapter 15 describes the fundamentals of solar heating).

This chapter provides information, procedures, and tips to make your conventional home central heating system more efficient. The topics included are: heat distribution systems, furnaces, distribution system and furnace maintenance, checking fuel consumption, heating for comfort and economy, use of outdoor air for combustion, heat reclamation, and additional air registers and ducts. Chapter 14 describes supplementary heating sources including fireplaces, wood stoves, and other supplementary heaters. Chapter 15 describes some ideas on solar heating.

12-1. HEAT DISTRIBUTION SYSTEMS

Heat is distributed by moving air or by pipes carrying water or steam. Hot air systems include gravity air systems and forced air systems; piping systems include gravity and forced hot water and steam. Radiant heating and zoned heating systems are also included. Each distribution system is described so that you have an idea of how your distribution system operates.

Hot Air Gravity Distribution System

Heat is produced in a hot air gravity system in the *plenum*, which is a chamber above the furnace heating area. The air in the plenum is heated and is then distributed. In the gravity system of distribution, heat rises through air ducts by *gravity*; that is, heated air in the plenum rises by convection through the hot air distribution ducts to the rooms. Since air cannot circulate without a return path, return outlets and ducts from the rooms allow the heavier, cooler air to return to the heating system by gravitational flow. *Dampers* and *registers* provide a means of controlling the amount of flow from each distribution duct. A damper or a register acts as an adjustable gate to partially or fully block the air from exiting a hot air duct; dampers are located within the duct and are adjustable in the basement whereas registers are located at the ends of ducts in the floors or walls of the rooms being heated.

Forced Air Distribution System

The *forced* air distribution system is the same as the gravitational hot air system except that a blower is added to force the air from the plenum through the distribution ducts to the rooms. The forced air system heats rooms more quickly than the gravitational system because of the forced delivery. The blower also provides somewhat of a suction pressure that pulls the cooler air from the return ducts. Although the forced air system causes more *drafts* in a room, it also provides more level heat because it reduces the tendency of the air to stratify in different temperature levels within rooms. Dampers and registers are used in the forced air system as in the gravity system. Cool returned air is passed through a filter in either the forced air or the gravitational system before it is reheated in the plenum.

The temperatures in different rooms can be varied by adjusting the dampers and/or registers to the rooms. Partially closing the damper or register to one room allows less air to flow into the room and therefore some additional airflow to the other rooms (and vice versa). Adjusting the dampers and/or registers until each room is as warm as desired is known as *balancing* the system (DIY 12-7).

Note that the same forced air ducting system used for heating a home can also be used for distributing air-conditioned air to cool the home. In most cases, an A-frame cooling coil is added into the heater plenum. Air conditioning is covered in Chapter 13.

Hot Water Distribution System

In a hot water distribution system, water in a boiler is raised to the boiler tank heating temperature. The hot water is then forwarded by *gravity* or by *force* to *radiators* in each of the rooms to be heated. Radiators may be the old *cast iron* style, a *convector*, or modern *baseboard* radiators. Cast iron radiators may have from 2 to 56 sections and can be heated by hot water or steam. They deliver heat mostly by convection as cool air near the floor is warmed by the radiator, flows by the radiator sections to the top, and then continues upward toward the ceiling; radiators also produce some heat via radiation from the hot cast iron.

Convectors are *built-in radiators*. Hot water or steam is circulated through a finned metal pipe in the radiator cabinet. Cooler air at the floor is heated as it passes by the finned tube and then rises toward the ceiling. The convector could be considered more decorative than the radiator, but it is not any more efficient as a heating device.

The most efficient type of radiator is the baseboard radiator which is located along the baseboards on the outside walls. Baseboard radiators may have radiant cast iron tubes or the more efficient finned tube construction. Heating is accomplished by convection currents. All radiators, whether of the cast iron, convector, or baseboard configuration, have air valves used to vent trapped air.

Hot water distribution is by *gravity* or by *force*. Heat is transferred from the boiler to the radiators and hence the rooms in a *gravity* hot water distribution system because of the principle that hot water expands as it is heated, becomes lighter, and therefore rises. The hot, lighter water rises through pipes to radiators where it cools (and therefore becomes heavier). The cooler water returns through another set of pipes to the boiler for reheating and recycling. The speed of gravitational cycling is dependent upon the temperature difference between the boiler water (about 180 degrees F) and the returning water (about 140 degrees F). A water *expansion* tank in the system takes care of increased water and pressure in the system due to the heat.

A *forced* hot water system is also known as a *hydronic* system. Water is circulated through the system from the boiler to the radiators by a *circulating pump* called a *circulator*. The circulator is usually located in the return pipeline and is turned on and off by a command from the thermostat. The forced water system gets hot water to the rooms more quickly than a gravity system. A flow control valve prevents the flow of water by gravity when the circulator is not operating.

In addition to supplying hot water to the radiators, the hydronic system also heats hot water for domestic use. The hot water for the faucets does not come directly from the boiler, but from a coil immersed in the boiler's hot water; the domestic supply water passes through the coil and is heated by the hot water in the boiler surrounding the coil.

Hot water system piping configurations may be two-pipe, one-pipe or series-loop. The two-pipe hydronic system uses one pipeline for the supply line to the radiators and the second pipeline returns the cooled water to the boiler. The radiators are connected to the lines in *parallel*, just as there are parallel electrical circuits. Each radiator in the two-pipe system receives very nearly the same temperature inlet water.

In the series-loop hot water piping system, hot water flows from the boiler to the first radiator, then the second, third, and so on. In this configuration, the first radiator receives the hottest water and the last radiator along the line receives the coolest water because the water has been continually cooling as it moves along from one radiator to the next. If a single radiator is shut off, all of the water stops flowing and all radiators cool off.

The one-pipe system is a series/parallel arrangement in which water moves serially through the system until it comes to a radiator; at the radiator, part of the water continues serially along the pipe and part of the water is forced through a T into the radiator. If one of the radiators is turned off, the others continue to heat.

Steam Distribution System

A steam heating system is similar to a hot water system except that the water temperature is higher, leaving the boiler at 212 degrees F. In the boiler, water is heated to the boiling point where it vaporizes into steam. The steam rises through pipes to the radiators where it then condenses to water because of contact with the cooler metal. The water returns to the boiler for reheating into steam.

Steam heating systems may have a two-pipe or a one-pipe system. In a two-pipe system, steam rises to the radiators through one set of pipes and returns as condensed water vapor through the second set of pipes to the boiler. In a one-pipe system, steam rises through the series of pipes, condenses upon contacting the radiators, and returns within the same pipes to the boiler. As the water is heated to steam, pressure is built up in the system; to prevent the steam pressure from increasing to a level that is unsafe for the system, a safety valve is included in the system. The valve opens up at about 15 pounds pressure, venting the excess pressure. The valve remains open until the pressure decreases to a safe limit.

Radiant Heat

Radiant heat is produced by rows of parallel hot water pipes or an electric-resistance heating cord (Fig. 12-1) placed in concrete floors or plastered ceilings or walls. In a hot water radiant system, hot water from a boiler is circulated by a circulating pump through pipes embedded in floors or ceilings. In an electric system, electric current flows through the wire heating cord which provides resistance to the flow of the electric current; in resisting the current flow, heat is generated. The heat, in either the hot water or the

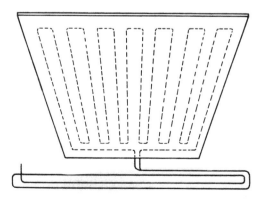

FIGURE 12-1. Radiant heating electric cables are an integral part of this gypsum drywall ceiling. A nonheating lead extends from the back of each panel for electrical connection. Panels produce about 15 watts per square foot with surface temperatures of about 100 degrees F.

electric resistance heating, is controlled by a thermostat. The piping or wires, as applicable, heats the embedding material; the material transmits the heat to the surrounding air.

Radiant heat has the advantages of being invisible (no registers, convector cabinets, radiators, or ducts) and no floor space needs to be sacrificed. Radiant heat is most effective in homes with no basements because it keeps the lower floor warm.

One type of radiant heater uses electricity to heat water (including antifreeze) contained in a hermetically sealed copper tube (manufactured by Intertherm Inc.). The heaters are self-contained and are directly connected to 240 volts housepower (with the exception of several portable baseboard units that plug directly into 117 volts). A thermostat controls the temperature of each unit. Different lengths of baseboard units are available and you can heat an individual room, an addition, or a whole house.

A radiant heater having electrically heated water in a sealed tube has several desirable features. When the water is heated, it circulates through the length of the oval tubing; a finned area along the top of the oval tubing lets the heat radiate quickly into the room. Since the top run of tubing is slightly cooler than the bottom run of tubing, water from the bottom circulates to the top by gravity causing continuous circulation throughout the tubing. When the thermostat contacts open, indicating that the room is sufficiently warm, the electrical heating is turned off; however, the water continues to circulate by gravity and continues to radiate heat providing a more constant type of heating than can be provided by off-on systems such as all-electric or hot air. This radiant heating method provides comfortable heat that is clean and healthful.

Zoned Heating Systems

A *zoned* heating system is the most practical and economical heating system for residential use. In a hydronic system, one boiler provides heat. Two or more piping systems (of one type or another) with individual circulating pumps and thermostats control various sections of the house. For example, the temperature of the bedroom section can be established by one thermostat and set several degrees lower than the living and dining sections; a recreation room might be on a third zone. Zoned heating systems are also possible with hot air systems; in this case a motorized damper can be used to cut off hot air to one duct section or the plenum can be split by a damper as directed by a second thermostat.

12-2. FURNACES

Now that you know how your heat distribution system operates, let's discuss how the heat is produced. The heat is produced in the *furnace* which is an enclosed structure. Heat can be produced by an oil burner, gas burner, electricity, heat pump, or by a coal furnace. Heat can also be produced by solar energy (Chapter 15); however, solar energy is not yet sufficiently developed to provide all of our heating, and, at this time, it is not economical nor practical for many homeowners to install it in homes already built. Heat from solar energy usually needs another source of heat to supplement the solar heat during periods of extreme cold, sunless days, and evenings.

Oil Burners (Fig. 12-2)

An oil burner system consists of an oil supply tank, a burner, and a combustion chamber. Upon a demand for heat initiated by a thermostat contact closure, fuel oil is admitted from the tank to the combustion chamber where it is ignited and burned. The heat of combustion heats the hot air plenum (heat exchanger) in hot air distribution systems or the hot water supply tank in hot water distribution systems. There are two types of oil burners: a *pressure* burner and a *vaporizing* burner.

There are also two types of *pressure* oil burners: high pressure (most often used in home furnaces) and low pressure. In the *high pressure burner*, oil is pumped from the oil tank through a filter to a pressure-regulating valve. When the pressure reaches about 100 psi, the valve opens and the oil proceeds to a burner nozzle where the oil is atomized. After the oil leaves the nozzle, it is mixed with air from a blower to produce a combustible mixture. A transformer is used to produce a high voltage (from the input house voltage of 117 volts to about 10,000 volts) which produces a spark between two electrical contacts (called electrodes). The spark ignites the combustible oil/air mixture and the furnace produces heat. Heating continues until the thermostat contacts open which causes the oil pump motor to shut off. As the motor reduces speed, the oil pressure is reduced to the pressure regulator valve. At about 90 psi, the pressure regulator valve closes, cutting off all fuel to the nozzle; the flame extinguishes. The oil filter between the oil supply tank and the pump is used to remove impurities from the oil. A stack-control relay (on most oil burners) shuts off the oil pump motor if the oil spray does not ignite within a short time. Without the relay, the pump could continue to pump oil into the burner without ignition. If no hot gases pass through the stack-control relay indicating combustion, the relay shuts the oil pump motor off. This relay can be manually reset but should *only be reset once*. If the burner fails to ignite, there is a malfunction.

The *low pressure oil burner* is similar to the high pressure burner except that the pump produces a low pressure oil flow that mixes with air inside the nozzle. The mixture is sprayed from the nozzle as a vapor into the combination chamber where ignition from a high voltage spark takes place, as in the high pressure system.

Vaporizing oil burners are used in small central heating systems and space heater furnaces. A pool of oil is formed in the combustion chamber

air flow

air filter

flue connection

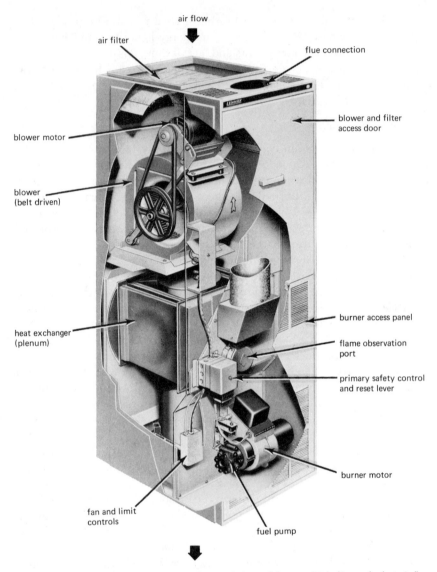

blower and filter
access door

blower motor

blower
(belt driven)

heat exchanger
(plenum)

burner access panel

flame observation
port

primary safety control
and reset lever

burner motor

fan and limit
controls

fuel pump

FIGURE 12-2. This forced air oil furnace has a belt driven blower which directs the hot air flow downward through the furnace. The air filter is normally in the air return inlet to the furnace as shown (*courtesy Lennox Industries Inc.*).

which is then lit by an electric spark or by hand. The heat of the burning oil causes additional oil to vaporize, and burning becomes hotter and more efficient. Air is provided by natural draft or by a small blower.

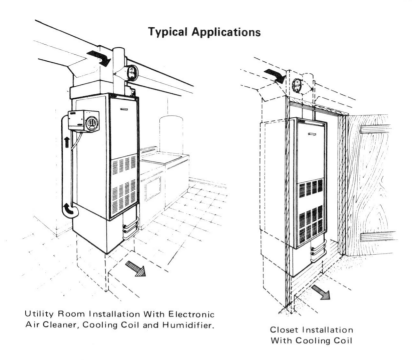

Typical Applications

Utility Room Installation With Electronic Air Cleaner, Cooling Coil and Humidifier.

Closet Installation With Cooling Coil

FIGURE 12-2. (Continued.)

Just as a fireplace needs a chimney to produce a draft, an oil burner needs a draft. This draft is provided by a vent pipe to the house chimney. Many oil burner systems use an automatic draft regulator in the vent pipe to control the amount of draft. The regulator closes when the burner is off, preventing residual heat in the burner from going up the chimney. Failure of the draft regulator to open during combustion periods can cause toxic fumes to build up within the furnace.

Gas Burners (Fig. 12-3 & 12-4)

A gas burner mixes gas (natural, artificial, or liquified petroleum gas) and air in a combustion chamber to produce heat. The gas is turned on by an automatic main gas valve on command from an electrical signal from the thermostat. The gas pressure is regulated to a low pressure by a pressure regulator and is ignited by a pilot flame that burns continuously, or by an electric spark. Electric spark igniters are more modern and more efficient than gas pilot flames which constantly consume gas even when the heater is off.

The pilot heats a *thermocouple* which is an electrical device that converts heat to an electrical current. The current from the thermocouple keeps a

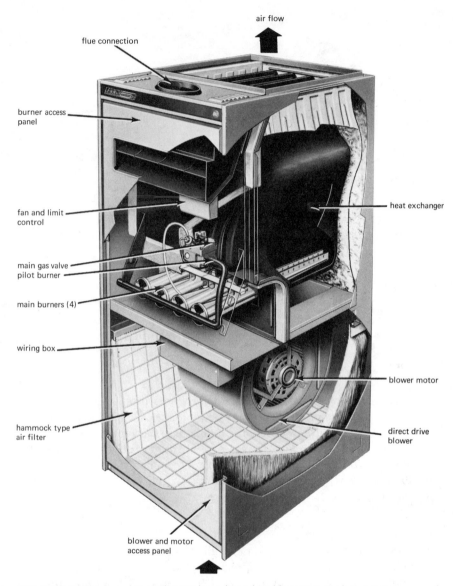

air flow

flue connection

burner access
panel

fan and limit
control

heat exchanger

main gas valve
pilot burner

main burners (4)

wiring box

blower motor

hammock type
air filter

direct drive
blower

blower and motor
access panel

FIGURE 12-3. This forced air gas furnace has a direct drive blower causing hot air to flow upward
(*courtesy Lennox Industries Inc.*).

safety control valve energized whenever the pilot is burning so that the main
gas valve can be turned on when commanded by the thermostat. If the pilot
flame goes off, there is no heat to the thermocouple. Subsequently, the
electrical current from the thermocouple is terminated so that the safety

Typical Applications

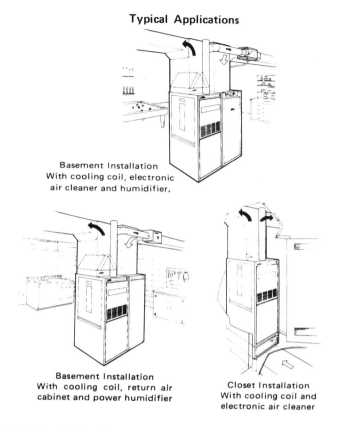

Basement Installation
With cooling coil, electronic
air cleaner and humidifier.

Basement Installation
With cooling coil, return air
cabinet and power humidifier

Closet Installation
With cooling coil and
electronic air cleaner

FIGURE 12-3. (Continued.)

control valve is closed preventing any gas from going to the pilot flame or to the main burner when a command is sent from the thermostat to the main gas valve.

The gas jet may be a ribbon type, multijet, or a single jet. The multijet type is similar in appearance to a gas range; the single jet type shoots a flame against a deflecting surface to spread the flame over the surface being heated.

Hot gases caused by the combustion of the gas and air pass through the furnace into the vent pipe and then out through the chimney. A draft diverter hood on the vent pipe maintains a small draft; it also prevents air currents from backing down the chimney and blowing out the pilot flame.

Studies have shown that about 5,400 cubic feet of gas normally burned by a pilot can be saved per year in the average home by using an electrode system that ignites the pilot just prior to lighting of the main burner. An electrode system can be easily installed by a trained serviceman in most furnaces in a

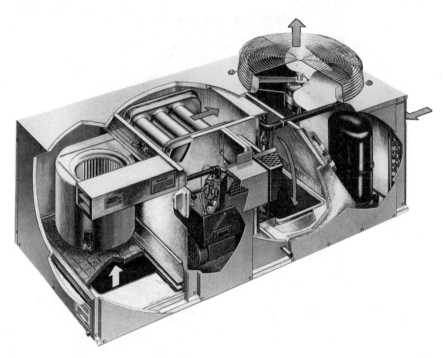

FIGURE 12-4. This all-season gas heating and cooling unit is designed for outdoor installation (*courtesy Lennox Industries Inc.*)

relatively short time. The system can reduce the operating costs of the furnace, save energy, and pay for itself in a relatively short time.

The electrode system consists of a solid-state igniter, electrode, and flame sensor and gas valve. When the thermostat calls for heat, it signals the solid-state igniter and allows gas to flow to the pilot. The solid-state igniter is activated and simultaneously sends pulses to the electrode which cause a sparking to light the pilot. When the flame sensor determines that the pilot flame is lit, the main gas valve is opened. The main gas valve will not open unless the pilot is lit and the mercury-filled flame sensor is satisfied. The expansion of the mercury must complete an electrical circuit before the main valve is opened. When the heating cycle is completed and the thermostat signal is removed, the main burners and the pilot are turned off. For more information, write White-Rodgers Division, Emerson Electric Co., 9797 Reavis Road, St. Louis, Missouri 63123.

Electric Furnace (Fig. 12-5)

An electric furnace utilizes a series of from two to five heavy-duty resistance heating element circuits. When electric current flows through the

Typical Applications

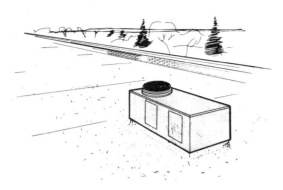

(Bottom supply and return air)

(Side supply and return air)

FIGURE 12-4. (Continued.)

elements, heat is produced because of the resistance of the elements to the flow of electricity. A fan blows air through the elements to a ducting network which distributes the heated air to the rooms. A thermostat calls for the heat and sets the furnace into action. A filter at the base of the furnace filters the incoming air to the furnace before it is heated.

Each heating element circuit contains a sequence-relay in addition to the resistance element. The purpose of the sequence-relay is to turn each element on in sequence 30 seconds after the previous element has turned on. This sequencing prevents extreme current surges (power drains) at the time of the initial call for heat by the thermostat. No chimney is needed for draft or for combustion gas venting; thus, the electric furnace can be located almost anywhere.

Electric duct heaters are sometimes used in conjunction with electric furnaces. Duct heaters are small resistance heaters mounted anywhere in a

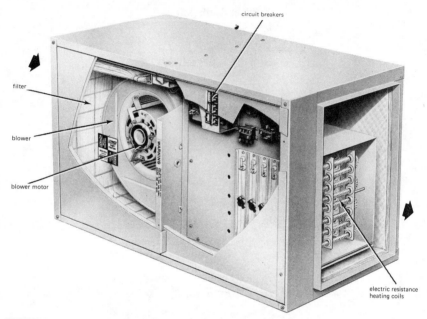

circuit breakers

filter

blower

blower motor

electric resistance
heating coils

FIGURE 12-5. A forced air electric furnace can be installed for an up, down, or horizontal airflow. No flue, chimney, nor combustion air are required for electric furnaces
(*courtesy Lennox Industries Inc.*).

ducting system. They are fitted with their own thermostats so that the temperature of the room served by the duct can be set as desired.

Heat Pumps (Figs. 12-6 & 12-7)

Although not brand new in concept, the heat pump has been redesigned
in recent years to be more efficient—experts say it's about four times as efficient
as oil and gas furnaces and it should become a major means of space *heating*
and *cooling* in the 1980s. A heat pump is powered by electricity and for cooling
runs just like an air conditioner; for heating, the heat pump still runs just like
an air conditioner, but in reverse. That is, the heat pump takes heat from the
outside air, even when the outside air temperature is below freezing, and uses
the heat to warm the home. When the outdoor temperature drops below
about 20 degrees F, the heat pump system uses resistance heaters (built in
integrally) to supplement the heat pump. Although it uses electricity, the heat
pump gets more heat out of a dollar's worth of electricity than do resistance
heaters in baseboard units or electric furnaces.

Regardless of outside temperature, the heat pump is always collecting
heat from one place and transferring it to another place. Switch-over between
heating and cooling is completely automatic. It is all electric; it cools without

Typical Applications

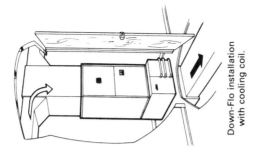

Down-Flo installation with cooling coil.

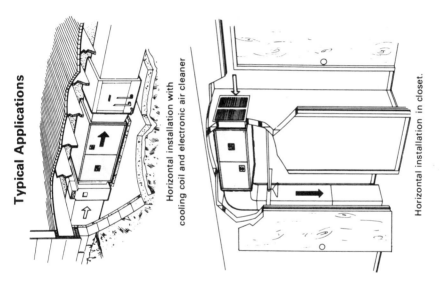

Horizontal installation with cooling coil and electronic air cleaner

Horizontal installation in closet.

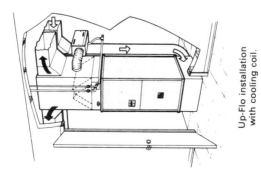

Up-Flo installation with cooling coil.

FIGURE 12-5. (Continued.)

water and heats without burning fuel of any kind. No flue is needed because there is no combustion. Thus, although the cost of a heat pump is a little more than a central heating system plus a separate cooling system, no chimney, gas connections, oil tanks, or cooling towers are required. If a satisfactory warm air duct already exists, it can be used with a heat pump installation.

If you already have a furnace with an air duct distribution system, you can install a heat pump into the ducting system. When the outside temperature drops below about 20 degrees F, the heat pump is supplemented by the existing furnace. Heat pumps come in house size and room size; the room size has its own electric resistance coil (similar to an electric baseboard heater) to supplement the heat pump as needed. In addition to being packaged with separate outdoor and indoor components (Figs. 12-6 and 12-7), some heat pumps are available in a single package for outdoor installation (Fig. 12-8).

Heat pump efficiency is referred to as coefficient of performance (COP). A straight electric resistance heat furnace operates at a COP of 1.00 or 100 percent efficiency. It returns 3,413 Btuh of heat for every kilowatt hour of electricity put in. (Gas and oil furnaces are normally rated at 70 to 80 percent efficient, but in some areas, the relatively low cost of these fuels makes them more economical to use than a heat pump with supplemental electric heat). As

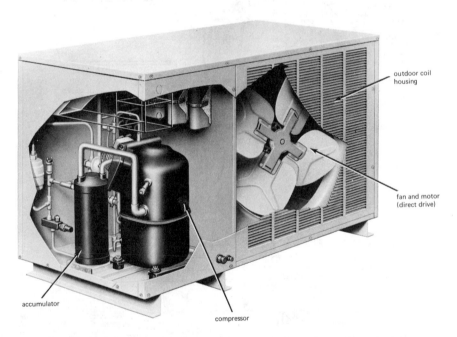

outdoor coil
housing

fan and motor
(direct drive)

accumulator

compressor

FIGURE 12-6. The outdoor unit of the heat pump can be located on a concrete slab or on a rooftop. The compressor and controls are mounted in a separate compartment from the coil and fan (*courtesy Lennox Industries Inc.*).

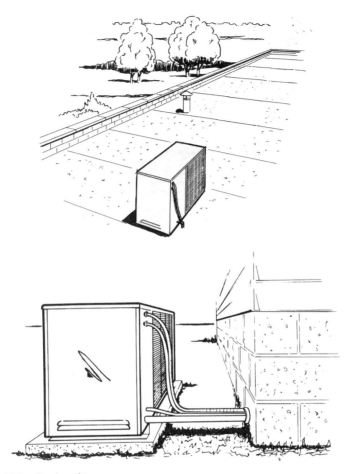

FIGURE 12-6. (Continued.)

the outdoor temperature drops, the heat pump COP also drops. When it reaches the point where the heat pump can no longer draw enough heat from the air to heat the home, the system switches to auxiliary electric heaters for its heating needs. These supplemental heaters keep the inside temperature at a comfortable level.

Figures 12-9 and 12-10 are schematic, refrigerant flow diagrams of one manufacturer's heat pump. The solid arrow indicates the flow of refrigerant for cooling operations and the dashed arrow indicates refrigerant flow during heating operations. Operation in the cooling, heating, and defrosting cycles are discussed in the following subsections.

For space cooling (Fig. 12-9), the compressor discharge gas flows through the four-way reversing valve to the outdoor coil where the gas

Typical Applications

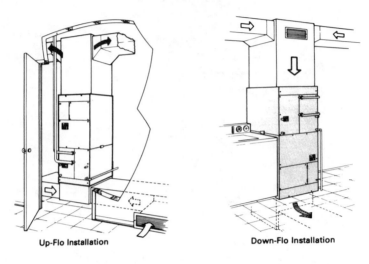

Up-Flo Installation   **Down-Flo Installation**

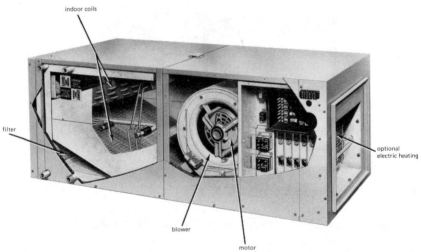

FIGURE 12-7. The indoor unit of the heat pump contains the indoor coil, the motor, and the blower. An optional electric heating unit supplements the heat pump, as required (*courtesy Lennox Industries Inc.*).

condenses to a liquid. Subcooled liquid leaves the outdoor coil and flows through a check valve to the indoor unit where it passes through a strainer and the capillary tube to the indoor coil.

The capillary tube restricts the flow as the refrigerant expands from a high to a low pressure. Evaporation takes place in the indoor coil and slightly super-heated gas flows through the vapor line to the outdoor unit. The gas flows through a parallel port in the four-way reversing valve to the suction-line

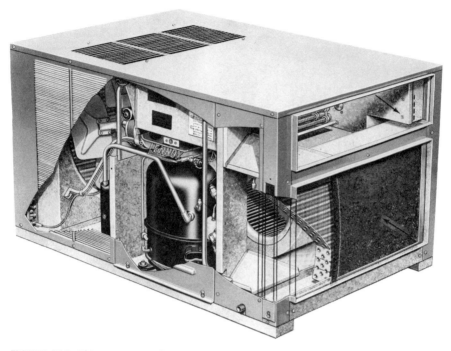

FIGURE 12-8. This compact outdoor single package heat pump contains all the refrigeration components, air movers, air filters, and optional electric heating unit in one complete package (*courtesy Lennox Industries Inc.*).

accumulator and the suction-line heat exchanger to the compressor suction to complete the cycle. Essentially no liquid flows through the liquid side of the suction-line heat exchanger during the cooling cycle, so no super-heat is added to the suction gas. For cooling operation, the accumulator does not store refrigerant liquid, so heat exchange is not needed, nor even desirable as it would tend to overheat the gas.

For heating operation (Fig. 12-10), the four-way reversing valve is de-energized so that compressor discharge gas flows through the vapor line to the indoor unit. As the vapor passes through the indoor coil, the vapor condenses to a liquid and becomes subcooled. It then flows through a check valve, bypassing the capillary tube, to the outdoor unit.

The check valve in the outdoor unit forces all the liquid refrigerant through the suction-line heat exchanger, the filter-drier and the capillary tube. The capillary tube restricts the flow as the refrigerant expands from a high to a low pressure. As the low-pressure refrigerant flows through the outdoor coil, it evaporates and flows as a vapor through the four-way valve to the suction-line accumulator, suction-line heat exchanger and suction line to the compressor to complete the cycle.

Typical Applications

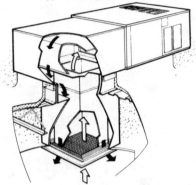

Rooftop installation with optional RT9 duct enclosure and combination ceiling supply and return air system.

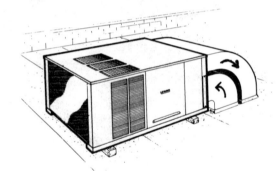

Rooftop
installation

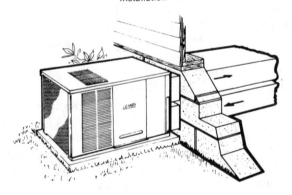

Unit on slab
at grade level

FIGURE 12-8. (Continued.)

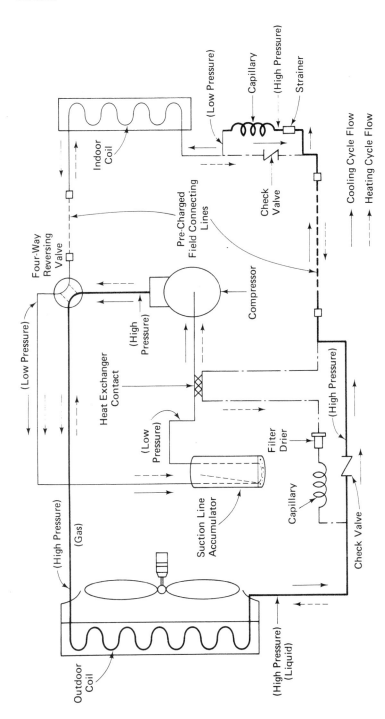

FIGURE 12-9: The heat pump is shown in the cooling and defrost cycles (*courtesy York Division of Borg-Warner*).

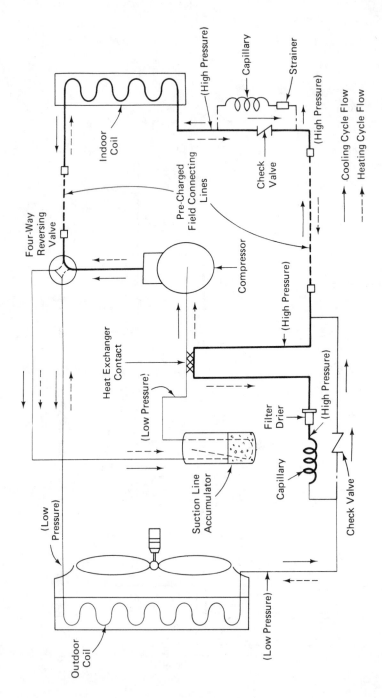

FIGURE 12-10. The heat pump heating cycle is similar to an air-conditioning cycle in reverse (*courtesy York Division of Borg-Warner*).

When the outside air temperature is below 50 to 55 degrees F, less refrigerant charge is required in the system than is needed for the cooling operation. Therefore, the excess refrigerant in the system ends up in the suction-line accumulator, since the capillary tube tends to keep the indoor coil, operating as a condenser, fairly well drained. As a consequence, the accumulator may be empty of refrigerant liquid at 55 degrees F outside air, and it may be nearly half full when operating on the heating cycle at −10 degrees F. In other words, the system charge in the heat exchangers adjusts to satisfy the load and the excess ends up in the accumulator.

Since the suction-line accumulator is a liquid-separation device, oil, as well as refrigerant liquid, will separate from the gas, with liquid laying in the lower portion of the accumulator. Since oil must be returned to the compressor to replenish its supply, a small metering orifice is located at the bottom of a U-tube inside the shell. The U-tube is opened at one end near the top of the shell so it can flow gas which has been separated from the liquid. The gravity head of the liquid in the shell forces small quantities of refrigerant liquid and oil into the suction tube, by way of the metering orifice, where it mixes with gas and returns to the compressor. The suction-line heat exchanger is provided to evaporate these droplets of refrigerant liquid to avoid crankcase dilution of the oil.

Frost and ice which form on the outdoor coil during the heating cycle must be defrosted when it blocks the airflow through the coil (Fig. 12-9). A defrost cycle is initiated when the defrost switch (air pressure differential) has been closed for 12 seconds more than it has been open (this delay eliminates the affect of wind gusts) and the liquid line temperature is 39 degrees F or lower. The defrost relay is energized which energizes the reversing valve to switch the refrigerant circuit to the cooling cycle, stops the outdoor fan and energizes at least one electric heater in the indoor unit to prevent drafts in the conditioned space.

The defrost cycle is terminated when the liquid-line temperature exceeds 75 degrees F or the liquid-line temperature exceeds 45 degrees F for 5 minutes. This permits defrost termination if the wind velocity does not permit liquid-line temperature reaching 75 degrees F. The defrost relay is de-energized to return the unit to the normal heating cycle.

Coal Burners

Coal burners and stokers are used in very old homes and in homes in areas where coal is abundant. In the furnace, coal is shoveled by hand through the input door onto an existing pile of hot coals. Air is introduced for draft and combustion through adjustable slotted doors. Burned ashes are "shaken down" through the grates by a mechanical lever and are shoveled out of the bottom door of the furnace.

A *stoker* feeds the coal automatically by means of a screw and many such burners also remove ashes from the burner into cans. There are two basic designs of stokers: the hopper-fed and the bin-fed. The hopper stoker *pushes* the coal from the coal bin to the firebox; the bin-fed *pulls* the coal from the bin. Hoppers require filling once a day. A fan forces air into the burner for combustion. Coal fires are started with kindling.

Furnace Sizing

If your house has been insulated after it was built, then your furnace may be too big for your home. In general that means it's inefficient and would use less fuel overall if it were smaller. Here's how to tell if it is too large; wait for one of the coldest nights of the year, and set your thermostat at 70 degrees F. If the furnace burner runs *less* than 40 minutes out of the next hour (time it only when it's running), once the house temperature reaches 70 degrees F, your furnace is too big. In the coldest weather, the furnace should run almost continually. A furnace that's too big turns on and off much more than it should, and that wastes energy. Call your service company; depending on your type of fuel burner, they may be able to cut down the size of your burner without replacing it (such as decreasing orifice and jet size with a proportionate decrease in combustion air).

12-3. DISTRIBUTION SYSTEM AND FURNACE MAINTENANCE

Maintenance should be performed periodically on your distribution system and furnace to maintain heating efficiency and ensure proper operation during the entire heating system. Perform the applicable maintenance procedures described in this section for the following units: thermostat, hot air distribution system, hot water distribution system, steam heating system, oil burner, gas burner, electric furnace, heat pump, coal furnace, and chimney flue and combustion chamber.

Thermostat Maintenance

No matter what type of heating (and cooling) system you may have, you should clean the thermostat contacts yearly. There is one exception; if you have a reasonably modern thermostat containing a mercury switch (you can usually see a "puddle" of mercury in a plastic capsule inside the thermostat housing), there is no need to clean it because the electrical contact (made by the mercury) is sealed. (You must insure that the mercury switch is level on the wall however. If the thermostat position is accidentally moved, place a level across the flat leveling surface of the thermostat. Loosen the mounting screws slightly, rotate the thermostat until level, and tighten the screws.

Failure to have the thermostat level causes inaccurate heater turn-on and turn-off temperatures.)

To clean the contacts of a thermostat, remove the thermostat cover. Use a soft artist's brush to brush dust out of the thermostat. Place a piece of clean bond paper between the thermostat contacts; advance the thermostat temperature setting higher, if necessary, to close the contacts on the paper. Pull the paper through the contacts several times to clean them; never use an abrasive on the contacts as it could damage the contacts.

Hot Air Distribution System Maintenance

With the exception of cleaning or replacing filters every 30 to 60 days during the heating season, hot air distribution system maintenance is performed once a year just prior to the start of the heating season. Clean filters as described in DIY 12-1 and perform the other maintenance procedures described in DIY 12-2. Remember that if filters are permitted to become clogged with dust and dirt, the circulation of air (heated or air-conditioned) will be diminished and the efficiency of the heating system reduced.

Do-It-Yourself 12-1
AIR FILTER MAINTENANCE

Materials: dishwashing detergent, filter coat (for metal filters; see procedure)
Tools: none

Procedure:

1. If the filter is fiberglass, replace it when it is dirty.
2. If the filter is made of a synthetic sponge-like material, remove the filter, shake it out, wash it *carefully* in a solution of dishwashing detergent and water, rinse with clear water, dry, and return it to the furnace. Do not squeeze the filter because this could tear the material.
3. If the filter is metal, flush clear water through the filter from the dirty side. Drain and dry the filter. Then spray the filter with a *filter coat* which aids in trapping dust particles. Filter coat is available from a heating supply dealer; failure to use the coating can severely limit the filtering ability of the filter.

Do-It-Yourself 12-2
HOT AIR DISTRIBUTION SYSTEM MAINTENANCE

Materials: detergent, cleaning rag, SAE-10 or SAE-20 motor oil, silicone lubricant (if rust is found), duct tape (if the duct joints leak)

Do-It-Yourself 12-2 (continued)
HOT AIR DISTRIBUTION SYSTEM MAINTENANCE

Tools: vacuum cleaner, wire brush, ruler, soft utility brush for dusting

Procedure:

CAUTION
Ensure that all electrical power is off to the system. Place the emergency switch to off, the thermostat to the lowest setting, and the power circuit breaker or fuse main switch at your house electrical distribution panel to off.

1. Open the blower cabinet at the base of the heater. Vacuum the dust out. Dust the blower blades with a soft brush. Using a detergent in water and a soft rag, clean the blower blades taking care not to cause any damage.
2. The motor that drives the fan (blower) has two oil cups (unless the motor is a sealed bearing unit that requires no lubrication), one at each end of the shaft. Lift the cup lids and lubricate the motor as directed by the manufacturer; if no directions are available, lubricate with several drops of SAE-10 or SAE-20 motor oil. A squirrel cage fan driven by the motor also has oil cups at each end of its axle; lubricate with as much SAE-10 or SAE-20 motor oil as necessary until the felt pads will not absorb any more oil without overflowing.
3. Check the belt between the motor and the fan for cracks and extensive fraying. Replace if necessary. Check the belt for proper tension; when the center of the belt is depressed with moderate pressure, the belt should deflect about ½ inch. Adjust the belt as necessary by relocating the position of the motor on its mount. Close the blower cabinet.
4. If a humidifier is used in the plenum, check the plenum for rust. If rust is present, brush it off with a wire brush. Spray the chamber with a silicone lubricant.

NOTE
Electrical power can be turned on as desired.

5. Turn the blower on, either manually or by setting the thermostat up for a heating cycle. Check duct junctions for leaks; stop leaks by wrapping the junctions with duct tape.
6. Clean registers and/or floor grilles with a vacuum cleaner. If possible, without damaging wall surfaces, remove registers and insert the vacuum hose with a soft utility brush on the end of the hose to remove dust from inside the duct (be sure the end *won't* come off of the vacuum.)

Hot Water Distribution System Maintenance

Perform the procedures in DIY 12-3 prior to the beginning of the heating season. Bleed air from the system (DIY 12-3) two months after the start of the season.

Do-It-Yourself 12-3
HOT WATER DISTRIBUTION SYSTEM MAINTENANCE

Materials: soda solution (available from plumbing suppliers) to clean out boilers, SAE-20 or SAE-30 motor oil

Tools: handle or key to open radiator valves, drain hose or bucket

Procedures:

NOTE

The following procedures describe what should be done to maintain a hot water distribution system. Follow the manufacturer's directions for performing these procedures. Since many hot water heating systems also provide for heating the domestic hot water supply, it may be necessary to remove power or set thermostats very low before beginning these procedures.

1. Lubricate the circulating pump motor (if not a sealed unit) by placing a few drops of SAE-20 or SAE-30 motor oil in the oil cup.
2. Check the operation of the pump, flow control valve, and piping for leaks.
3. Drain and flush the boiler. This removes most rust, mineral deposits, and sediment from the boiler. If the boiler has a clean-out (filler) opening, the boiler can be cleaned with a soda solution; buy the powder from a plumbing supply house. Be sure to follow manufacturer's procedures or have your serviceman show you how this year; next year you can do it yourself. Draining and flushing is performed basically as follows. Turn the burner off. Attach a hose from the drain outlet and run it to a drain or have several pails ready for draining out the water. Shut off the water supply valve to the boiler. Now open the drain valve. Open the venting valves on the highest radiators in the house to admit air into the radiators and piping so that the water can drain out. After the sediment has drained out, open the water supply valve letting fresh water into the boiler to flush it out. After flushing, refill the system by closing the drain valve (keep the water supply valve open) and turning the boiler on. Keep the venting valves open at the highest radiators *until* you hear water nearly filling the pipes of the radiators to the top; then close the radiator venting valves. Later in the day after the system has heated and water has circulated through the system, bleed the air from the system (step 4).
4. Bleed air from the system at the start of the season and about two months thereafter. Air rises to the top of the system and keeps out the hot water. Some radiators contain automatic venting valves that make this hand venting unnecessary; automatic venting valves are inexpensive and easy to install. Using a handle or key in the vent valve and a cup under the valve, open each radiator vent valve on the top floor and draw out air until water comes out; take care since the water is hot.
5. Drain the expansion tank at the beginning of the heating season and at periodic intervals to get rid of rust and other sediment in the tank. To drain the tank, turn the supply valve to the tank off. Open the tank drain valve at the bottom of the tank and drain the water into a bucket (or attach a small hose and run it to a drain). When the tank is drained, close the drain valve and open the supply valve.

Steam Heating System Maintenance

Steam heating system maintenance is limited to draining a bucket of water from the drain valve of the boiler every three to four weeks and checking the boiler water level. Sediment falls to the bottom and draining a bucket of water removes the sediment. Sediment insulates the boiler from the flame. Keep the boiler filled to the center of the water gauge indicator (or to the level recommended by the manufacturer).

Oil Burner Maintenance

An oil burner should be serviced yearly to improve efficiency and performance, and in the interest of safety. Your oil dealer normally performs this task for a fee prior to the start of the heating season. In addition to preseason service, the fee also normally includes emergency service and the replacement of any malfunctioning part during the subsequent year; thus, the maintenance agreement is also an insurance policy that you may need during the peak heating season.

You can perform some of the oil burner maintenance such as inspecting for oil leaks, changing air filters (QRC 12-1), and adjusting dampers and registers (DIY 12-7). Have your oil burner serviceman adjust and clean the burner unit, adjust the fuel-to-air ratio for maximum efficiency, change the oil filter, and check the oil pump and oil burner nozzle. (You can clean the oil filter by removing its attaching bolts or unscrewing it. Don't damage the gasket. Wash the filter in kerosene and replace it into the housing.) The serviceman should perform four tests and get satisfactory results; if satisfactory results are not obtained, the proper adjustments must be made and the tests repeated. The following tests should be performed using test instruments (a small diameter hole is made in the stack to insert the test instruments):

1. *Draft test*: determines if the draft of the chimney is sufficient to properly burn the oil or if excessive heat is being lost up the chimney. A draft gauge is inserted through a ¼-inch hole in the stack to make this test; it gives a continuous indication of the draft and instantly shows the slightest change as adjustments are made.

2. *Smoke test*: determines if oil is being burned cleanly and completely. A smoke gauge is used to take a sample of the combustion smoke passing through the stack. The color of the sample paper is compared to a standard smoke shade chart.

3. *Carbon dioxide* (CO_2) *test*: determines if the fuel is being burned completely (an efficiency test). The level of CO_2 should run 8 to 14 percent (below these percentages indicates an inefficient burner). A sample of the stack flow is taken in a sampling device.

4. *Stack temperature test*: determines if the stack gases are too hot or not hot enough. A thermometer (400- to 1200-degree F range) is inserted in the stack; the temperature should usually be less than 450 degrees F (an excessively high stack temperature indicates an inefficient furnace).

When the furnace is all checked out, the CO_2 content will have been raised as high as possible without causing a smoky fire, the draft will be set to provide 0.02 to 0.03 inch water draft over the fire, and the stack temperature will be known. The CO_2 content and the stack temperature are transferred to a chart to read the efficiency rating of the furnace.

For safety, it is always best to let a trained serviceman familiar with your furnace make all equipment changes, alterations, or adjustments. Furnace testing instruments are manufactured by Dwyer Instruments, Inc., P.O. Box 373, Michigan City, Indiana 46360, and Bacharach Instrument Company, 625 Alpha Drive, Pittsburgh, Pennsylvania 15238.

Gas Burner Maintenance

A gas burner should be serviced every three years to improve efficiency and performance and in the interest of safety. You can perform some of the maintenance when the system is completely shut off including removing the side panels that have handles (don't take out screws) for dusting and vacuuming, setting the pilot by adjusting the pilot adjustment screw until the flame is ½ inch high and adjusting the heater to a blue flame by rotating the air shutter on the main burner. Have a serviceman check the operation of the main gas valve, pressure regulator, and safety control valve. He should also set the primary air supply nozzle for proper combustion and make the following tests using test instruments (a small diameter hole is made in the stack to insert the test instruments):

1. *Draft test*: determines if the draft of the chimney is sufficient to properly burn the gas or if excess heat is being lost up the chimney. A draft gauge is inserted through a ¼-inch hole in the stack to make this test; it gives a continuous indication of the draft and instantly shows the slightest change as adjustments are made.
2. *Stack temperature test:* determines if the stack gases are too hot or not hot enough. A thermometer (300- to 1,200-degree F range) is inserted in the stack; the temperature should usually be less than 330 degrees F (an excessively high stack temperature indicates an inefficient furnace).
3. *Carbon dioxide (CO_2) test:* determines if the fuel is being burned completely. The CO_2 level should run 8 to 14 percent (below these percentages indicates an inefficient burner). A sample of the stack flow is taken in a sampling device.

When the furnace is completely checked out, the CO_2 content will have been raised as high as possible without causing a smoking fire; the draft will be set to provide 0.02 to 0.03 inch water draft over the fire and the stack temperature will be known. The CO_2 content and the stack temperature are transferred to a chart to read the efficiency rating of the furnace.

For safety, it is always best to let a trained serviceman who is familiar with your furnace make all equipment changes, alterations, and adjustments.

Electric Furnace Maintenance

There is very little, if any, maintenance required on an electric furnace. Check the manufacturer's instructions. Do not forget to perform maintenance on the *heat distribution* system, however, such as cleaning air filters, thermostats, ducts, and so on.

Heat Pump Maintenance

Heat pump system maintenance is simple because no combustion takes place; thus there is no soot to accumulate in a combustion chamber, smoke pipe, or flue, and no tests for draft, smoke, flue temperature, or CO_2 content are required. Maintenance consists of cleaning the outside coil, checking condensate drain, filter cleaning or replacement, blower lubrication, and waxing (DIY 12-4).

Do-It-Yourself 12-4
HEAT PUMP MAINTENANCE

Materials: SAE-10 or SAE-20 nondetergent oil for motor lubrication, automobile wax for waxing
Tools: garden hose

Procedure:

NOTE

Follow the manufacturer's specific maintenance procedures; if none are available, use the following as a guide.

1. Clean and inspect the outside coil every few months (every month during lawn care season). If the coil is dirty, wash it with a garden hose.
2. Heat pumps contain automatic defrost systems to keep the outdoor coil from icing up, even during cold weather. If ice builds up excessively, contact your serviceman.
3. During the cooling season, check the condensate drain for a free and running condition. If the water does not run freely, check and remove any blockage in the drainpipe.

4. Check the filters monthly; if dirty, refer to DIY 12-1 for cleaning procedures.
5. Lubricate fan and blower motors annually.
   a. If there are no oiling ports (cups), the motors are prelubricated and sealed and no lubrication is required.
   b. If there are oiling ports (cups), add a few drops of SAE-10 or SAE-20 nondetergent oil.
6. Wax the cabinet with automobile wax semiannually. This helps protect the unit from the weather.

Air ducts may be balanced as described in DIY 12-7; the position of register dampers may be different during the heating season than during the cooling season.

Heat pump manufacturers have warned that duct systems cannot be separately controlled by zones. Heat pumps require a large volume of air blowing over the indoor coil during winter heating, or else the compressor outside the house can be seriously damaged. Reducing the airflow into the rooms either by zone control (dampers in main trunk lines) or by closing down some of the dampers in the registers may reduce airflow to a point where damage could occur. Consult your heat pump operation and maintenance manual or contact the manufacturer before you appreciably restrict the flow of the circulating air.

Coal Furnace Maintenance

Coal furnace maintenance is accomplished by the do-it-yourselfer at the end of the heating season. Maintenance consists of cleaning the furnace and then oiling to prevent rust (DIY 12-5).

Do-It-Yourself 12-5
COAL FURNACE MAINTENANCE

Materials: SAE-10 or SAE-20 motor oil
Tools: broom, brush, shovel

Procedure:

1. Make sure the fire is completely out. Work on a warm day.
2. Use a broom or brush to knock soot off the top and sides of the heater.
3. Clean the soot from the top of the grate and the ashpan.
4. Clean the flue pipe. Carefully remove and number each section of the flue so you can put it back together easily. Tap each section to loosen the soot and then brush the sections out. Put the flue back together. If there are any leaks at the section junctions, wrap the sections with duct tape.
5. Oil the inside of the coal screw and hopper to prevent rusting.

Chimney Flue and Combustion Chamber Cleaning

For efficient operation, maximum warmth, and to prevent smoke from backing up into your home, your chimney flues (one for the furnace and one for the fireplace) and combustion chamber should be cleaned when an inspection shows a heavy accumulation of soot. A buildup of ⅛ inch soot decreases efficiency by about 8 percent. Follow the procedures in DIY 12-6.

If you spot a leak in the chimney (smoke coming out of the hole), repair the leak with a mixture of one part cement to three parts sand. Clean out loose mortar around the hole and adequately moisten the hole and surrounding area with water before applying new mortar.

Do-It-Yourself 12-6
CHIMNEY AND COMBUSTION CHAMBER CLEANING

Materials: newspaper or plastic, masking tape, burlap or canvas bag, bricks (or snow tire chains; see procedure), water hose
Tools: vacuum cleaner, brush, dust pan, soft wire brush

Procedure:

Chimney Flue

1. Disconnect the flue at the furnace.
2. Seal the chimney flue with paper and tape or plastic and tape at the duct opening, at fireplace openings, and at any other openings so that soot and dirt will collect against the paper or plastic for easy removal.
3. As you seal off, check for holes and rust spots. Repair the holes and remove rust.
4. Fill a burlap or canvas bag with rags and bricks until it fits the chimney. The rags provide bulk and the bricks provide weight. If the chimney is capped, use an old snow tire chain instead of the bag.
5. Tie the top of the bag with a strong rope.
6. Raise and lower the bag in the chimney to knock off the soot. Start with the first few feet of area and gradually work your way down the chimney. If using the snow tire chain, lower it down and rattle it around to break the soot loose.
7. After all the loose soot is removed, *lightly* spray water against the walls of the flue. Then raise and lower the bag some more.
8. Remove the paper or plastic from the flue and fireplace openings and clean out the soot into a trash can. Clean the final soot out at the openings with a shop vacuum cleaner. Reconnect the flue at the chimney.

Combustion Chamber

1. Shut off all power to the furnace by placing the emergency switch to "Off," the thermostat to minimum setting, and the circuit breaker to "Off." If your system has a fuse rather than a circuit breaker, remove the fuse.

Do-It-Yourself 12-6 (continued)
CHIMNEY AND COMBUSTION CHAMBER CLEANING

2. Remove the smoke pipe.
3. If the lining of the combustion chamber is firebrick or steel, vacuum the soot off (don't use any attachments; the metal end of the vacuum will help remove the soot). If the lining of the combustion chamber is a clothlike material, don't vacuum, scrub, or do anything else to it; you'll do more damage than good.
4. Vacuum out the soot in the smoke pipe too. A soft wire brush should help.
5. Attach the smoke pipe to the furnace. Reapply power.

12-4. HEATING FOR COMFORT AND ECONOMY

The ultimate goal is to heat your home economically, yet feel comfortable. There are a number of techniques, tips, hints, and other steps that will help you meet these objectives. Try to utilize all of the general tips in QRC 12-1 and the specific tips for your type of heating system in QRC 12-2 through 12-6.

12-1. General Tips and Techniques for Heating Comfortably and Economically.

12-2. Hot Air Distribution System.

12-3. Hot Water and Steam Distribution System.

12-4. Radiant Heat Distribution System.

12-5. Oil Burners.

12-6. Gas Burners.

Quick Reference Chart 12-1
GENERAL TIPS AND TECHNIQUES FOR HEATING COMFORTABLY AND ECONOMICALLY

1. Be sure all windows, doors, and other places where air can escape or infiltrate, such as attic access doors and light fixtures, are caulked (Chapter 5), weather stripped (Chapter 6), or otherwise sealed.
2. In the interest of safety, as well as possible improvement in efficiency and performance, routine cleaning and adjustment of furnaces should be performed annually. Refer to the applicable QRC in this chapter.
3. Close off any rooms not heated. Also keep closet doors closed.
4. Don't overheat any particular room. Balance the heating to the desired temperature in each room by adjusting the distribution system (hot air duct dampers or registers or hot water radiator valves). Refer to DIY 12-7.
5. If your heating system is zoned, set the thermostats in bedrooms and areas of high activity at temperatures lower than the living areas.

Quick Reference Chart ,12-1 (continued)
GENERAL TIPS AND TECHNIQUES FOR HEATING COMFORTABLY
AND ECONOMICALLY

---

6. Keep fireplace dampers closed.

7. Clean air filters monthly; do it on the first day of each month so it's easy to remember when the maintenance is required.

8. Perform each of the maintenance procedures described in Section 12-3.

9. After you've completed most of the weatherizing tips and additions up to this chapter, you may find that your furnace is too large. Your furnace serviceman can effectively reduce the size by changing the nozzle size of the oil burner or orifice size of a gas burner.

10. If a part of the heating distribution system is located directly beneath a window enclosed by a drapery, much of the warm air will rise and become trapped within the drapery/window area. Deflect the heat away from the drapery and window and out into the room by one of the following methods:

    a. Hot air system: if possible, change the position of the heat duct deflectors to cause the warm air to be blown outward instead of upward.

    b. Hot air system: add a plastic or metal deflector to the heat duct.

    c. Radiant heat: form and place sheet aluminum behind the radiator or other radiant heat device.

Do-It-Yourself 12-7
BALANCING HEATING AND COOLING SYSTEMS

Materials: none
Tools: thermometers

Procedure:

NOTE

This procedure is for setting the levels of temperature in your heated rooms. The procedure must be repeated if another room is placed into use or taken out of use during the season. The procedure is also used to balance air conditioning systems (Chapter 13). This same procedure can also be used to balance a hot water or steam heating system; instead of adjusting dampers or registers, adjust the radiator shut-off valves.

1. Leave the thermostat at one setting for several hours before going on to Step 2. Of course, the blower is running, and all dampers in ducts and registers should be open.

2. Using several thermometers, check the temperatures in all rooms. Be sure the thermometers register equally; lay them out together so they can equalize. If there are differences in their readings, take this into account when measuring temperatures in the various rooms. Take temperatures in each room 2 or 3 feet

---

Do-It-Yourself 12-7 (continued)
BALANCING HEATING AND COOLING SYSTEMS

---

off the floor and near the center. Doors to rooms should be left in their normal positions, closed or open. Let the system operate about 30 minutes before taking readings.

3. If you now find a room too warm or cool, partially close or open the damper to that room. Do one room at a time. It's best to start with the room that contains the thermostat if that room tends to be uncomfortable (Fig. 12-11).

Adjusting Damper in Supply Air Duct.

Floor or Wall Register has Lever-Controlled Damper.

FIGURE 12-11. Adjust dampers in supply ducts; leave register dampers wide open. If there are no dampers in the ducts, adjust the dampers in the registers (*courtesy Lennox Industries Inc.*).

4. As air delivery is reduced at some outlets, it automatically increases at others. After air has been reduced to rooms that received an excess amount, allow the system to run for 30 minutes or more. Then check temperatures again; the formerly uncomfortable rooms will have greater comfort.

5. Keep adjusting dampers until rooms reach the temperature balance you want. Be sure to allow enough time for temperatures to stabilize after you make each adjustment. Also, check temperatures in each room every time, because as you cut delivery to one room, you can never be sure which other rooms will receive the resulting gain.

6. Once the correct adjustment is established, place a small mark on the damper or register to indicate the lever position.

Quick Reference Chart 12-2
HOT AIR DISTRIBUTION SYSTEM

1. Tape all duct junctions with insulating tape to prevent the escapement of warm air.
2. Insulate all ducts in unheated areas such as attics, crawl spaces, and so on (Section 7-14).
3. The distribution fan switch should be set to extract all of the useful heat from the plenum. Set the fan "Off" switch at about 75 degrees F and set the fan "On" switch as close to the fan "Off" as possible (about 80 degrees F). This resetting of both switches can increase the system's efficiency by about 8 percent.
4. Set the distribution fan to "On" manually and let it operate continuously. This helps prevent stratification of layers of different temperature air.

NOTE

Many new hot air distribution systems have a *continuous air circulation* system incorporated in them; included are a two-speed motor and a fan-speed control (thermostat). When the furnace is on, the fan operates at high speed; when the furnace is off, the low-speed fan operates to provide continuous air circulation. You can update your furnace if the fan is belt-driven by replacing the existing motor with a two-speed motor of the same horsepower, shaft size, direction of rotation, mounting size, and with the high speed equal to the old motor speed. Replace the existing fan control switch with a control designed for two-speed operation. Set the control for high-speed fan operation above 80 degrees F. Be sure all electrical power is off at the fuse box or circuit breaker; if you are unfamiliar with electricity, have a serviceman make this modern installation.

5. Do not obstruct registers with furniture or curtains. If the registers are high on the wall, consider adding plastic deflectors which will deflect the warm air downward.
6. The hot air ducts should either have adjustable dampers in the ducts or in the registers at the output of the ducts. If there are no adjustments, replace the existing registers with adjustable registers or add dampers to the ducts. To add a damper, measure the duct size and buy the proper size damper from a plumbing supply house. Open the duct at the branch-main junction. Drill holes in the damper shaft opposite each other in the branch duct. Attach the damper and then the shaft handle. Join the ducts back together.
7. Balance the system (DIY 12-7).
8. For problem rooms such as a room having a lot of glass, consider installing an automatic damper on the duct leading to the room. The automatic damper is relay-controlled and opens and closes on thermostat demand. The automatic damper can also be effectively placed in a main duct if the duct supplies all rooms, such as bedrooms, which would be set at the same temperature (but lower than the other rooms on other main ducts).
9. If you have a two-story house, put most of the heat in the lower level. Since heat rises, the upper level may receive sufficient heat without the necessity of additional ducts or supplementary heating.

Quick Reference Chart 12-3
HOT WATER & STEAM DISTRIBUTION SYSTEM

1. Insulate all pipes that run through unheated spaces (Section 7-16).
2. If radiators are to be painted, use a metal finish or an enamel in preference to a flat paint.
3. Reflect heat from radiators into the room; place a sheet of aluminum or aluminum foil against the wall behind each radiator.
4. Bleed air out of the radiators on the top floor at the start of each heating season and two months thereafter (DIY 12-3).
5. Adjust the valves on radiators to set the room temperature. If the room is not in use, close the valve to turn the radiator off.
6. Valves of radiators in steam heating systems are normally all on or all off. Buy valves that let you set the flow; in this manner, you can control the temperature in a room.
7. Balance the heating system (DIY 12-7).

Quick Reference Chart 12-4
RADIANT HEAT DISTRIBUTION SYSTEM

1. Locate portable, radiant heaters so that they radiate heat to the most advantageous location.
2. Don't locate a portable, radiant heater near a home-heater thermostat.
3. All portable, electric, radiant heaters should have a tip-over switch which removes all electrical power to the heater if the heater falls over.
4. Don't block the radiant heat by placing furniture, draperies, or similar items in front of the heater.
5. Radiant heaters having water-filled tubing provide a more constant heat because when the heat source is eliminated, the water remains warm in the tubing and continues to provide heat.

Quick Reference Chart 12-5
OIL BURNERS

1. Have a serviceman provide a yearly tune-up and inspection. This cuts fuel bills as well as the chances of costly breakdowns.
2. Installation of a baffle in the firebox will spread the flame from the oil burner to the side walls causing the medium to heat more quickly. Have a specialist make the installation to prevent hazards.

Quick Reference Chart 12-6
GAS BURNERS

1. Turn the burner pilot permanently off during the nonheating season. This can save about 750 to 1,000 cubic feet of gas per month during the nonheating season. Refer to your gas burner manual for shutoff procedures (and turn-on procedures for the heating season).
2. New burners should be equipped with electric ignition rather than a pilot. This saves about 10 percent in per-year costs and pays for itself in a year's time.

12-5. USE OF OUTDOOR AIR FOR COMBUSTION

When oil and gas burners were discussed (Section 12-2), you learned that air is mixed with the oil or gas for combustion in the combustion chamber. In reality, about 2,000 cubic feet of air are required for each gallon of fuel oil burned and about 1,250 cubic feet of air are needed for safe, complete combustion of each 100 cubic feet of natural gas per hour. The heat warms the plenum above it where air circulates to the rooms via supply air ducts and returns to the plenum via return ducts. The air in the ducts does not enter the combustion chamber. The combustion gases and air rise up the chimney. Where, then, does the air come from to supply the combustion chamber?

Combustion air is drawn into the combustion chamber by a blower which gets its air from its surrounding environment. If the furnace is in a heated basement or utility room, the air drawn in for combustion air is heated air at 68 to 70 degrees F. If the furnace is in an enclosed, unheated area, warm heated air drifts through the unheated area where it cools off and is then used for the combustion process. As this heated air is drawn in for combustion and then sent up the chimney, the "used" air must be replaced; the replacement air comes in through cracks around windows and doors and through the walls of your home. The influx of this replacement air causes cold drafts across the room and, of course, more cold air that must be heated. Since the outside air is cold and low in humidity, the influx of air also drops the humidity level in the house.

How can air be provided for combustion without providing the warm interior air that must be replaced with cold air that infiltrates to replace the warm air used for combustion? The answer is to provide the furnace with outside air for combustion. This has been proven effective in a number of studies and has reduced oil and gas consumption by 8 to 15 percent. Not only has consumption been significantly reduced, but drafts of cold air that were coming through and around windows and doors and through walls to replenish interior air used in combustion were virtually eliminated. Because much less cold air with low humidity was drawn into the homes, the interior humidity level increased. Thus, by providing outside air to your furnace for combustion, you can decrease fuel consumption, decrease cold air drafts, and increase the humidity level in your home.

Outside air is easily and inexpensively supplied to the furnace for combustion by running a duct (about 4 inches) from an exterior wall to the heater. You can use pieces of old rainspouting or clothes dryer flexible-vent tubing. Cut a hole through the wall to the exterior and install a clothes dryer vent (remove the one-way flapper valve inside it) or a breather vent of similar design and size. An even simpler way to install the vent is to remove a pane of glass from a cellar window and replace it with a piece of ¼-inch plywood with the vent mounted in its center. Run the ducting from the wall or window across the ceiling to the heater; then run it down and over the top of (or along side of) the furnace smoke pipe (the pipe from the furnace to the chimney). Running the pipe in this manner will heat the air to some extent. Terminate the air duct at the furnace burner air intake; you can simply leave the duct terminated or you can make an enclosed aluminum "boot" to mate with the furnace air intake making a sealed, air combustion chamber.

Now that you have probably changed the fuel-to-air ratio because of the ducting, it is important that the furnace be tested for the proper stack temperature and CO_2 content. Have your furnace serviceman make these tests (refer to Section 12-3).

12-6. HEAT RECLAMATION

With the furnace cleaned and tuned, there is still another method to reduce the cost of heating your home: heat reclamation. This involves capturing some of the heat flowing from the combustion chamber through the metal stack pipe to the chimney. You can't capture all of this wasted heat nor can you even capture most of the heat because some heat must pass up the chimney to cause a sufficient draft to vent the gases of combustion (a temperature of about 450 degrees F for fuel oil furnaces and about 330 degrees F for gas furnaces is considered correct).

There are a number of different types of heat exchangers that can reclaim heat from the smoke stack pipe. Several of these can be built easily at home and others can be purchased and installed easily by the do-it-yourselfer. The more heat extracted, the higher is the efficiency of the heat-reclaiming device. The reclaimed heat can be used to heat the immediate area around the furnace or with the internally mounted heat exchangers, the reclaimed heat can be blown through a duct into another room such as a recreation room or shop (Fig. 12-12).

The simplest method of extracting heat from the smoke stack pipe is to simply circulate air over it. Use a small blower that operates on 117 volts and

DID YOU KNOW. . .?

As Much as Half the Heat Your Furnace Produces Actually Goes Up the Chimney.

But Now, Much of This Wasted Heat Can be Saved with a Simple Air-to-Air Heat Exchanger.

"Free Heat"

Furnace

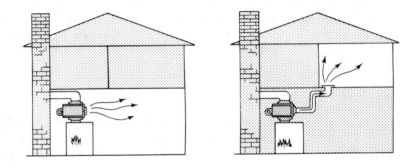

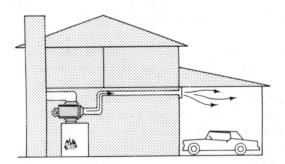

FIGURE 12-12. Heat reclaimed from the smoke stack pipe can be used to heat the surrounding area or directed through ducting to other rooms (*courtesy Dolin Metal Products, Inc.*).

direct the air into the area where more heat is desired. Connect the blower electrical terminals in parallel with the oil pump motor or (as a second choice) in parallel with the furnace blower.

You can extract heat more efficiently if the radiating area of the heat exchanger can be increased. Using aluminum sheet, flashing, or grass edging, cut exchangers as shown in Fig. 12-13. Measure the circumference of the stack

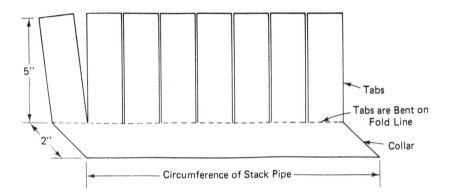

FIGURE 12-13. A simple smoke stack pipe heat reclaimer can be constructed from aluminum sheet, flashing, or grass edging.

pipe by wrapping a piece of string around it. You'll need one exchanger for each 4 inches of stack pipe length. Wrap each exchanger around the flue so that close contact is made between the stack and the exchanger collar; hold the collar on with wire (do not use any material that could catch on fire). Space the exchangers every 4 inches along the stack pipe. Bend and twist the flaps alternately left and right. To increase efficiency further, direct the air of a small blower toward the collars.

The exchangers discussed thus far have been external to the stack pipe. Several manufacturers produce internal heat extractors that insert into the stack pipe and thereby extract more heat than can be extracted by external means only (Fig. 12-14). To be effective, the stack pipe temperature needs to be higher than 450 degrees F but less than 800 degrees F; too low a temperature will not provide sufficient heat to justify the installation; too high a heat can damage the heat exchanger. Once installed, a serviceman should check for proper stack pipe temperature and CO_2 content. The amount of heat that can be recovered is a function of stack pipe temperature and runs from about 4 percent at 450 degrees F to 8 percent at 800 degrees F. You can determine the desirability of retrofitting a stack pipe with an internal heat exchanger by first measuring the stack temperature and then estimating the percentage recovery and the annual saving on the total fuel bill.

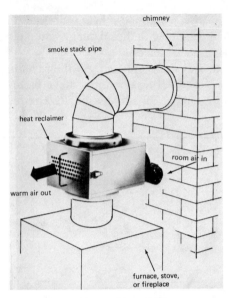

FIGURE 12-14. The air passing through any heat reclaimer must be fresh air. **Combustion** gases that pass through the smoke stack pipe are toxic and cannot be vented into the home. A heat reclaimer is a heat exchanger (*courtesy Dolin Metal Products, Inc.*).

Removing too much heat from the exhaust gases can decrease their buoyancy and result in a reverse flow of gases. Lack of a proper draft in the chimney can cause toxic gases in the house and can ruin the furnace.

Internal heat exchangers are placed into the stack pipe (either vertically or horizontally) after about 10 inches of stack pipe are first removed ahead of the draft diverter on a gas fired furnace or the barometric damper in an oil fired furnace. Heat pipes extend into the path of hot, escaping flue gases and the heat conducts along the heat pipes to the reclaimer section of the heat exchanger. A small blower in the exchanger directs the heat through a duct into the immediate area or through an extension duct into another room or into the house ducting itself. An adjustable 117-volt thermostat is connected as directed by the manufacturer (some plug into a regular convenience outlet while others connect in parallel with the heater blower or pump motor).

A *heat pipe* is used to transfer heat from one location to another with very little temperature differential (loss). The heat pipe (Fig. 12-15) has three basic parts: a hollow metal container, a wire screen or cloth capillary wick structure, and a working fluid (water, refrigerant, or liquid metal). The wick is secured to the inside wall of the container and is saturated with the working fluid. The container is evacuated and is hermetically sealed.

The boiling temperature of the working fluid is reduced to the point where it is in equilibrium with its own vapor because of the strong vacuum

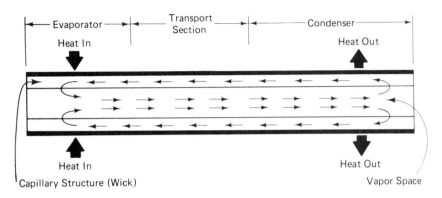

FIGURE 12-15. The heat pipe is a by-product of the space age; it is used to transfer heat from one place to another with very little temperature loss.

inside the tube. When heat is added to one end of the tube (evaporator), the liquid vaporizes and expands to fill the tube. The rapid generation of vapor creates a pressure gradient that forces excess vapor to move to the colder end (condenser). The lower temperature causes the vapor to condense; the condensate is returned to the evaporator end by the capillary pumping action of the wick. The cycle continues to repeat.

The heat reclaimer in Fig. 12-16 uses a number of heat pipes placed in the stack pipe to transfer heat into another duct. The transfer of heat is very effective.

FIGURE 12-16. This heat reclaimer utilizes a number of heat pipes that transfer heat very efficiently (*courtesy Isothermics, Inc.*).

Do-It-Yourself 12-8
AIR REGISTER INSTALLATION

Materials: register
Tools: tin snips or saber saw with metal cutting blade, screwdriver, drill and bits

Procedure:

1. Determine the location of the register along the hot (output) air duct from the furnace. Measure the duct width and purchase the desired register with adjustable damper (approximate outside dimensions available are 10 × 6, 12 × 4, 12 × 6, 18 × 4½ inches).
2. Measure the inside of the register where the louvered openings are located; the width and length will be the size of the opening to be cut in the duct. Mark the rectangular pattern on the duct just big enough so the damper will fit through.
3. Drill a ¼-inch starter hole in each corner of the pattern.
4. Use tin snips or a saber saw to cut out the rectangular pattern.
5. Place the register in place in the opening. Mark the location of the two screw holes.
6. Drill two screw holes, insert the register in place, and tighten the attachment screws.

12-7. ADDITIONAL AIR REGISTERS AND DUCTS

Perhaps your home was built a number of years ago and finally you've built a recreation room, a hobby room, or an office into your basement. If you have forced air heat, and if the new room feels chilly, it's easy to install additional registers into the existing ducts or to add a duct into the existing duct work. (Registers for air conditioning are installed in exactly the same way). Heating the lower level provides a second benefit; as the heat rises, it heats the upstairs rooms. Do-It-Yourself 12-8 describes how to add a register into an existing duct. Adding another duct is similar to adding a register except you'll need to see your heating supplier for ducts, boots, and registers.

Incidentally, you can't just pour hot air into a room. You have to have a complete *path* for air circulation which means that you must provide a return air duct in the cold air return to the furnace from the room where you add in a hot air register. Check first. Your furnace may already contain a return register near its base; if it doesn't, add a register with a damper near the bottom (below the filter).

If adding duct work to the end of a long run of existing duct does not provide sufficient heat, you can add a *duct booster* which is an electric motor and fan. The motor and fan are installed in the ducting in such a way as to draw warm air from the furnace to the needed location. A thermostat can be located in the ducting to automatically turn the fan on and off, or the fan motor can be wired directly to an on-off wall switch for direct control.

chapter 13

MORE EFFICIENT
AIR CONDITIONING

In some areas of the country, the heat and humidity of summer are too great to offset with natural air circulation. Some of us become accustomed to air-conditioned offices and cars and therefore feel that we need an air-conditioned home because we "feel" the heat so much when we are home from the office. Air-conditioned air also helps us to function more efficiently, sleep better, and work in the home with less loss of energy.

Not only does air conditioning cool the living area, but it also removes moisture, filters dust, and circulates the air. Reduced humidity provides more personal comfort; it allows perspiration to evaporate from the skin, making us feel cooler. Reduced humidity also helps prevent mildew and the swelling of furniture joints, drawers, and sticking doors.

An air-conditioner filter traps irritating and odor-carrying dirt, dust, and pollen. Windows and exterior doors are kept closed which reduces the amount of dirt blown into the home, reducing dust on furniture, and decreases the need to clean furniture and draperies so often. The air conditioner circulates cooled, dried, and cleaned air gently; you don't have to depend upon unpredictable breezes or cross ventilation for cooling.

This chapter discusses procedures for operating air conditioners efficiently and maintenance procedures for getting more efficient air conditioning from your room or house air conditioner. It also provides information on the selection of the correct sizing and installation of air conditioners.

13-1. AIR CONDITIONERS

Air conditioners, whether a single room air conditioner (Fig. 13-1) or a central air conditioner (Figs. 13-2 & 13-3), have the same basic parts and operate in a similar manner. Parts include a filter, fresh air control, controls, evaporator, condenser, evaporator fan, condenser fan, and motor compressor. In a window or through-the-wall room air conditioner, all of the parts are located in a single housing. In a central air conditioner, some parts are usually located outdoors and others are located indoors, although some systems have all parts outdoors; if the heating system is a hot air system, part of the air-conditioner components are located in the heater plenum chamber and the air ducts branch to the living space.

FIGURE 13-1. Room air conditioners are most easily installed through a window, although this prevents the window from being opened to let in fresh air on cooler days. Room air conditioners can also be mounted through the wall (*courtesy Amana Refrigeration, Inc.*).

Hot moist air is drawn through a filter (which removes dust and pollen) by means of an evaporator fan. The air passes through a cooling (evaporator) coil that absorbs the heat of the air. The absorbed heat is then passed onto an outdoor condensing coil via refrigerant tubes where the heat is blown into the outside air by a condenser fan. The refrigerant tubes that carry the heat from

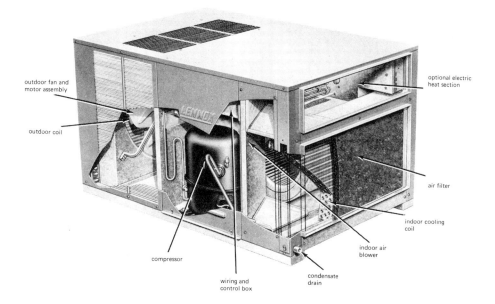

outdoor fan and
motor assembly

outdoor coil

compressor

wiring and
control box

condensate
drain

indoor air
blower

indoor cooling
coil

air filter

optional electric
heat section

FIGURE 13-2. This central air conditioner is a one-piece unit mounted outdoors on a concrete slab. A duct extends from the unit through a wall in a crawl space, basement, utility room, or attic (*courtesy Lennox Industries Inc.*).

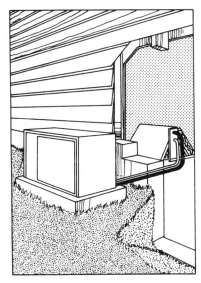

Outdoor Cooling Unit with
Indoor Coil Set on a Furnace

FIGURE 13-3. Many central air-conditioning systems are in two parts (*courtesy Amana Refrigeration, Inc.*).

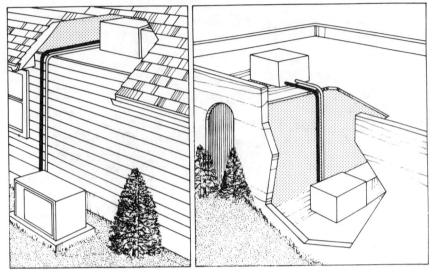

Cooling Unit Connected to
Indoor Coil-Blower that Contains
Cooling Coil and Blower. Perfect
System for Non-Ducted Heating
Systems.

Cooling Unit Installed on a Roof.
Connects to Indoor Furnace or
Coil-Blower.

FIGURE 13-3. (Continued.)

the interior to the exterior are filled with a refrigerant called Freon which
changes from a liquid to a gas and back to a liquid state as it cycles through the
refrigerant lines, evaporator coil, condenser coil, and motor compressor. The
motor compressor is a pump which draws a low pressure on the cooling side of
the refrigerant cycle and squeezes or compresses the gas into the high pressure or condensing side of the cycle. The fresh air control admits outside air
into the system, if desired.

At the same time that the heat is being removed from the circulating air,
the evaporator (cooling) coil also condenses moisture from the warm,
moisture-laden air that passes over the cold tubing. The condensate is drained
away. The clean, cool, dehumidified air is then circulated. This process is
continuous and is regulated by a thermostat that selects the comfort level
desired (Fig. 13-4).

13-2. EFFICIENT USE OF AIR CONDITIONERS

Efficiency in air conditioning is measured by the energy savings, or dollars,
that you can save by following some simple procedures. Basically, you want to
prevent the loss of cool air from the room, the infiltration of hot air, cooling of

HOW YOUR COOLING SYSTEM OPERATES*

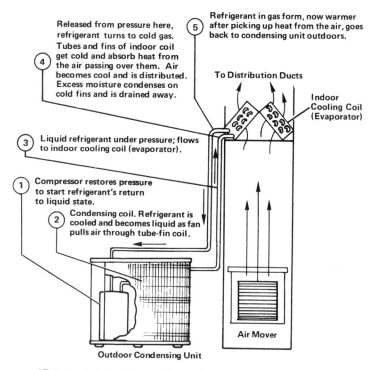

Released from pressure here, refrigerant turns to cold gas. ⑤

Tubes and fins of indoor coil get cold and absorb heat from the air passing over them. Air becomes cool and is distributed. Excess moisture condenses on cold fins and is drained away. ④

Refrigerant in gas form, now warmer after picking up heat from the air, goes back to condensing unit outdoors.

To Distribution Ducts

Indoor Cooling Coil (Evaporator)

③ Liquid refrigerant under pressure; flows to indoor cooling coil (evaporator).

① Compressor restores pressure to start refrigerant's return to liquid state.

② Condensing coil. Refrigerant is cooled and becomes liquid as fan pulls air through tube-fin coil.

Air Mover

Outdoor Condensing Unit

*This sketch shows indoor cooling coil mounted on a warm air furnace, using the filter and air mover in the furnace. Other coils may be installed with their own blowers, all in a separate cabinet.

FIGURE 13-4. Room and central air conditioners have similar parts and operation (*courtesy Lennox Industries Inc.*).

unused space, proper control of the air conditioner and distribution system, and blocking the inlet and outlet louvers of the air conditioner and distribution system. Quick Reference Chart 13-1 describes tips on increasing room air conditioner efficiency and QRC 13-2 describes tips on increasing central air conditioner efficiency; read the applicable QRC and take measures to increase your air conditioning efficiency.

13-3. AIR CONDITIONER MAINTENANCE

Your air conditioner will last longer and operate more efficiently if you perform preventive maintenance on it. This consists of monthly filter cleaning or replacement, cleaning of louvers and coils, and yearly lubrication. Do-It-Yourself 13-1 describes the procedures.

Quick Reference Chart 13-1
INCREASING ROOM AIR CONDITIONER EFFICIENCY

The following procedures will enable you to cool a single room at a lower cost.

1. If possible, locate a window air conditioner on the north side (followed in preference by east and west). Also shade the unit if possible with an awning or trees, but do not let either block the airflow to the air conditioner. The unit should tilt slightly toward the rear (outside) to allow condensate to run out. Insure that no hot outside air can enter the house from around the sides of the air conditioner.

2. Keep storm windows in place. Latch the window securely. If a window is partially raised for air conditioner installation, place insulation at the junction of the upper and lower window sashes.

3. Overhangs, awnings, shades, draperies, reflecting glass, and so on, greatly reduce heat gain through windows. Remember, it is about seven times more effective to block heat from coming in than it is to remove it. Refer to Chapter 8.

4. Keep shrubbery trimmed around the condenser to allow free air circulation.

5. Do not let curtains, shades, furniture, etc., block the output of the air conditioner. Direct the louvers upward and position them so that the air is directed evenly across the room.

6. Set the air conditioner blower to maximum and the thermostat to a setting that keeps the room temperature at 78 to 80 degrees F. When the outside temperature rises above 95 degrees F, keep the inside temperature 15 degrees F below the outside temperature. (Keeping the room at 80 degrees F rather than 75 degrees F reduces the air-conditioning load by about 15 percent.) Do not set the thermostat low in an attempt to speed up the cooling process; the air conditioner puts out the same amount of coolness regardless of the thermostat setting.

7. The air conditioner "ventilate-circulate" control should be kept in the circulate position. The ventilate (air damper) position brings in outside air; outside air is not required because there is sufficient air leakage through windows and doors to provide fresh air. Use the ventilate position only when necessary to rid the room of odors.

8. Enter and leave the room quickly through the door. Do not leave the door partway open.

9. Be sure the room door fits and latches tightly. Weather stripping can be added around the door.

10. Keep closet doors tightly closed. There is no need to cool unused space.

11. Keep lights and other appliances off as much as possible. Lights and appliances generate heat that opposes the air conditioner cooling.

12. Do not operate the room air conditioner in mild weather because this can cause frosting of the cooling coil. When the frost melts, the water can cause damage to the window frames, walls, and furnishings below the unit.

13. Clean the filter every 30 days (DIY 13-1).

Quick Reference Chart 13-2
INCREASING CENTRAL AIR CONDITIONING EFFICIENCY

The following procedures will enable you to cool the rooms of your home at a lower cost.

1. Close all air ducts to nonair-conditioned rooms.

2. Close the doors to all rooms, closets, and storage areas that are not air-conditioned. If the rooms are never used in the summer, consider weather stripping the doors.

3. Use snap-on diffusers on registers to deflect cool air upward. Provide a high wall air return. Don't let such things as furniture and draperies block the air ducts.

4. Set the air conditioner blower to maximum, continuous circulation and the thermostat to a setting that keeps the temperature at 78 to 80 degrees F. When the outside temperature rises above 95 degrees F, keep the inside temperature 15 degrees F below the outside temperature. (Keeping the home at 80 degrees F rather than 75 degrees F reduces the air-conditioning load by about 15 percent.) Do not set the thermostat low in an attempt to speed up cooling; the air conditioner puts out the same coolness regardless of the thermostat setting.

5. Set registers and/or dampers in the airflow ducts to balance and control the temperature and flow of air (DIY 12-7).

6. The outside damper of the air conditioner should be closed for greater effectiveness and economy. Most houses have enough natural air leaks through windows and exterior door openings and closures to provide sufficient fresh air into the home. The damper should be opened, as necessary, to rid the house of unpleasant odors.

7. Clean the filter every 30 days (DIY 13-1).

8. Unless the room is to be unused for an extended period of time, do not close the register in the room since the entire system would have to be rebalanced when use of the room is resumed.

9. Don't turn off the system just because you'll be away for a day. Heat and moisture build up in the house. It takes quite a while (and energy) to restore comfort, but costs relatively little to maintain it.

10. Don't open windows after dark unless the outdoor temperature *and* humidity are low. Night air may seem cool but it is also filled with moisture. This increases the work your system must do the next day.

11. When entertaining a large group, lower the thermostat two to four degrees a few hours before the guests arrive.

12. The most effective way to reduce solar heat gain (up to 80 percent) is through the use of external shading devices such as awnings, overhangs, louvered sun screens, and trees. Shades, blinds, draperies, etc., reduce heat gains by up to 50 percent, their effectiveness depending upon the amount of solar energy reflected back through the window (Chapter 8).

13. Keep lights off during the day and the TV off when it is not being used. If possible, don't use heat-producing appliances during the day. Use dishwashers, clothes washers, driers, and irons in the early morning or evening hours. Be sure the drier is vented outside.

14. Use vent fans in the bathroom and kitchen to rid the house of heat and moisture.
15. Keep shrubbery trimmed around the condenser to allow free air circulation. Shade the condenser with an awning or a tree to help the air conditioner operate more efficiently, but don't let them interfere with air circulation.
16. Air-conditioning ducts in nonair-conditioned areas (such as basements or attics) should be insulated (Chapter 7).
17. The thermostat should be located on an inside wall where comfort is of greatest significance or in a hallway where it will sense the air temperature from several rooms. The thermostat should not be located near electrical applicances, warm pipes, in direct sunshine, behind draperies, or in any location where it is shielded from air circulation.
18. Add additional insulation to your home (Chapter 7). Add a vent fan to the attic (Chapter 10).

Do-It-Yourself 13-1
AIR CONDITIONER MAINTENANCE

Materials: dishwashing detergent, filter coat (for metal filters [see procedure]), SAE-10 or SAE-20 nondetergent motor oil
Tools: vacuum cleaner with small dust brush attachment, wrench (for adjusting belt tension)

Procedure:

NOTE

When available, use the manufacturer's air conditioner maintenance procedures. If unavailable, or to supplement the manufacturer's procedures, follow these procedures.

Filters (the first day of the season and every 30 days thereafter)

1. If the filter is fiberglass, replace it when it is dirty.
2. If the filter is made of a synthetic sponge-like material, remove the filter, shake it out, wash it *carefully* in a solution of dishwasher detergent and water, rinse with clear water, dry, and return it to the air conditioner. Do not squeeze the filter because this could tear the material.
3. If the filter is metal, flush clean water through the filter from the dirty side. Drain and dry the filter. Then spray the filter with a *filter coat* which aids in trapping dust particles. A spray can of filter coat is available from an air-conditioning dealer; failure to use the coating can cause extensive damage.

Air Ducts and Louvers (yearly)

1. Use a vacuum cleaner with a dusting brush attachment to draw dust away from the air duct registers or louvers.

2. On extremely dirty ducts, remove the register or louver and insert the vacuum hose into the register to remove as much dust as possible.

Clean Evaporator and Condenser Coils (yearly)

1. Turn unit power switch to "Off" and remove power plug or place circuit breaker to "Off."
2. Remove access panels.
3. Clean blowers, coils, motors, and other parts with a vacuum cleaner with a dusting brush attachment to remove loose dust and dirt.
4. Wipe off any greasy residue with a soft cloth dampened with dishwasher detergent. Take care not to bend or otherwise damage any parts.
5. Clean the finned surfaces of the condenser to remove leaves, mud, etc.

Lubrication (yearly) (use manufacturer's instructions, if available)

1. For motor *without* oiling ports: this type of motor is permanently lubricated and has sealed bearings. No lubrication is necessary.
2. For motor *with* oiling ports: add a *few* drops of SAE-10 or SAE-20 nondetergent motor oil.
3. If blower bearings require lubrication, the bearings are equipped with grease cups. Once a year, turn the cups down a full turn to lubricate the bearings.

Belt Tension (yearly)

1. On belt drive blowers, check the belt annually for wear and proper tension. Replace frayed or damaged belts.
2. Belt tension should be as loose as possible without slippage; this is usually about a ½-inch deflection at the center of the belt with a couple of pounds of pressure applied. Belt tension is adjusted by loosening a nut(s) on the motor mount, sliding the motor, and retightening the nut(s).

Evaporator Drain Line

1. Check that the evaporator drain line is flowing by pouring a little water into the drain.
2. If the drain line is clogged, clean the line.

Dampers (at beginning of season)

If the air-conditioning system uses the same ducts as the heating system, different settings are usually required for summer cooling than for winter heating. Adjust the dampers (DIY 12-5).

It is *important* to keep the filter clean. Dirt blocks airflow, reducing cooling capacity. Dirt will also accumulate on the evaporator coil which will reduce its cooling ability and can cause a buildup of frost and ice that further impedes the flow of air.

13-4. SIZING AND BUYING AIR CONDITIONERS

Room air conditioners are frequently undersized when they are used to cool large living rooms or combined living and dining areas. When used in bedrooms, room air conditioners are frequently oversized. For the most efficient and effective use, a room air conditioner should be slightly undersized so that it runs almost constantly. When the air conditioner is not running, there is no circulation of air nor resulting dehumidification as the air passes over the chilled evaporator. Dry air increases evaporation of moisture from the skin which makes you feel cooler.

Central air-conditioning systems generally provide the most effective and economical means of total house cooling, especially when the air distribution ducts are the same ones used for heating. In many cases, hot air heating systems are easily modified for dual use with air conditioning systems; all you need to do is add a cooling coil, a condensing unit, and a combination heating-cooling thermostat. If you presently heat with a system other than forced hot air, then you have to install a duct system for distribution of air from the central air-conditioning system.

An air conditioner's cooling capacity is rated in Btu per hour and cooling efficiency (energy efficiency ratio [EER]) in Btu per watt-hour. A British thermal unit of air-condition cooling is the energy required to decrease the temperature of one pound of water by one degree Fahrenheit. Air conditioners are measured in Btu per hour or in tons of air conditioning; one ton equals 12,000 Btu per hour.

It is desirable to have an air conditioner that can produce the maximum number of Btu per hour of cooling at the least amount of electrical energy (watts) utilized. Thus, an efficient air conditioner is one having a high energy efficiency ratio (EER). The EER is derived by dividing the number of Btu cooling capacity per hour by the number of watts. This is the amount of cooling per given amount of electricity; the higher the EER for a given cooling capacity (in Btu), the greater is the efficiency of the air conditioner. Thus, if two 7,000 Btu air conditioners are considered having an EER of 5.6 and 9.0, the air conditioner with the EER of 9.0 is the most efficient. In selecting an air conditioner, buy one with at least an EER of 8.5; it may cost more at the outset, but you'll save in the long run on electricity. Most air conditioner manufacturers label their air conditioners with the EER which ranges from 5.5 to 10.2. Note that EER is not cooling capacity. Cooling capacity is measured in Btu per hour; for room air conditioners, the Btu rating ranges from 4,500 to 36,000. Don't guess what cooling capacity you need for your application. For the required capacity of a single room, use the Cooling Load Estimate Form in Fig. 13-5; for a central air-conditioning system, have at least two professional installers make estimates for you.

COOLING LOAD ESTIMATE FORM

Customer _____ Estimate by _____ Date _____

HEAT GAIN FROM	QUANTITY	FACTORS						Btu/Hr (Quantity x Factor)
		NIGHT	DAY					
1. WINDOWS: Heat gain from sun.			No Shades*	Inside Shades*	Outside Awnings*	(Area x Factor)		
Northeast	___ sq ft	0	60	25	20			
East	___ sq ft	0	80	40	25		Use only the largest load	
Southeast	___ sq ft	0	75	30	20			
South	___ sq ft	0	75	35	20			
Southwest	___ sq ft	0	110	45	30			
West	24 sq ft	0	(150)	65	45			3600
Northwest	___ sq ft	0	120	50	35			
North	___ sq ft	0	0	0	0			

*These factors are for single glass only. For glass block, multiply the above factors by 0.5; for double-glass or storm windows, multiply the above factors by 0.8.

HEAT GAIN FROM	QUANTITY	NIGHT	DAY	Btu/Hr
2. WINDOWS: Heat gain by conduction. (Total of all windows.)				
Single glass	24 sq ft	14	.............(14).............	336
Double glass or glass block ..	___ sq ft	7	.... 7	
3. WALLS: (Based on linear feet of wall.)			Light Construction Heavy Construction	
a. Outside walls				
North exposure...........	___ ft	30	...... 30.............. 20	
Other than North exposure....	20 ft	30	......(60).............. 30	1200
b. Inside Walls (between conditioned and unconditioned spaces only) (5+20+15.5...	50 ft	30	............(30).............	1500
4. ROOF OR CEILING: (Use one only.)				
a. Roof, uninsulated	___ sq ft	5	.... 19	
b. Roof, 1 inch or more insulation.	___ sq ft	3	.... 8	
c. Ceiling, occupied space above.	___ sq ft	3	.... 3	
d. Ceiling, insulated with attic space above	___ sq ft	4	.... 5	
e. Ceiling, uninsulated, with attic space above	300 sq ft	7	............(12).............	3600
5. FLOOR: (Disregard if floor is directly on ground or over basement.)	___ sq ft	3	.... 3	
6. NUMBER OF PEOPLE:	2	600	............(600).............	1200
7. LIGHTS AND ELECTRICAL EQUIPMENT IN USE	200 watts	3	............(3).............	600
8. DOORS AND ARCHES CONTINUOUSLY OPEN TO UNCONDITIONED SPACE: (Linear feet of width.)	___ ft	200	.... 300	
9. SUB-TOTAL	x x x x x	x x x x x	x x x x x	12036
10. TOTAL COOLING LOAD: (Btu per hour to be used for selection of room air-conditioner(s).)	12036 (Item 9) X		1.00 (Factor from Map) =	12036

NOTE: See Reverse side for instructions on use of this form.

Notes: West exposure
under attic
1 - outside wall
2 - 3x4 windows, single glass, no shade
2 - people
200 watts - lights

FIGURE 13-5. Use this cooling load estimate form to determine the needed capacity of a room air conditioner; an example is worked out. For central air conditioning, have at least two experts give you an estimate (*Source: Association of Home Appliance Manufacturers*).

INSTRUCTIONS FOR USING COOLING LOAD ESTIMATE FORM
FOR ROOM AIR CONDITIONERS

A. This cooling load estimate form is suitable for estimating the cooling load for comfort air-conditioning installations which do not require specific conditions of inside temperature and humidity.

B. The form is based on an outside design temperature of 95 F dry bulb and 75 F wet bulb. It can be used for areas in the continental United States having other outside design temperatures by applying a correction factor for the particular locality as determined from the map.

C. The form includes "day" factors for calculating cooling loads in rooms where daytime comfort is desired (such as living rooms, offices, etc.), as well as "night" factors for calculating cooling loads in rooms where only nighttime comfort is desired (such as bedrooms). "Night" factors should be used only for those applications where comfort air-conditioning is desired during the period from sunset to sunrise.

D. The numbers of the following paragraphs refer to the correspondingly numbered Item on the form:

1. Multiply the square feet of window area for each exposure by the applicable factor. The window area is the area of the wall opening in which the window is installed. For windows shaded by inside shades or venetian blinds, use the factor for "Inside Shades." For windows shaded by outside awnings or by both outside awnings and inside shades (or venetian blinds), use the factor for "Outside Awnings." "Single Glass" includes all types of single-thickness windows, and "Double Glass" includes sealed air-space types, storm windows, and glass block. Only one number should be entered in the right-hand column for item 1, and this number should represent *only the exposure with the largest load.*

2. Multiply the total square feet of *all* windows in the room by the applicable factor.

3a. Multiply the total length (linear feet) of all walls exposed to the outside by the applicable factor. Doors should be considered as being part of the wall. Outside walls facing due north should be calculated separately from outside walls facing other directions. Walls which are permanently shaded by adjacent structures should be considered as being "North Exposure." Do not consider trees and shrubbery as providing permanent shading. An uninsulated frame wall or a masonry wall 8 inches or less in thickness is considered "Light Construction." An insulated frame wall or a masonry wall over 8 inches in thickness is considered "Heavy Construction."

3b. Multiply the total length (linear feet) of all inside walls between the space to be conditioned and any unconditioned spaces by the given factor. Do not include inside walls which separate other air-conditioned rooms.

4. Multiply the total square feet of roof or ceiling area by the factor given for the type of construction most nearly describing the particular application. (Use one line only.)

5. Multiply the total square feet of floor area by the factor given. Disregard this item if the floor is directly on the ground or over a basement.

6. Multiply the number of people who normally occupy the space to be air-conditioned by the factor given. Use a minimum of 2 people.

7. Determine the total number of watts for lights and electrical equipment, except the air conditioner itself, that will be *in use* when the room air-conditioning is operating. Multiply the total wattage by the factor given.

8. Multiply the total width (linear feet) of any doors or arches which are continually open to an unconditioned space by the applicable factor.

   NOTE—Where the width of the doors or arches is more than 5 feet, the actual load may exceed the the calculated value. In such cases, both adjoining rooms should be considered as a single large room, and the room air-conditioner unit or units should be selected according to a calculation made on this new basis.

9. Total the loads estimated for the foregoing 8 items.

10. Multiply the sub-total obtained in Item 9 by the proper correction factor, selected from the map, for the particular locality. The result is the total estimated design cooling load in Btu per hour.

E. For best results a room air-conditioner unit or units having a cooling capacity rating (determined in accordance with the AHAM Standards Publication for Room Air Conditioners, RAC-1) as close as possible to the estimated load should be selected. In general, a greatly oversized unit which would operate intermittently will be much less satisfactory than one which is slightly undersized and which would operate more nearly continuously.

F. Intermittent loads such as kitchen and laundry equipment are not included in this form.

FIGURE 13-5. (Continued.)

It is important to correctly size the air conditioner to ensure energy efficiency. Oversized air conditioners not only waste energy, they are also less effective because the humidity is not decreased and held at a constant level unless the air conditioner is running. If the air conditioner is oversized, it will cycle off frequently, decreasing its ability to dehumidify. A properly sized air conditioner will run practically full time on the hottest days. Slight undersizing is better than oversizing.

With the air conditioner running almost constantly, the air in the room stratifies into layers of cool and warm air. Continuous circulation together with properly placed air supply ducts provide the most constant temperature air.

If you own central air conditioning, you may want to add an air economizer built by Honeywell, Inc. An air economizer turns off the part of the air conditioner that uses a lot of electricity and it then circulates outside air through the house when the outside air is cooler than the inside air. The air economizer consists of a duct running from the return air plenum or a return air duct of your furnace to an outside grille, a motorized damper, an enthalpy (a control that senses outside temperature and humidity), and a two-stage thermostat.

13-5. INSTALLING AIR CONDITIONING

Room air conditioners can either mount in a window frame or mount through the wall. The window-mounted models are easily installed but have the disadvantages of blocking light, view, and natural fresh airflow when you don't want the air conditioner turned on. Mount as directed in DIY 13-2.

Do-It-Yourself 13-2
WINDOW AIR CONDITIONER INSTALLATION

Materials: weather stripping, wood screws
Tools: hammer, screwdriver

Procedure:

Air conditioners mount in window frames either on a mounting frame that comes with the unit or by setting the unit on the window ledge and securing the lower window sash.

Mounting Frame Provided

1. With the lower sash raised, set the air conditioner frame in the center of the window.
2. Secure the frame to the window sill with wood screws.
3. Secure outside support brackets between the air conditioner frame and the window frame.
4. Slide the air conditioner into place in the frame. The air conditioner should tip slightly to the rear to let condensate drain out.
5. Lower the window to the air conditioner case.
6. Slide air conditioner side panels outward to the sides of the window. These usually remain in place by finger tightening knurled screws.

Do-It-Yourself 13-2 *(continued)*
WINDOW AIR CONDITIONER INSTALLATION

---

7. Place weather stripping or gasket material around the air conditioner at the window frame to seal the space between the frame and the unit.

No Mounting Frame

1. With the lower window sash raised, set the air conditioner on the window sill so that the lip along the air conditioner base fits down into the sash.
2. Lower the window sash against the air conditioner so that the lip along the air conditioner top fits along the window bottom rail.
3. Secure an L-shaped bracket with a wood screw in the jamb directly above the window sash to prevent the sash from sliding upward.
4. Slide the air conditioner side panels outward to the sides of the window. The side panels are usually secured in place with thumb screws.
5. Place weather stripping (if necessary) around the air conditioner at the window frame to seal the space between the frame and the unit.

Room air conditioners can be installed through the wall. This procedure involves removing external siding, removing internal plaster or plasterboard, removing one or more studs, installing framework, installing the air conditioner, reinstalling some external siding, and installing molding on the interior. Only experienced do-it-yourselfers should attempt through-the-wall room air conditioner installations. Refer to a home maintenance manual such as the *Reader's Digest Complete Do-It-Yourself Manual* for specific procedures.

Central air conditioners are difficult to install and you should contact several contractors for estimates on installation. If you already have a forced hot air furnace system and you are an experienced do-it-yourselfer, you might consider installing a do-it-yourself central air-conditioning system. One manufacturer is the McGraw-Edison Company, Air Comfort Division, Albion, Michigan 49224. Write them for a brochure.

chapter 14

HEATING WITH FIREPLACES, WOOD-BURNING STOVES AND FURNACES, AND OTHER SUPPLEMENTARY HEATERS

At the outset of this book, it was stated that the book was written for those who have a home and expect to stay in it for many more years. It was then suggested that a little time and money should be invested in weatherizing your home now so that you could save money that was being and could be needlessly spent on energy consumption.

There are two possibilities that exist within your home: one, you already have that American dream fireplace that you enjoy cozily sitting by on cold winter evenings as the heat warmed by your central heating system is sucked up the chimney; or, you don't have a fireplace and now the all-American nostalgic fireplace can be added as a supplementary source of heat and you'll save energy. This is probably wrong unless you've been doing some reading on recent fireplace designs and alternate energy designs.

In this chapter we'll first assume that you have a fireplace; we'll discuss how the inefficient fireplace works and what you can do to make the existing one more efficient. Secondly, we'll assume that you don't have a fireplace, but you'd like to install one (or better yet a wood-burning stove or furnace) to supplement your existing heating, to reduce your total fuel bill, and to prepare for an emergency in the event you temporarily lose central heating power for one reason or another. This chapter covers existing home fireplaces, wood, fire-building, adding a fireplace, wood-burning stoves and furnaces, and other supplementary heating.

14-1. EXISTING HOME FIREPLACES

If your home was built at least several years ago, the fireplace is probably one of the older, inefficient designs, perhaps even located where you don't want it, and, certainly located where it doesn't provide much warmth to the home; its heating efficiency is about 10 percent. It is installed only for aesthetics and coziness, not for the heat it can deliver (Fig. 14-1).

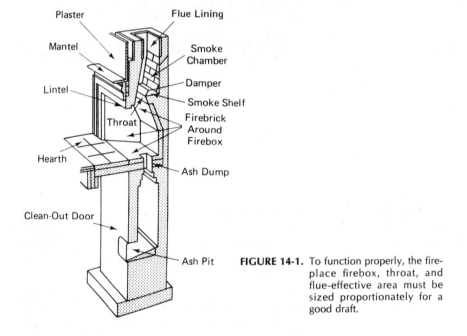

FIGURE 14-1. To function properly, the fireplace firebox, throat, and flue-effective area must be sized proportionately for a good draft.

    For a fireplace to burn, it needs an ample supply of fuel, heat, air, and a properly sized flue through which the gases of combustion are vented to the outside by a draft (flow of air). Where does this air supply come from? It comes from the room air which has *already* been heated by the central heating system. Thus, air at 68 degrees F is supplied (drawn) from the room to the fireplace for combustion. As the flue is drawing air because the hot air in the fireplace is rising, room air too is sucked along with it up the chimney. The 68-degree F air from the room to the fireplace has to be replaced and this air comes in through the cracks around windows and doors and through the walls of your home at the outside air temperature causing drafts.

    In all fairness to the fireplace, it does provide a little heat into the living area by radiation and convection currents and provides that cozy atmosphere. As mentioned, its efficiency is about 10 percent.

But, you say, you've *weatherized* your home as has been outlined in this book so that cold air just cannot get in and therefore you won't have drafts. This is partly true; while you've reduced the infiltration to a large degree, you'll probably never be able to eliminate it completely. If you've done a really good weatherizing job, you may find that the fireplace smokes; it cannot get enough air for combustion and draft. In this case, you'll need to open a window slightly so that 10 percent efficient, cozy little fireplace can provide a pleasant fire without smoking, which really is rather absurd.

14-2. MAKING YOUR FIREPLACE MORE EFFICIENT

What can you do about the inefficiency of your fireplace? How can you provide sufficient air for combustion without using heated room air? How can you prevent the warm air in the room from being drawn up the fireplace? How can you control the rate of burning, and hence heat, of the firewood?

To reduce the use of room air for combustion, provide outdoor air to the front of the fireplace; this is similar to providing outside air to the furnace (Section 12-5). Run a 3- by-10-inch (minimum) duct (see your plumbing supplier) from an outside wall to the base. Make the duct as short as possible and pitch it slightly downward toward the exterior so water cannot enter; as the air travels through the duct, the air will be preheated somewhat by the surrounding internal air. The grille at the floor should have a damper in it to close when the fireplace is not being used and the grille should be as long (and narrow) as possible. Also place a throw rug over the grille when the fireplace is not in use.

Existing fireplace efficiency can be increased somewhat by placing hardened (tempered) glass doors across the front of the fireplace. The glass allows the beauty of the fire to be seen and does not significantly decrease the amount of radiated heat from the fire. The major attribute of the glass doors is that they prevent heated room air from being drawn into the fire for combustion or simply drawn up the chimney as wasted heat. This is particularly important when the fire is put out or slowly burns out. The doors also prevent sparks from jumping from the fire onto the rug and prevent smoke from backing into the house as the fire goes out. The doors also prevent children or pets from touching the flame or coals. An outside air duct and grille placed just inside the glass doors provide a combination that greatly reduces the loss of heated room air up the chimney.

In the event that you can't afford glass doors (or don't want them), cut a piece of sheet metal (or other noncombustible material) a little larger than the fireplace opening and place it over the opening as the fire dies down for the evening. If you let the fire burn out naturally without closing off the opening, your heated room air will go up the chimney all night and cold drafts will blow down the chimney. You can't close the flue (if you have one) until the fire is completely out or you'll have smoke in the room. Closing off the fireplace at the opening cuts off the room air supply but the flue is still open for smoke to rise. If the damper doesn't close tightly, you absolutely need glass doors, metal, or some other noncombustible material to close the front of the fireplace when the fire is out.

14-3. WOOD

Your objective is to obtain wood that burns longest and produces the largest number of Btu per cord of wood. Woods that are dense, heavy, and have been adequately dried (seasoned) meet the objective by burning slowly, cleanly, and hot. They produce more heat and less creosote. Quick Reference Chart 14-1 lists the burning characteristics of the more common woods.

Pines and other resinous soft woods are not very good for fires because they burn too fast. Pitch pockets in these woods cause the fire to spit; smoke from resinous trees leaves flammable deposits in chimneys. The deposits can catch fire causing chimney fires.

Quick Reference Chart 14-1
WOODS FOR BURNING

Wood	Cord Weight (Dry) (lb)	Splitting Difficulty	Relative Amount of Heat	Equivalent Fuel Oil (gal.)	Available Heat for Cord Air-Dry (Million Btu)
Ash	3,300	moderate to easy	high	154	20.0
Aspen	1,900	easy	low	97*	12.5
Basswood	1,900	easy	low	93*	
Beech	3,500	very hard	high	165	21.8
Birch, yellow	3,500	moderate to easy	high	160	21.3
Cherry	2,550	moderate to easy	medium	129	
Elm	3,000	very difficult	medium	135	17.2
Hemlock	2,100	moderate	low	98*	
Hickory	3,900	hard	high	191	24.6
Maple, hard	3,500	moderate	high	160	21.3
Maple, soft	2,800	moderate to easy	medium	132	18.6
Oak, white	3,600	moderate	high	175	22.7
Pine, white	1,800	easy	low	94*	13.3
Spruce	2,100	easy	low	99*	
Tamarack	2,500	easy	medium	132	

* Not efficient wood for heating. Use, if necessary, for fire starting.

If you'll be doing a lot of woodcutting yourself, purchase a chainsaw. Cut trees into lengths that will fit into your fireplace, stove, or furnace. Usually, logs under 6 inches in diameter can be immediately stacked for drying; the logs should be split if they are greater than 6 inches in diameter. Unsplit logs can retain a fire all night but it takes an unsplit log about three times as long to dry out as it does a split log which has more surface area to increase the drying rate.

Be sure to use the proper tools for splitting wood: wedges (two), splitting hammer, maul, or sledge. Put the wedge on a natural crack of the log. The second wedge is used to free the first wedge as necessary. Be sure to wear safety glasses when splitting wood, especially when metal is pounded against metal.

Green wood splits easier than seasoned wood, frozen wood splits easier than unfrozen wood, and lighter woods with straight grains are easy to split. After cutting and splitting, the wood is stacked for seasoning (drying). Six months of drying time is adequate, but a year is better. So split your wood now for next year. Wood having 12 to 20 percent moisture is good for burning; more than 25 percent makes for poor burning. *Green* (undried) wood needs a lot of heat in a fire to drive out the moisture before the wood catches for burning.

For seasoning wood quickly, stack it in chimney fashion; two logs in one direction, the next two on top at 90-degree angles, another two on top aligned with the first logs, and so on. Stack in an open-air shed, barn, garage, or build a lean-to. You can place sheet plastic over the top to keep rain off, but be sure there is ample room for air circulation. If seasoning must be done inside, stack the wood in a cabinet-like storage area where a fan can be used to circulate air through and around the logs.

Check that logs are dried by banging two pieces together; a ringing sound indicates proper seasoning.

14-4. FIRE-BUILDING

The amount of heat and the rate of burning of wood in your fire depends largely on the type of wood (Section 14-3) and how you build the fire. The most effective fire is built by placing a large log at the back and at the front of the fireplace on a set of andirons or a grate that raises the fire 4 to 6 inches off the floor. Roll paper tightly into balls and place them into the center of the logs. Add kindling. *Do not add any kind of starters such as charcoal lighter, kerosene, or gasoline to the fire.* Place a third log on top. Before lighting the fire, get a draft moving up the chimney. Open the damper, insert three sheets of newspaper in the chimney and burn them completely; the heat rising starts the draft. Light the paper at the base of the logs and keep the fire between the logs. Let the ashes build up under the grate and move them to control the draft of the fire.

Burn wood slowly rather than allow a roaring fire. A slow-burning fire burns more completely. A roaring fire draws more air and thus may be drawing more cold air into your home via infiltration. Keep the flue damper closed as much as possible without causing the fire to smoke. This will enable the fireplace to give off as much heat as possible and will decrease the loss of heat up the chimney. Whenever the fireplace is not in use, the flue must be closed. Otherwise, the room heat will rise through the chimney and be lost outdoors.

With glass doors installed across the front of the fireplace and with outdoor air ducted to the front of the hearth, adjust the air louver so there is a good draft. The air should flow smoothly through the fire to produce even burning and should carry smoke and flue gases up the chimney; the fire still radiates heat into the room through the glass doors. The draft should not be strong enough to make the fire roar and it should be in front of and under the fire. If the draft is too high above the fire, the fire will smoke.

Quick Reference Chart 14-2 lists fireplace troubles, probable causes, and remedies.

Several fireplace tubular grates are available that are used to catch some heat before it goes up the flue. These grates consist of a number of C-shaped pipes about 1 inch in diameter spaced 3 to 4 inches apart and welded to a base (Fig. 14-2). The base of the grate extends slightly out in front along the hearth and the top extends just out of the fireplace. Room air is heated by convection; cooler air is drawn into the lower ends of the pipes, is heated as it passes through the pipes, and exits through the tops into the room. Although seemingly sound in theory, actual use shows only a slight amount of increased

Quick Reference Chart 14-2
FIREPLACE TROUBLES

Trouble	Cause	Remedy
Smokes on windy days	Downdraft caused by:	
	1. Tall trees close to chimney.	1. Remove close branches, remove tree.
	2. Wind changes direction quickly and often.	2. Add revolving chimney cap.
	3. Chimney height inadequate. It should be at least 2 feet above any point of the roof within a radius of 10 feet, or at least 3 feet above a flat roof.	3. Raise chimney. You can check it by adding a length of flue liner on top of chimney.

Quick Reference Chart 14-2 (continued)
FIREPLACE TROUBLES

Trouble	Cause	Remedy
	4. Chimney diameter insufficient; soot and tar (from burning green or resinous woods) have accumulated on inside of chimney.	4. Sweep chimney (DIY 12-6).
	5. Damper in the flue not opening properly.	5. Dirty. Be sure to clean it out after chimney sweep. Reach up chimney from fireplace and clean smoke shelf off with a stick and a vacuum cleaner.
Insufficient draft	Fire not high enough in fireplace.	1. Use a grate that raises the fire 4 to 6 in. This will give better draft and decrease smoking.
	Room sealed too well. If you seal all the windows and doors well, you may not get a good draft because outside air leakage is necessary for combustion.	1. Crack a window. Better still, run an air duct to the front of the fireplace from the outside; this will prevent cold drafts. Can run the air to the ash chute.
	Opening in front of fireplace is too high (usually the height is $2/3$ to ¾ the width).	1. Reduce the height by adding a piece of angle iron across width at the front top of the fireplace.
		2. Can also raise height of fireplace by adding additional firebrick on the fireplace base. Use mortar containing fire clay.
Firewood burning too quickly; fire roaring	Too much draft.	1. Partially close off the flue damper, but not so much that the fire starts to smoke.
		2. Partially close off the damper in the floor or fireplace glass door through which air is provided to the fire.

FIGURE 14-2. A grate consisting of a number of C-shaped tubular steel tubes heats room air flowing through the tubes by convection (*courtesy Thermograte, Enterprises Inc.*).

efficiency to the fireplace. Instead of buying one, you should consider building one.

An improved model of the tubular *static* grate is a *dynamic* grate which uses a blower to aid in circulation of room air through the tubular grate (Fig. 14-3). Air is drawn in through the base of the grate and exits through the top into the room. Use of the blower increases the amount of air circulated because of the increased flow rate. It also keeps the tubes from getting excessively hot which will decrease their life expectancy.

By combining tempered glass doors with a blower-powered, C-shaped tubular grate, fireplace efficiency can be somewhat increased (Fig. 14-4). The blower draws room air through the tubular firegrate and gently blows the heated air back into the room. The door prevents room air from being drawn up the chimney and it assists the flue damper in preventing cold air drafts from coming down the chimney and into the room when the fireplace is not in use. Small, louvered doors in the base of the front of the glass doors are opened as necessary to admit air into the fireplace for combustion. You can increase fireplace efficiency and decrease drafts in the room by venting outside air through a duct to a louvered grille placed in the hearth directly in front of the louvered doors.

At least one manufacturer makes a tubular fireplace grate through which water is circulated in a closed loop to heat water for either a hot water heating

FIGURE 14-3. Several companies also make dynamic firegrates that use a small blower to increase the flow of room air through the firegrate tubes (*courtesy Woodmack Products, Inc.*).

FIGURE 14-4. Tempered glass doors block heated room air from rising up the chimney. The blower assists the heating of room air that is heated in the C-shaped tubular firegrate (*courtesy Thermograte Enterprises Inc.*).

Hot Water System

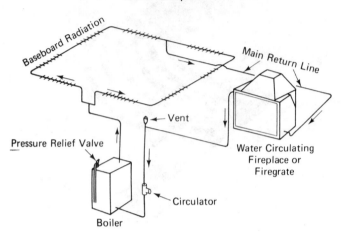

Forced Air or Heat Pump System

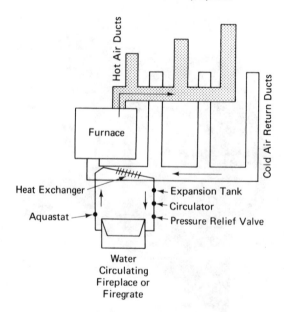

FIGURE 14-5. A water-circulating fireplace or firegrate can preheat water for your hot water system or preheat air via a heat exchanger for your hot air heating system (*courtesy Ridgway Steel Fabricators Inc.*).

system, forced air, or heat pump system (Fig. 14-5). A complete fireplace unit is also available in which water circulates through a grate and the double walls of the fireplace as well. In hot water systems, the furnace or boiler water is preheated, thus lowering the fuel consumption of the heating system. A heat exchanger mounted in the cold air return duct is used in a forced hot air system.

14-5. ADDING A FIREPLACE

If you want to add a fireplace to your home to really *heat* it, skip this section and go on to Section 14-6, wood-burning stoves and furnaces. A fireplace without any heat-circulating device is only about 10 percent efficient; with heat circulating devices, the efficiency is raised somewhat to maybe 20 to 30 percent efficient. Good wood-burning stoves are about 50 percent thermally efficient (fuel oil furnaces are approximately 65 percent efficient). If you are interested in designing and *building* a fireplace on your own, consult your local library for books of design information and plans. If you are interested in buying a prefabricated fireplace that you can easily install yourself and add a chimney and interior trim, then continue reading this section.

You can buy fireplaces in numerous sizes and designs to fit any size room and decor. You can select from wood-burning fireplaces, gas log fireplaces, and electric fireplaces. Models can be built in, hung from a ceiling, may be freestanding, or may even be on wheels for wheeling around a patio.

Before making any purchase, you must decide exactly on why you want a fireplace and what physical limitations your home offers. Hopefully, with your mind turned to weatherizing your home, you'll want a wood-burning fireplace that can provide some atmosphere, and in addition can somewhat supplement your central heating system. Do you have gas connections to your home? Can you use an existing chimney? Is gas or electricity the cheaper energy in your region? Are you high atop a condominium where it is impossible to add a chimney for a wood-burning or a gas log fireplace? Do you want only the atmosphere of a fireplace without it producing any heat? Subsequent subsections discuss fireplaces that the do-it-yourselfer can install: wood-burning fireplaces, gas fireplaces, and electric fireplaces.

Wood-burning Fireplaces

To design and build your own fireplace, you must study fireplace design, particularly flue size, fireplace opening size, relationship of flue size to fireplace opening, introduction of air for combustion, combustion gases and local building codes. You'll probably need to break into your existing interior and possibly exterior walls, build a concrete footing, and work with bricks or stone. As mentioned, this is beyond the scope of this book and probably beyond the capability of the average do-it-yourselfer.

But, you don't need concrete forms, concrete, or brick masonry to have a wood-burning fireplace. You can buy a factory-built fireplace of correct design that complies with local building codes and is approved by the Underwriters' Laboratories. These prefabricated fireplaces cost less than conventional, contractor-built fireplaces (about half as much) and can be installed by the do-it-yourselfer during a weekend. Many of the fireplaces can be placed right against a wall and on the floor; no insulation between them is required. A 2- by 4-inch lumber framework is then built around the fireplace and the frame is finished off in brick, stone, or tile as desired.

The easiest types of wood-burning fireplaces to install are the freestanding (Fig. 14-6) or wall-hung (Fig. 14-7) fireplaces. You don't need to add house support framing or do any masonry work. Existing flues may be utilized, or prefabricated metal flue components which interlock easily may be extended upward from the fireplace to a rooftop termination; all parts are available at a fireplace dealership. Hearths may be of crushed stone. Chimney-top designs are from contemporary to traditional styles in simulated brick. The freestanding fireplaces act like giant radiators and throw off heat on every side including the exposed stovepipe leading to the flue.

FIGURE 14-6. Freestanding fireplaces are available in many shapes and decorator colors to accent any room. The preassembled woodburning fireplace comes with a screen to prevent sparks from jumping to the floor. It can be vented to an existing chimney or completed with all-metal chimney components (*courtesy The Majestic Co.–An American Standard Co.*).

FIGURE 14-7. This wall-hanging fireplace installs without masonry, framing, or finishing. Its triple-wall construction and triple-wall chimney allow placement against combustible materials (*courtesy The Majestic Co.—An American Standard Co.*).

The Franklin stove fireplaces (Fig. 14-8) are somewhat improved from the day of Benjamin Franklin, their inventor. The Franklin stove is higher in efficiency than the freestanding and wall-hung fireplaces; its heavy iron construction radiates a lot of heat, especially the models with iron doors that close. Glass doors allow heat to radiate out, block the flow of room air up the chimney, and protect against flying sparks; they also keep children's hands from entering the fireplace.

The development of the prefabricated, wood-burning fireplace, scientifically designed and factory built, allows greater flexibility of placement at sharply reduced cost (Fig. 14-9). The stringent proportioning problems required in all masonry construction are solved at the factory. In addition, modern prefabs are more efficient fireplaces, with better fuel combustion and without smoking, as so many ill-conceived masonry fireplaces have been known to do.

The prefabricated, wood-burning fireplace is basically a steel box (Fig. 14-10) which is placed in the desired location on any floor, in any room, against a wall or within a wall, in a corner or in the center of the room. These units employ a thermosiphoning principle using multiple wall construction which allows the exterior of the fireplace to remain cool even with a fire blazing in the hearth. Because of this, the fireplace may be safely placed against any combus-

FIGURE 14-8. This Franklin fireplace heats rooms 750 to 1,000 square feet. A louvered top aids
in quick heat dissipation. An optional thermostatically controlled fan improves
circulation (*courtesy Martin Industries*).

tible material, thus requiring no more elaborate construction than framing it
with two-by-fours and finishing in a choice of exterior enclosure materials.
There is no need for masonry, footers or special foundations. All the preinstal-
lation designing and proportioning already have been done, including placement of a damper, smoke dome and fire screens (Fig. 14-11).

The erection of a chimney and flue is also a simple proposition with the
prefabricated system (Fig. 14-12). Round, multiwalled, interlocking chimney
components are fastened to the top of the fireplace unit, then extended
upward through the ceiling and the roof where they may be terminated with a
choice of chimney housings, from contemporary designs to the traditional
styles in simulated brick patterns. Again, these flue components are designed
to be placed near or adjacent to combustible materials, no matter where the
location. As with the basic unit, no masonry is required.

FIGURE 14-9. The prefabricated fireplace eliminates the problem and risk of the do-it-yourselfer designing an inefficient or smoky fireplace (*courtesy The Majestic Co.—An American Standard Co.*).

The homeowner is still free to choose any style preferred to finish the exterior of a fireplace (Fig. 14-13). Once the unit has been framed in, whether within or against a wall, the trim may be applied with any kind of material without a concern for combustibility. Traditional mantels, extended or raised hearths, incorporated cabinetry, shelving, or any other treatment is possible with the modern, prefabricated fireplace.

A wide variety of available models allows you to select the shape and size which will best suit your room decor. Most units installed are of the front-opening variety, but models with left or right side opening for island installation are also available (Fig. 14-14). Some are designed for flat wall placement or with tapered sides for corner locations.

For multiple installations, these modern units allow "stacking"; that is, placing one fireplace above the other on subsequent floors. This is possible

FIGURE 14-10. Prefabricated models come with all parts for quick assembly. Interior and exterior trims are unlimited. The triple-wall construction allows location anywhere in any room with zero clearance required between the fireplace and combustible materials (*courtesy The Majestic Co.—An American Standard Co.*)

FIGURE 14-11. Wood-burning fireplaces are even made and approved for mobile home installations. It can be faced with any decorative material to match the decor (*courtesy Dyna Corp.*).

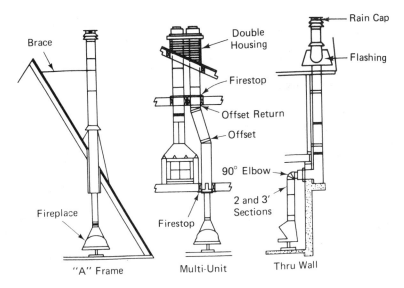

FIGURE 14-12. Typical chimney installations are shown for wood-burning and gas-burning fireplaces (*courtesy Heatilator Fireplace–Div. of Vega Industries, Inc.*).

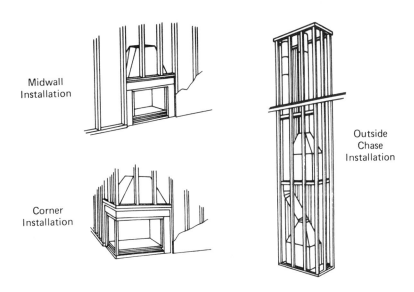

FIGURE 14-13. When the placement of the fireplace and chimney (and the air ducts of heat insulating fireplaces) are in place, framing begins as in normal wall construction; (a) midwall installation, (b) corner installation, (c) outside chase installation (*courtesy Heatilator Fireplace—Div. of Vega Industries, Inc.*).

FIGURE 14-14. A side-opening wood-burning fireplace serves as a room divider (*courtesy The Majestic Co.—An American Standard Co.*).

with a variety of available chimney components, allowing the metal flue to be angled to the same area and terminated together at a common top housing. An even more dramatic and economical possibility is back-to-back units.

The controlled *heat circulating fireplace* gives you improved heat efficiency and the advantage of a factory-engineered design (Fig. 14-15). Room air enters the front of the fireplace through individually controlled air intakes on the front sides of the fireplace opening. By adding optional fans, this airflow efficiency may be increased. Once inside the heat exchanger chambers of the fireplace walls, the air is heated by the fire. It is then released from the top sides of the fireplace back out into the same or an adjoining room through the air ducts and grilles. When no heat circulation is desired, the air ducts may be individually shut off, using the pull chains over the air intakes.

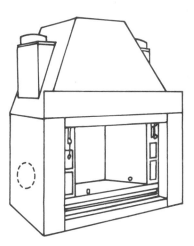

FIGURE 14-15. This fireplace lets you enjoy the cozy fire and receive the additional benefit of air heated and circulated to the same or adjoining rooms (*courtesy Heatilator Fireplace—Div. of Vega Industries, Inc.*).

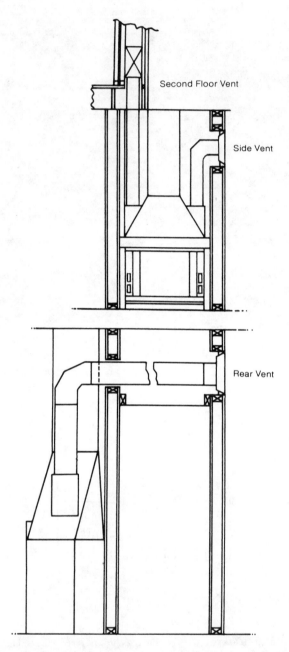

Second Floor Vent

Side Vent

Rear Vent

FIGURE 14-16. Heat-circulating fireplaces can duct heated air into the same room as the fireplace or into other rooms (*courtesy Heatilator Fireplace—Div. of Vega Industries, Inc.*).

The ways in which you can install the heat circulation ducts are numerous (Fig. 14-16). Since there are two ducts from the fireplaces, you can direct the heat into two different rooms. These rooms may be on the same floor or a floor above. The elbows for the ducts that come with the fireplace are adjustable from zero to 90 degrees. This allows offset runs of the ducting around existing obstacles when used with additional straight and elbow ducting. The heat ducting is of double-wall construction. It should also be kept in mind that a 1-inch minimum clearance from combustibles should be maintained at all times for heat ducting, and a 2-inch clearance for chimney sections.

You can add an optional fan kit to the heat circulating fireplace to increase the airflow through the circulation system. This can be especially helpful if you are planning to run an extension of duct work to an adjoining room. The fan kit includes covers that seal the standard air intake openings, and two fans, duct and grille assemblies. The fireplace has alternate openings in the side which accept this feature

Gas Log Fireplaces

Gas log fireplaces (Fig. 14-17) provide an easy way to bring the traditional glow of an open fire into any room setting. Although it is less economical to burn gas than wood, gas log fireplaces are ideal in locations where wood is not easily available or is not preferred because of the inconvenience, accumulation of ash, or danger of sparking from the fireplace. Designs have been advanced to where the fire in an artificial gas log very closely resembles a real wood-burning fire. They are especially designed to meet safety and building requirements and offer an answer to homeowners who seek traditional fireplace charm with the conveniences of a modern appliance.

Prefabricated gas log fireplaces employ realistic gas log sets which closely resemble the dancing flames of a real wood-burning fire. Prebuilt models are provided with an automatic pilot, control valve and pressure regulator with a 100 percent safety shut-off feature. Most fireplaces utilize pushbutton convenience to start the fire and have an automatically controlled damper which is opened when the fireplace is operating but is closed when the unit is not being used, thus closing the flue and avoiding loss of warm air in winter or cool air in summer. Many models also feature circulating fans which help disperse the warm air of the fireplace throughout the room.

Like the prefabricated wood-burning fireplace, gas-fired fireplaces are factory built and designed for installation in any location and against any combustible material. This is because multiple wall construction of the steel unit keeps the outside cool, even when a fire is blazing in the hearth. The fireplace is set in its desired location, framed in with two-by-fours, and may be vented through existing or prefabricated, approved gas vents and flues, in some cases supplied with the basic unit. The final exterior trim may be of any kind of material to suit the demands of room decor; no masonry is required.

FIGURE 14-17. A realistic gas log fireplace has pushbutton firing without the inconvenience of log toting and ash dumping. The fireplace can be installed anywhere a gas supply can be furnished; however, gas is not as economical as wood (*courtesy The Majestic Co.—An American Standard Co.*).

Electric Fireplaces

There are instances in which a fireplace adds a special degree of cozy charm to complete a room decor. Conventional fireplaces, whether all masonry or prefabricated units, are not always possible or practical, particularly in this age of high-rise apartments, mobile homes, and condominiums. In such instances, the electric fireplace takes on special importance.

The wall-hanging electric fireplace (Fig. 14-18) is for persons desiring the atmosphere of a fireplace without a real fire. Installation merely involves hanging the unit on a wall like a picture and plugging it into a nearby outlet. A flip of a switch activates the flame-flickering electric logs and, with no building or laying of brick, with no toting of logs or dumping of ashes, the room is bathed in the warm glow of a natural-looking fire. Since there are no require-

FIGURE 14-18. An electric fireplace that simulates a flickering flame is the answer for apartments, condominiums, and offices where installation of a wood-burning fireplace to a chimney is impractical. Room heating units can be added (*courtesy The Majestic Co.—An American Standard Co.*).

ments for a permanent flue or built-in construction of any kind, the fireplace is as easy to move as an average piece of furniture.

For homes in mild climates, or where a room has been added on and the central heating does not adequately handle it, electric fireplaces may be equipped with heating units. These fireplaces feature circulating and thermostat control and are available for 120-volt or 240-volt circuits. Styles available in electric fireplaces make them adaptable to almost any decorative scheme.

14-6. WOOD-BURNING STOVES AND FURNACES

Fireplaces can provide some heat to parts of a house. But, if you're really serious about heating with wood, choose a thermostatically controlled stove or furnace instead of a fireplace. If you already have a fireplace, consider blocking it off, then add a stove in front of the fireplace and run the stove chimney up the fireplace chimney. If you have a choice of chimneys in the house, use the chimney on the inside wall because it will draw better and give off more heat to the interior.

Wood-burning stoves and furnaces are more efficient than fireplaces because they burn wood more completely. Proper combustion of the wood depends upon proper regulation of fuel, air, and heat. Wood actually burns in three stages: removal of moisture, burning of combustible gases (principally methane and carbon monoxide), and burning of the remaining solid carbon or charcoal. Good wood-burning stoves and furnaces are designed to burn the wood completely in all three stages. The designs include airtight doors and the introduction of both *primary* and *secondary* air into the firebox. Airtightness allows the stove or furnace burning to be controlled by the admittance of air as needed to keep the fire burning at the same controlled rate.

In the first stage of wood burning, moisture is driven from the wood. All wood, even dried, contains some moisture. As the wood heats, the moisture is vaporized; obviously, for wetter wood, more heat is needed to vaporize the moisture and thus dry the wood out in the fire. The most efficient firewood providing the greatest amount of heat is dried and seasoned wood. Wood should be air-dried for 6 months to a year which will leave it with a moisture content of about 20 percent.

In stage two, the wood begins to break down into gas, charcoal, and volatile liquids. If air is supplied properly at the top of the burning wood, then the gases and volatile liquids will burn completely, providing heat and reducing the wood to charcoal. This is where regular fireplaces without a source of *secondary* air fall down in efficiency; air is not provided at the wood surface where the gases and volatile liquids are produced and without oxygen from the air (and at the correct temperature of about 1,100 degrees F), the gases and volatile liquids do not burn. Wood-burning stoves and furnaces of proper design provide heated secondary air to the proper location to provide efficient and complete burning of the gases and volatile liquids that increase the heat of the fire by about 50 percent. Ineffectively designed woodburners allow the gases and volatiles to escape up the chimney unburned.

Finally, the charcoal burns, providing most of the useful heat. Primary, preheated air is introduced to the coals under the burning wood in effective wood-burning stoves and furnaces.

The rate of burning of the wood can be controlled if the stove or furnace is airtight and has an adjustable draft intake in the front of it. As the fire burns, it burns at a faster and faster rate because everything is getting hotter. Here is where you have control; as the fire burns faster, you begin to close the draft intake cutting off the supply of oxygen; this is much more effective than partially closing the flue damper. As the oxygen supply is cut off, the fire burning rate decreases. If the damper(s) is closed completely, the fire will go out; if the stove or furnace is not airtight, however, the fire will not go out. If the stove or furnace is not airtight, you have less control of draft and therefore of the fire.

Some wood-burning stoves and furnaces have thermostatically controlled intake draft controls. As the fire burns more rapidly, the thermostat moves the draft intake(s) to cut off input air, reducing the burning rate; as the fire burning rate decreases, the thermostat opens the draft intake letting in more air to increase the burning rate.

Thus, the keys to efficient, long-duration burning are well-seasoned firewood and control of the draft. With no control, fast burning occurs, with excessive heat wasted up the chimney. You can see just how inefficient the old home fireplace can be with no control at all; the fire gets excessive air, making it burn faster and faster. This causes more and more room heat to rise up the chimney (of course the added air makes the fire burn even faster and we have a vicious cycle of uncontrolled burning).

To get complete combustion, or complete burning of the wood (charcoal, gases, and volatiles), additional preheated air must be supplied to the correct location. In effectively designed fireplaces, secondary preheated air is supplied from a secondary air inlet to the area just above the fuel bed. The secondary air supplies oxygen to the gases and volatiles to afford complete burning resulting in about 50 percent more heat.

Figures 14-19 and 14-20 illustrate a downdraft console, wood-burning heater capable of heating four to five average rooms. The heater is thermostatically controlled so that slow, controlled burning of about 4.5 cubic feet (about 100 pounds) of wood lasts for 12 hours or more. An optional, automatic blower is used to keep the floors warm and to circulate heat more evenly throughout the rooms. Refer to the callouts in Fig. 14-20 for a discussion of the parts of a downdraft console:

(a)

FIGURE 14-19. Two downdraft woodburners are shown. The console (a) is more decorative; its top can be lifted off for emergency cooling (*courtesy Ashley, Martin Ind.*)

(b)

FIGURE 14-19. (continued) The stove model (b) is ideal for homes, cabins, and utility buildings where heating comfort and economy are more important than styling (*courtesy Ashley, Martin Ind.*).

1. *Automatic thermostat damper*, controlled by a highly sensitive bimetal helix coil, regulates the amount of combustion air needed to maintain the selected comfort level.

2. *Downdraft system* preheats primary air to further insure complete combustion of wood gases, thereby increasing efficiency.

3. *Dual primary air intakes* above firebox grates insure even burning along the total length. This feature also eliminates hot spots and extends the life of the firebox.

4. Cast iron *firebox liners* assure maximum heat transfer and the life of the heater.

5. Cast iron *grates* hold the wood.

6. Cast iron *flue collar* is used to prevent burnout since it is subjected to extremely high temperatures.

7. Cast iron *feed* and *ash door frames* prevent warping and maintain an airtight seal with firebox and matching doors.

8. Cast iron *feed* and *ash doors* do not warp from high temperatures and maintain an airtight seal with the use of soft woven asbestos gaskets.

9. *Secondary air intake* responds automatically to temperature changes and admits air into the firebox above the fire zone, increasing the efficiency by burning off wood gases which would go unburned in other heaters.

10. *Ash drawer.*

11. *Double wall cabinet door* maintains its rigidity and is cool to the touch.

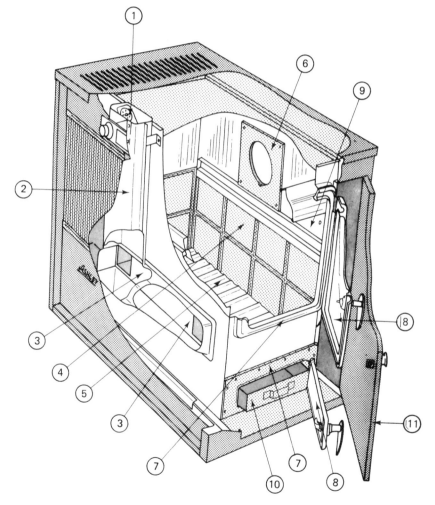

FIGURE 14-20. A downdraft wood-burning heater or stove gets more heat out of every piece of wood than other updraft models. Refer to the text for an explanation of the keyed parts callouts *(courtesy Ashley, Martin Ind.).*

The downdraft system shown in Fig. 14-20 works as follows: the primary air for combustion is regulated by means of the thermostatically controlled damper (1). The primary air is drawn down the draft manifold (2) where it is preheated. The air is then evenly distributed along the entire firebox above the fire by means of the air intakes (3). Secondary intake air (9) responds automatically to temperature changes and admits air into the firebox above the fire zone increasing the efficiency by burning off wood gases which go unburned in other heaters not supplying secondary air. This system of air supply promotes even burning over the length of the firebox, eliminates hot spots

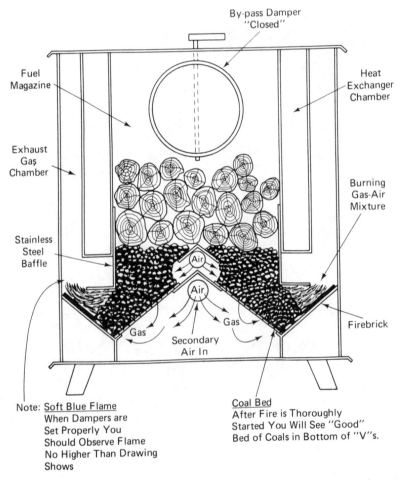

By-pass Damper "Closed"

Fuel Magazine

Heat Exchanger Chamber

Exhaust Gas Chamber

Burning Gas-Air Mixture

Stainless Steel Baffle

Air

Air

Gas

Firebrick

Gas

Secondary Air In

Note: <u>Soft Blue Flame</u>
When Dampers are
Set Properly You
Should Observe Flame
No Higher Than Drawing
Shows

<u>Coal Bed</u>
After Fire is Thoroughly
Started You Will See "Good"
Bed of Coals in Bottom of "V"'s.

FIGURE 14-21. This downdrafter stove introduces secondary air under the coals that causes clean and thorough burning of the volatiles which are forced down through the coals (*courtesy Vermont Woodstove Co.*).

that cause burnouts, and increases efficiency of burning of wood gases for increased heat output.

Another type of wood-burning stove is shown in Fig. 14-21. This is truly a downdraft stove in that secondary air is introduced under the bed of coals. The volatiles are forced through the heat of the coals, superheating them to well above their ignition point of about 1,100 degrees F. The air inserted causes complete combustion of the volatiles.

Figure 14-22 illustrates another wood-burning heater. The desired temperature is set on the regulator dial (1). This causes a bimetal coil inside the regulator to raise a chain connected to a magnetic damper (2) which automatically admits primary air into the heater. The primary air travels below the grate (3) which is protected at all times by 2 to 3 inches of ash. The air is then passed behind the liners (4) and deflected into the combustion chamber by the liner retainers (5). Item 6 is the gas combustion flue; smoke is forced downward, back into the fire before it can enter the combustion flue and exit out the chimney. Secondary air enters through the duct (7) where it is preheated before entering the combustion flue. The secondary air provides oxygen necessary for complete combustion. The woodburner in Fig. 14-23 also burns coal.

Before you buy any fireplace, stove, or furnace, take a good look around your area. Do you have wood available on your land in sufficient quantity? What are your neighbors using? What have your neighbors found to be efficient or inefficient? Finally, visit dealers in your area and compare the

FIGURE 14-22. This wood-burning heater is rated at 50,000 Btu/h to heat 4 to 6 rooms. It is thermostatically controlled and introduces secondary air to ensure complete, efficient combustion (*courtesy Riteway Manufacturing Co.*).

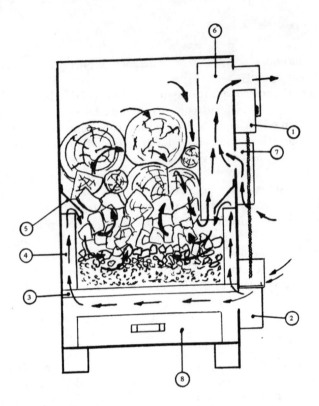

FIGURE 14-22. (Continued.)

FIGURE 14-23. The wood and coal heater is rated at 73,000 Btuh and takes up to 12-inch thick logs without splitting (*courtesy Riteway Manufacturing Co.*).

features and designs of the latest models. Models are available which you can connect into your central hot air or central hot water systems. Perhaps these are the solutions to your heating situation.

14-7. OTHER SUPPLEMENTARY HEATING

Sometimes you'll find that you need just a little supplementary heat in some areas of the home for comfort. This is particularly true in a bathroom or perhaps in a study, office, or hobby room on the north side of the house.

Electrically powered supplemental heating is the easiest and most convenient to install (Fig. 14-24). The small units operate on 117 volts and can be plugged into a wall outlet. Many designs are available including those which are portable or those that are installed in floors, baseboards, walls, or ceilings. Units that are thermostatically controlled provide heat when needed and turn

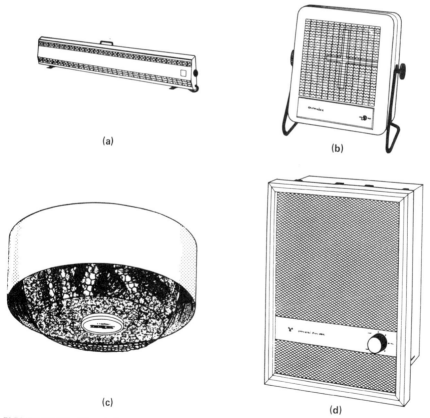

(a)

(b)

(c)

(d)

FIGURE 14-24. Electric heaters provide supplementary heating; (a) portable baseboard unit, (b) portable, (c) ceiling mounted, (d) wall mounted.

off automatically when the desired temperature is reached. Units having a blower circulate the air more effectively.

Gas-burning space heaters also provide supplementary heat. These are connected by a professional to a gas supply already within the home; an adequate air supply must be provided for combustion and a chimney must be provided for the exhausting of combustion gases. Obviously this type of supplementary heating is not as convenient as electric supplementary heating and since gas lines are no longer being installed in many areas of the U.S., it may not even be a possible means of heat.

In using supplementary heat, keep the tips in QRC 14-3 firmly implanted in your mind.

Quick Reference Chart 14-3
HINTS FOR SUPPLEMENTARY HEATING

1. Use a heater with a thermostat to automatically turn the heater on when the temperature drops and off when the temperature rises.
2. If a portable electric heater is used, *ensure* that it has a *tip-over* switch. This is a safety switch which cuts off all electric power to the heater if the heater falls over (an overturned electric heater could cause a fire).
3. A heater with a fan circulates warm air more completely than simply a radiant heater. A two- or three-speed fan control allows air to be circulated further and a two- or three-element electric heater lets you vary the amount of heat from the heater.
4. Remember that hot air rises; therefore, locate the heater to take advantage of this fact.
5. Don't place a supplementary heater in a position where its heat output can be felt by a room or house thermostat; while locally heating one area with supplementary heat, you could "fool" the thermostat and prevent the central heating system from turning on.

chapter 15

SOLAR HEATING

Solar energy has captivated the imagination of many of us, especially the do-it-yourselfers. Here is a plentiful source of free energy that can provide much of our heating, cooling, and power needs; it is virtually an untapped, unlimited, unpolluted supply. Enough solar energy reaches our earth each 15 minutes to supply the entire world's energy needs for a year. Little wonder then that man is beginning to look to solar energy as an alternative source of energy to replace fuel oil, natural gas, and electricity.

Perhaps you are thinking that you should spend all of your weatherization budget on solar heating and cooling apparatus instead of *little* items like caulk, weather strip, insulation, and storm equipment. Then, you think, you could get rid of your present heating and cooling equipment, substitute solar, and let the sun *do it all*. Unfortunately, in today's state of solar technology, this is a dream rather than a reality.

Solar energy can *help* us meet our objectives of weatherization. Solar energy, properly harnessed, can make us feel warmer in the winter and cooler in the summer. It can reduce the cost of our monthly heating and cooling bills, decrease the use of fossil fuels, make us more comfortable, and it is free of pollution.

But, before you contemplate solar heating, you've got to "button up" first, as described in the previous chapters: all paths of heat escapement and infiltration must be repaired; all windows and doors caulked and weather stripped; storm equipment in place; attics, crawl spaces, and hopefully exterior walls insulated; heater tuned and maintenance performed.

In today's solar technology, solar energy provides less than 100 percent of the energy needed to heat our homes and hot water. More precisely, solar energy can supply only about 40 to 70 percent of the total energy needed for most space heating and water heating applications. The percent of effectiveness of the solar energy depends upon your climate, house site location, severity of the winter, and how well your home is weatherized. For all practical purposes, solar air conditioning is not sufficiently advanced for practical, economical home use and therefore won't be discussed any further in this book. We can, however, further discuss solar energy as used in home space heating (the heating of the air within the home) and domestic hot water heating, because systems are sufficiently developed to be effectively utilized.

This chapter provides you with a basic understanding of solar heating systems. Specifically, the following topics are covered: passive and active solar heating systems, solar heating system components, solar collector and storage locations, typical solar heating systems, and things to come in solar heating and cooling.

15-1. PASSIVE AND ACTIVE SOLAR HEATING SYSTEMS

Solar energy is available to us only during daylight hours; on very cloudy days, the sunlight is blocked or severely subdued. Thus, adequate, direct solar energy is only available during fairly clear daylight hours; very little solar energy is available on cloudy days and none at nighttime. If heat generated by solar energy is to be used on cloudy days and during nights, a means of storing surplus thermal energy produced on clear days must be used.

The living space within our homes and our domestic hot water can be heated by either *passive* or *active* systems. A *passive* solar system is an assembly of *natural* and *architectural* components including collectors, a thermal storage device(s), and transfer fluid which converts solar energy into thermal energy in a controlled manner and in which no pumps, blowers, or controls are used to accomplish the transfer of thermal energy. On the other hand, an active solar system is an assembly of collectors, a thermal storage device(s) and transfer fluid which converts solar energy into thermal energy, and in which energy in addition to solar is used to accomplish the transfer of thermal energy; the active system uses pumps or blowers and controls. In most cases, the active solar system is able to reduce the use of supplementary fuels more than the passive system.

15-2. SOLAR HEATING SYSTEM COMPONENTS

As a minimum, a solar heating system consists of a *collector*, *transfer fluid*, *storage device*, and a *distribution system*. Additionally, it may consist of *controls*, a *heat exchanger*, and an *auxiliary energy* system. Our discussion focuses primarily on components that are effective today and are being commercially produced today for residential systems; components that are on the drawing board or in prototype, unproven systems, are not discussed. However, in the near future, these new ideas may make present components obsolete.

The *collector* gathers in the solar energy. Usually, the collector is a *flat-plate* design (Figs. 15-1 & 15-2) meaning that its top is flat; other collectors may have a slightly convex or a concave shape to aid in collection. The collector usually consists of a glass (or plastic top), a black, metal absorber, a transfer fluid, insulation, and a frame. The glass or plastic top allows the radiant energy of the sun to pass through onto the absorber plate. Radiant energy consists of short waves of radiation which readily pass through the glass. The absorber soaks up the short waves of radiation which warm the absorber plate. When the short waves hit the absorber, some are reflected back off but, in being reflected, become long waves. Glass does not allow long waves to pass through; thus, the long waves are trapped inside the glass by the collector.

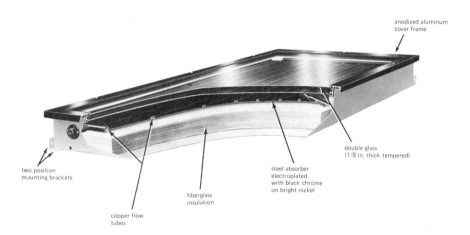

anodized aluminum cover frame

two position mounting brackets

copper flow tubes

fiberglass insulation

steel absorber electroplated with black chrome on bright nickel

double glass (1/8 in. thick tempered)

FIGURE 15-1. This commercial flat-plate collector is designed for use in systems for heating residences, commercial areas, domestic hot water, swimming pool water and so on. It can be installed on a roof or at ground level individually or in multiple banks end to end and/or side by side (*courtesy Lennox Industries Inc.*).

The absorber temperature rises as the sun rays pass through the collector glass onto the absorber. Solar energy is transferred into heat energy which must be moved from the collector to either the home for direct use or to a storage device for later use. A transfer fluid is passed over or through the absorber to carry the heat away from the absorber to the home distribution system or to storage. The transfer fluid can be water, water with additives, some other liquid, or air.

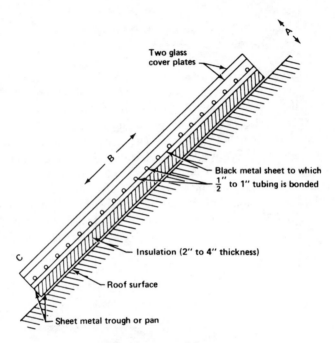

Two glass
cover plates

Black metal sheet to which
$\frac{1}{2}''$ to 1'' tubing is bonded

Insulation (2'' to 4'' thickness)

Roof surface

Sheet metal trough or pan

Notes: Ends of tubes manifolded together
one to three glass covers depending
on conditions

Dimensions: Thickness (A direction) 3 inches to 6 inches
Length (B direction) 4 feet to 20 feet
Width (C direction) 10 feet to 50 feet
Slope dependent on location and on winter-
summer load comparison

FIGURE 15-2. This liquid solar collector is used for residential heating and cooling. It is possible for the do-it-yourselfer to make collectors similar to this one.

Another type of collector is the *focusing* collector (Fig. 15-3). This collector is a specially concave-shaped mirrored surface that concentrates the solar rays along a straight line. This type of collector produces very high temperatures of up to several thousand degrees, but requires a very expensive tracking system to maintain its orientation relative to the incoming sunlight. With slightly overcast skies, however, the operation ceases.

The *transfer fluid* passes over or through the absorber and then leaves the collector through the distribution system. If the transfer fluid is liquid, the fluid flows through pipes from the collector to the heating system or to storage; if the transfer fluid is air, the fluid (air) travels through insulated air ducts.

Heat may be transferred from a solar collector either by a liquid or by air; each has its own advantages and disadvantages. A liquid collector can be made smaller than an air collector because it takes less *volume* of liquid passing over

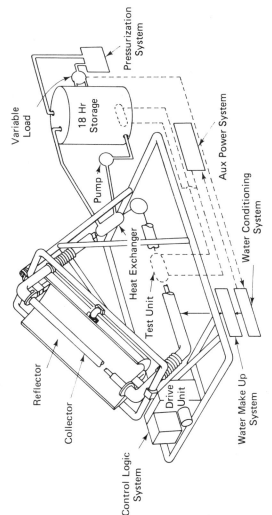

FIGURE 15-3. The highly reflective curved metal plates on this solar collector cause the sun's rays to converge on the glass tube in the center. Water or other suitable fluid in the tube is heated by the sun, circulated through the tubes of the heat exchanger, and recirculated to pick up more heat. The heat is transferred in the heat exchanger to another fluid that is pumped to the storage tank where it can be used to produce electricity, provide air conditioning, and furnish hot water and heating for homes and other buildings. The work at ERDA's Sandia Laboratories in New Mexico is typical of research under way in university, industry, and national laboratories throughout the United States.

Reflector

Collector

Control Logic System

Drive Unit

Water Make Up System

Test Unit

Heat Exchanger

Water Conditioning System

Aux Power System

Pump

18 Hr Storage

Variable Load

Pressurization System

the absorber plate to collect the heat than it takes for air to collect the heat. Likewise, liquid can be transported to and from the collector via pipes (of 3/4 to 1 inch diameter) whereas air must be transported via air ducts. Liquid systems are subject to freezing; if water without an antifreeze is used, the water must be drained from the system whenever a freeze is expected; this is done automatically in most systems because the system stops operation whenever the collector temperature decreases to about 5 degrees F above the storage device temperature. Pumps are required for liquid systems whereas blowers are required for air systems; both are powered by electricity. Liquid systems can develop leaks which can cause damage to the house structure. Liquid systems also may corrode the collector absorber plate eventually requiring collector replacement.

The transfer fluid may be circulated directly through the heating system or heat exchanger or it may be circulated into storage. For example, an air system may duct the hot air directly into a room or into the ducting system of a hot air heating system for distribution into several rooms; cold air returns through the return ducting system to pass over the collector absorber again. If a liquid solar collector system is used with a hot air heating distribution system, then a heat exchanger is needed; for example, a long coil of tubing may be placed inside a heat duct; heated fluid circulates through the coil and the air in the forced air heated system is blown past the coil which cools the coil and heats the air–the heat is *exchanged* from the coil to the air in the heat exchanger.

The heat exchanger shown in Fig. 15-4 is used to transfer heat from a circulating transfer medium to another medium used in storage or in distribution. Shell and tube heat exchangers consist of an outer casing or shell surrounding a bundle of tubes. The water to be heated is normally circulated in the tubes and the hot liquid is circulated in the shell. Tubes are usually metal such as steel, copper or stainless steel. A single shell and tube heat exchanger cannot be used for heat transfer from a toxic liquid to potable water

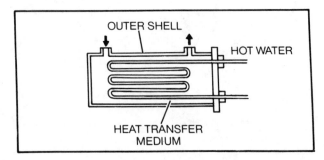

FIGURE 15-4. In this typical heat exchanger, liquid from the collectors is cycled through the shell. The water passing through the tubing is heated by the liquid in the shell. This type of heat exchanger is *not* used for potable (drinkable) water.

because double separation is not provided and the toxic liquid may enter the potable water supply such as in a case of tube failure.

Figure 15-5 illustrates a shell and double-tube heat exchanger that is similar to the previous one except that a secondary chamber is located within the shell to surround the potable water tube. The heated toxic liquid then circulates inside the shell but around this second tube. An intermediary, nontoxic, heat transfer liquid is then located between the two-tube circuits. As the toxic, heat transfer medium circulates through the shell, the intermediary liquid is heated, which in turn heats the potable water supply circulating through the innermost tube. This heat exchanger can be equipped with a sight glass to detect leaks by a change in color (toxic liquid often contains a dye) or by a change in the liquid level in the intermediary chamber, which would indicate a failure in either the outer shell or intermediary tube lining.

A *storage device* is necessary to hold thermal energy once it has been collected for use during the nighttime and on cloudy days. A good storage device is capable of storing enough heat to supply the system for four or five cloudy days. Storage devices for liquid systems may consist of a large insulated tank or a large tank surrounded by rock and then insulation; in this configuration, the heated tank heats the rocks which retain heat. A typical airflow collector system uses a rock bin for heat storage (Fig. 15-6). Insulation around the rocks prevents heat from escaping from the storage device. The distribution system removes the heat from the storage device and distributes it via ducts (air system) or pipes and radiators (water system).

Controls are valves, sensors, comparison circuits, thermostats, and relays that are used to control the transfer fluid and the distribution system. For example, the sensors sense the temperature of the fluid in storage and of the

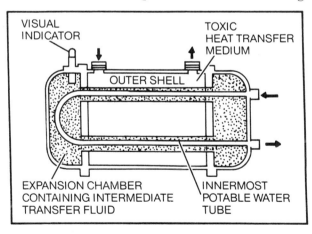

FIGURE 15-5. This is a shell and double tube heat exchanger. Potable water can be heated by this exchanger.

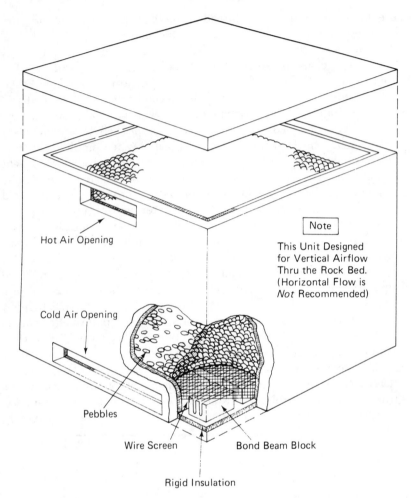

FIGURE 15-6. One way to store hot air produced by solar energy is to heat a bin of rocks or pebbles with the warm air (*courtesy Solaron Corporation*).

temperature of the collector on the roof and send electrical signals to the comparison circuits. If the collector temperature is approximately 10 degrees F higher than the storage temperature, the sensor automatically starts a pump or blower and positions valves to start the collection system into operation. If the temperature decreases to a 5-degree F difference, then the controls turn the collection system off.

The *auxiliary energy* system is the heating system that supplements the solar heating system as necessary and replaces it in the event of some malfunction. The auxiliary system is usually an oil- or gas-fired burner or electric heating element; wood stoves or furnaces comprise the auxiliary system in some homes. Home space-heating systems are set via thermostats; if the solar heating system proves inadequate to keep the home at the desired temperature, then the auxiliary energy system will be turned on to provide the necessary heat to raise the temperature to the desired level.

15-3. SOLAR COLLECTOR AND STORAGE LOCATIONS

Before you get too excited and ready to run off and buy some solar heating equipment, let's determine if your home is situated satisfactorily to receive solar energy. Remember, we are interested in installing solar heating equipment in your existing home which is fixed in place and design, unless you can afford to make an addition to your home which could aid in your placement of the collector and storage device.

To determine if you can install a solar heating system in your existing home, you need to determine a location for your collectors and storage device. Ideally, the collectors are mounted on the roof of the house, garage, porch, patio, or shed. On the roof, the collectors are out of the way, do not occupy yard space, are less likely to have something block the path of the sun, are safe from children, and do not present an unsafe area where someone could fall against the glass. If a roof is not suitable, the next best location is on top of a fence. As a last resort, the collector can be located on the ground. No matter what the location, be sure that no evergreen trees, fences, buildings, or other houses will block out the sun. Think into the future too. Will any structure be built on the adjacent property that could block the sun from your collector? Free air space that may later be blocked by a building is becoming a legal problem.

You will need to place your collectors so that, ideally, they face directly *true* south. This is not *magnetic* south. To determine true south, you have to subtract or add the magnetic *variation* of your part of the country from the reading on your compass; for example, in Baltimore, Maryland, approximately 9 degrees west variation is subtracted from the compass reading. You can obtain the magnetic variation from a current marine navigational chart or similar chart. (If your variation is west, you *subtract* the variation from the compass reading: if east, you *add* the variation to the compass reading.) If you can't locate the collector directly south, the next best location is from south to 15 degrees west of south (180 to 195 degrees). You'll get a 5 percent loss in efficiency between 157 and 203 degrees, and a 10 percent loss between 152 and 210 degrees; beyond the 10 percent loss, find a method to relocate the collector toward a more true direction of 180 degrees.

Once the physical location of the collector is established so that it faces as near true south as possible, the angle, with respect to the horizontal, must be established to place the collector perpendicular to the sun rays during midday. The ideal angle up from the horizontal for a collector for space heating is the latitude of your location (see Fig. 2-1) plus 15 to 20 degrees; you can set the collector at any angle within ±25 degrees of the ideal angle without much loss. For example, the ideal angle for a collector in Baltimore, Maryland having a latitude of 39 degrees is about 57 degrees from the horizontal. The angle for a collector for a domestic hot water system should be equal to the latitude for best year-round performance; the collector angle could also be made adjustable so that the pitch angle could be varied during the year.

You also need to know the approximate size of the collector area. For a space heating system, the maximum collector area is about 25 percent to 30 percent of the living area; thus, for a home with 1,500 square feet of living area, the collector area would be approximately 375 to 450 square feet. The collector area for a domestic hot water system is approximately 50 square feet. Solar flat-plate collectors weigh about 2.5 pounds per square foot, about the same weight as a layer of asbestos shingles.

The best place to locate a storage device is inside your basement; any heat that bypasses the storage device remains within the living space anyway. Other storage device locations include underground or inside a garage or shed. When the device is located outside underground, the construction and insulation must be able to withstand the elements of cold, frost, rain, and insects. The storage tank size (of a liquid system) is approximately one and one-half gallons for each square foot of collector area plus one gallon for expansion for each 50 gallons of liquid storage. Thus, for a 1,500-square-foot living area with a 375-square-foot collector area, a storage tank of 618 gallons capacity is needed. When rocks, insulation, and a block frame are placed around the storage tank (you can probably only buy a 750-, 1,000-, 1,500-, or 2,000-gallon storage tank), the size will be something like 8 feet wide, 12 feet long, and 6 feet high to as large as 8 feet wide, 25 feet long, and 7 feet high. For a solar domestic hot water tank, a storage tank of 150 gallons is usually adequate.

15-4. TYPICAL SOLAR HEATING SYSTEMS

Now that you know something about the components in a solar heating system, passive systems, and active systems, let's take a look at some typical systems that are being used successfully today. Included are a solar heating and hot water system, a solar hot air space heating system, a solar domestic hot water system, and passive solar heating.

Solar Heating and Hot Water System

Figure 15-7 illustrates the basic elements of a solar heating and hot water system. The system as shown provides two basic functions: capturing the sun's radiant energy, converting it into heat energy and storing this heat in an insulated energy storage tank; and delivering the stored energy as needed to

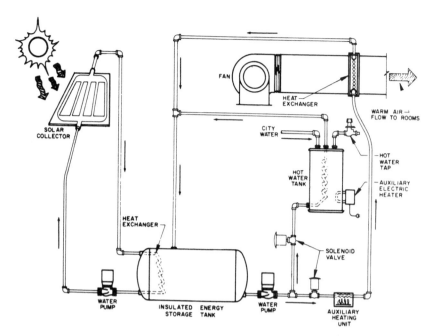

FIGURE 15-7. This schematic diagram of a solar space heating and hot water system indicates that if you already have a forced hot air heating system and a domestic hot water tank, you need to install collectors, a storage device, pipes, and controls to "go solar."

either the domestic hot water or heating system. The parts of the system which provide these two functions are referred to as the collection and delivery subsystems.

The key component in the collection subsystem is the collector. The basic function of the collector is to trap the sun's energy. Transparent cover plates made of glass or a suitable plastic material allow the rays of the sun to pass into the collector. Once inside the collector, the sun's rays are absorbed by a blackened metal absorber plate and transformed into heat energy. On sunny days absorber plate temperatures can reach well over 200 degrees F. Insulation is placed under the absorber plate to prevent heat loss out of the back side of the collector. Cover plates prevent the loss of heat from the top side of the collector. They perform the same function as the glass in a greenhouse. Heat energy is removed from the collector by circulating a fluid, such as water, through tubes in the absorber plate or by passing air over or under the absorber plate. The heat energy so removed is carried by means of insulated pipes or ducts to an insulated energy storage tank near or in the building.

The energy storage tank, which may be full of a liquid such as water, stores the heat so that it may be withdrawn upon demand. The transfer of energy to the storage tank is accomplished by means of a heat exchanger. Examples of heat exchangers include coils of metal tubing, the radiators in automobiles, and the finned radiators in electric baseboard heating units. Storage tanks and collectors are usually sized to enable the system to supply enough energy to support the building's heating and hot water requirements for a few consecutive cloudy days.

The delivery subsystem is divided into two parts: one for providing heat to the hot water tank and another for providing heat to the building's heating system. When either the domestic hot water or heating system requires heat, hot water from the energy storage tank is pumped to a heat exchanger in the domestic hot water tank or in the ducts.

On occasions when the temperature in the energy storage tank falls below the required temperature, perhaps due to increased demand for heat during a cold spell, an auxiliary heating unit such as a furnace or electrical resistance heater is turned on to provide the needed energy. A hot water tank, conventional heating unit, duct work, and some of the piping are already a part of many homes. Just the addition of collectors, an insulated energy storage tank, and associated piping are required for a solar system. The actual placement of collectors and energy storage tanks in a given case depends upon such factors as building orientation, available space, and roof angle, in addition to aesthetic factors. Figures 15-8 and 15-9 show solar collectors placed on the roof of homes.

FIGURE 15-8. This duplex has solar space and domestic hot water heating (*courtesy Solaron Corporation*).

FIGURE 15-9. Forced air space heating, pool water heating, and domestic hot water heating are all provided by solar energy in this residential home. The vertical glass on the southern exposure is permitting the sun to heat the inside passively (*courtesy Solaron Corporation*).

Solar Hot Air Space Heating System

A solar hot air space heating system is shown in Fig. 15-10. This system can be bought as a complete package ready to install in your backyard (International Solarthermics Corporation, Box 397, Nederland, Colorado 80466). No water is used; the aluminum collectors absorb, convert, and emit usable heat energy to a stream of air that is directed across their faces by a blower. The heat from the collectors is moved by air to the thermal storage bin containing small rocks. Insulated hot air ducts connect the heat storage bin with the hot air ducts of the home heating system. When heat is needed in the home, cooler house air is directed through the heat storage bin where the air is heated for distribution into the home.

Solar Domestic Hot Water Systems: Liquid

Domestic hot water systems operate on the principle of preheating water before it enters the home's main hot water heating tank. In the system in Fig. 15-11, hot liquid from the collector passes through the combined storage device and heat exchanger. Cold water from the supply comes into the storage tank as necessary. The water in the storage tank is heated by the hot liquid of the collector flowing through the coil of the heat exchanger. The preheated

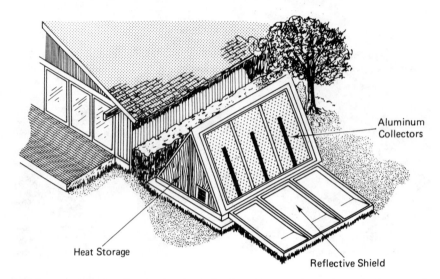

Aluminum
Collectors

Heat Storage

Reflective Shield

FIGURE 15-10. Designed as a solar *plug-in furnace*, this unit is located near the home on the ground and connects to a forced air heating system via insulated air ducts. No liquid is used.

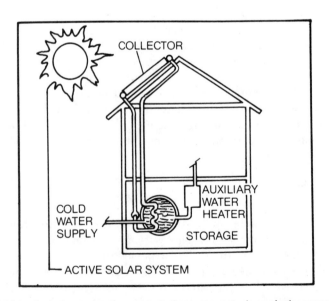

COLLECTOR

AUXILIARY
WATER
HEATER

COLD
WATER
SUPPLY

STORAGE

ACTIVE SOLAR SYSTEM

FIGURE 15-11. A solar hot water heating system is the most practical use of solar energy today.

water from the storage tank is sent to the auxiliary water heater; this heater is really the domestic hot water heater but it does not need to heat the water unless the water coming into it drops below about 140 degrees F (depending upon the thermostat setting). Thus, energy is saved because the cold, incoming water is preheated and little if any energy needs to be used in the auxiliary water heater to bring the water to the desired temperature (120–140 degrees F).

Solar Domestic Hot Water System: Air

Figure 15-12 illustrates a solar domestic hot water system utilizing an air collector. Solar energy is collected by the south-facing collector (1). These collectors may be mounted as shown, remote from the residence or on the roof. Air is circulated by the heat exchange unit (2) where the solar energy is transferred to the water being circulated by the domestic water-circulating pump (6). Solar-heated water is continuously circulated into the storage tank (3) as long as the control unit indicates that solar energy is available at the collectors (1) and until tank temperature in the storage tank (3) reaches 160 degrees F. At 160 degrees F the tank control stops the circulating pump (6) and heat exchange unit (2) operation. When hot water is required in the residence, hot water is drawn from the conventional domestic hot water heater (4) and preheated water is drawn from the storage tank (3) into the conventional domestic water heater (4). If the temperature in the storage tank

FIGURE 15-12. This is a domestic hot water system using an air type solar collector (*courtesy Solaron Corporation*).

(3) is above 120 degrees F (the suggested set-point for the conventional domestic hot water heater) no conventional energy will be used by the domestic hot water heater. Depending on local codes, a mixing valve (5) set at 120 degrees F may be installed for the mixing of water when tank storage temperatures above 120 degrees F can be obtained from solar. This eliminates the requirement for mixing at the faucets, increases the life of the conventional heater, and reduces conventional energy usage.

Passive Solar Heating

In passive solar heating, one side of your home needs to face almost directly south. About 25 percent of the south wall should be glass and if possible, part of the roof should also be glass (the glass must be insulated glass or be covered with storm windows). Insulating panels or shutters should be put over the glass at night to prevent heating losses.

The interior rooms of the south-facing glass must be designed to accept, absorb, store, and then reradiate the heat gained from the sun into the home during the night. Numerous techniques have been and are being used to perform these functions. For example:

1. A cinder block wall can be placed several feet behind the glass to absorb the sun's energy; if the wall is painted flat black, it will do a better job of absorbing solar energy. If air is circulated with a fan or ducted through the hollow cores of the block, the warm air can be distributed more effectively.
2. Oil drums (55 gallons) can be filled with water. The incoming sun heats the water in the daytime; at nighttime, the stored heat in the water is radiated into the home. (Plastic bottles can also be used.)
3. A heat absorbant floor will retain the sun's heat and give off the heat later.
4. A greenhouse can be used as a solar collector.

15-5. THINGS TO COME IN SOLAR HEATING AND COOLING

Our technology in solar systems has just begun. Successful systems are currently on the market. Private industry, government, universities, and do-it-yourselfers are researching, designing, innovating and developing new ways to effectively and economically collect and store solar energy for heating and cooling. Solar energy is a fascinating field of study and endeavor, and one in which we hope you will be involved too.

One such project under development by the Energy Research and Development Administration (ERDA) is the annual cycle energy system (ACES) (Fig. 15-13). The principal component of the ACES is an insulated

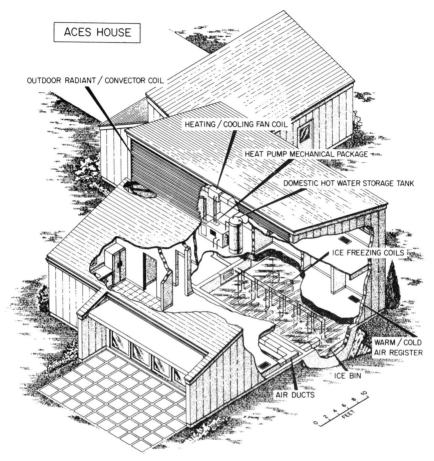

ACES HOUSE

OUTDOOR RADIANT / CONVECTOR COIL

HEATING / COOLING FAN COIL

HEAT PUMP MECHANICAL PACKAGE

DOMESTIC HOT WATER STORAGE TANK

ICE FREEZING COILS

WARM / COLD AIR REGISTER

ICE BIN

AIR DUCTS

0 2 4 6 8 10 FEET

FIGURE 15-13. This is an annual cycle energy system (ACES) home being developed by ERDA. The principal component of the system is an insulated tank of water which serves as an energy storage bin.

tank of water which serves as an energy storage bin. For well-insulated homes within the applicable zone (the geographic area between Atlanta, Georgia and Minneapolis, Minnesota), this bin need not exceed two cubic feet of water for each square foot of living space. Thus, a home with 1,500 square feet of living space would require a 3,000-cubic-foot tank of water. This would be equal in size to one-fourth to one-third the size of a basement. The energy bin could be located in the basement or could be built under a driveway, carport or patio of a home under construction.

Heat is obtained from the bin during the winter by a heat pump which draws the heat from the water in much the same manner that the conventional home heat pump draws heat from air. Heat drawn from the water is used to warm the building and to provide domestic hot water. This removal of the heat from the water gradually turns the water into ice over a period of months. In the summer months, the chilled water from the bin is used to provide air conditioning for the building without the operation of the heat pump compressor. This action causes a gradual melting of the ice over a period of months and thus stores heat for use in the winter.

You have been introduced to several concepts of solar heating and cooling systems and components in this chapter. There are an undefined number of designs, prototypes, models, and actual products being developed or currently on the market. Your best bet is to contact all of the solar energy system distributors in your area, read additional books on solar heating and cooling, and read current magazines such as *Popular Science* and *Popular Mechanics* which periodically update solar energy progress. Know before you buy or build.

GLOSSARY

This glossary lists terms used in weatherizing, heating, cooling, and in home construction, maintenance, and repair. The glossary provides a quick reference to homeowners who may be unfamiliar with some of the terminology.

Active solar system.   An assembly of collectors, thermal storage device(s), and transfer fluid which converts solar energy into thermal energy, and in which energy in addition to solar is used to accomplish the transfer of thermal energy.

Air duct.   Pipes that carry warm air and cold air to rooms and back to the furnace and/or air-conditioning system.

Ambient.   Surrounding on all sides; an encompassing atmosphere; environment.

Auxiliary energy subsystem.   Equipment utilizing energy other than solar both to supplement the output provided by the solar energy system, and to provide full energy backup during periods when the solar systems are not operating.

Barometer.   An instrument for determining the pressure of the atmosphere and hence for assisting in judgment as to probable weather changes (and for determining the height of ascent).

Batt.   Insulation in the form of a blanket, rather than loose filling.

British thermal unit (Btu).   The energy required to increase the temperature of one pound of water by one degree Fahrenheit. One ton (of air conditioning) equals 12,000 Btu per hour.

Building paper.   Heavy paper used in walls or roofs to damp-proof.

C.   Thermal conductance. C is similar to k, but applies to the actual thickness of the material. Thus, it is the measure of the rate of heat flow for the actual thickness of the material of an area of one square foot at a one degree F temperature difference between the inner and outer surfaces. If the k of a material is known, the C can be found by dividing k by the material thickness.

Casement.   A window sash that opens on hinges at the vertical edges.

Casing.   Door and window framing.

Caulk. To stop up seams and make them airtight and watertight by filling with a compound or material.

Chase. A groove in a masonry wall or through a floor to accommodate pipes or ducts.

Coefficient of heat transmission. Refer to U.

Coefficient of performance (COP). A measure of heat pump efficiency; it is the ratio of the useful heat energy transferred, to the amount of energy put into the system. A straight electric resistance heat furnace operates at a COP of 1.00, or 100 percent efficiency; it returns 3,413 Btu of heat for every kilowatt (1,000 watts) hour of electricity put in.

Collector. A device for collecting solar energy and transforming it into thermal energy.

Collector subsystem. The assembly used for absorbing solar radiation, converting it into useful thermal energy, and transferring the thermal energy to a heat transfer fluid.

Combination storm-screen. Two glass storm panels and a half screen enclosed in a frame which is installed on the exterior of a double-hung or horizontal sliding window. Adjustment of the panel and screen positions allows for a full storm panel during winter and a half screen and half storm panel during the summer.

Condensate. A product of condensation; a liquid obtained by condensation of a gas or vapor.

Condensation. The act or process of condensing; a reduction to a denser form (as from vapor to water).

Conductive heat. Air or material heated through contact with hotter surroundings, such as air touching hot iron.

Control subsystem. The assembly of devices used to regulate the processes of collecting, transporting, storing, and utilizing solar energy.

Convection current. The movement of heat (air), such as heated air or liquid, or moving by force such as by electric fan.

Cooling capacity (Btu per hour). The quantity of heat in Btu a room air conditioner is capable of removing from a room in one hour.

Cornice. Horizontal projection at the top of a wall or under the overhanging part of a wall.

Crawl space. A shallow, unfinished space beneath the first floor of a house which has no basement, used for visual inspection and access to pipes and ducts. Also, a shallow space in the attic, immediately under the roof.

Cripple. A vertical support under a window.

Damper. A device used to vary the volume of air passing through a duct by varying the cross-sectional area. Small door between the fireplace firebox and flue throat to regulate the draft.

Degree-day. The difference of the average temperature of the day from a standard of 65 degrees F. Example: If the average temperature during 24 hours is 45 degrees, the degree-day would be 20 (a heating day). The degree-day numbers for a complete heating season are added together to get the number of degree-days for a particular climate.

Dehumidifier. A device for removing moisture from the air.

Dehumidify. To remove moisture from the air.

Dew point. The temperature at which a vapor begins to condense.

Diffused. Not concentrated or localized; scattered.

Diffuser. An outlet discharging supply air in various directions and planes.

Distribution system. The pipes or ducts used to move the transfer liquid from subsystem to subsystem in a heating system.

Dormer. The projecting frame of a recess in a sloping roof.

Double glazing. An insulating window pane formed of two thicknesses of glass with a sealed air space between them.

Double-hung windows. Windows with an upper and lower sash, each supported by cords and weights.

Downdraft. Air coming down the chimney.

Draft. Air movement. Can be caused by breeze, difference in air pressures on opposite sides of house, or by a difference in temperature, as warm air in a fireplace rising up the chimney.

Dry wall. A wall surface of plasterboard, or material other than plaster.

Duct. A channel, pipe, or tube that conveys air or a substance such as water.

Eaves. The extension of a roof beyond house walls.

Energy efficiency ratio (EER). The unit's power using efficiency (how much cooling you get for a given amount of electricity). The EER is obtained by dividing the unit's cooling capacity in Btu/hour by its wattage. This value represents the relative electrical efficiency of room air conditioners.

ERDA. Energy Research and Development Administration.

Evaporate. To pass off in vapor or in invisible minute particles; to convert into vapor; to expel moisture from.

Fascia. A flat, horizontal member of a cornice placed in a vertical position.

Fiberboard. A material made by compressing fibers (as of wood) into stiff sheets; Masonite™.

Fill-type insulation. Loose insulating material which is applied by hand or blown into wall spaces mechanically.

Flashing. Noncorrosive metal used around angles or junctions in roofs and exterior walls to prevent leaks.

Floor joists. Framing pieces which rest on outer foundation walls and interior beams or girders.

Flora. Plant life; the plant life characteristic of a region, period, or special environment. Trees, shrubs, and vines.

Flue. A passageway in a chimney for passing smoke, gases, or fumes to the outside air.

Footing. Concrete base on which a foundation sits.

Fossil fuel. Fuel derived from the remains of carbonaceous fossils, including petroleum, natural gas, coal, oil shale (a fine-grained laminated sedimentary rock that contains an oil-yielding material called kerogen), and tar sands.

Foundation. Lower parts of walls on which the structure is built. Foundation walls of masonry or concrete are mainly below ground level.

Framing. The rough lumber of a house: joists, studs, rafters, and beams.

Free area. The total area of the *openings* in an air outlet or inlet through which air can pass.

Frost. A covering of minute ice crystals on a cold surface: the process of freezing.

Furnace. An enclosed structure in which heat is produced.

Furring.   Thin wood or metal applied to a wall to level the surface for lathing, boarding, or plastering, to create an insulating air space, and to damp-proof the wall.

Gable.   The triangular part of a wall under the inverted "V" of the roof line.

Geothermal energy.   Energy extracted from the heat of the earth's interior.

Girder.   A main member in a framed floor supporting the joists which carry the flooring boards. It carries the weight of a floor or partition.

Glass, insulating.   Two sheets of glass bonded together in a unit to enclose a captive air space. Edges of sheets are often melted together with an inert gas between the sheets.

Glazing.   Fitting glass into windows or doors: the glass panes or lights in the sash of a window.

Glazing bead.   A plastic or wood strip applied to the window sash around the perimeter of the glass on the outside to hold the glass in place.

Glazing compound.   A pliable substance applied between the window sash and the lights of glass to seal against the elements and sometimes to adhere the glass to the sash.

Glazing double.   A single-glazed sash with an additional glass panel installed on the sash to provide an air space between the two lights of glass. The second glass can either be movable or fixed and can be installed on either the inside or outside of the sash. Double glazing differs from insulating glass in that there is no positive seal around the edges of the two lights of glass to provide a true, dead air space and there is no desiccant within the unit to absorb and hold moisture.

Glazing, single.   A single sheet of glass installed in a window sash.

Glazing, triple.   A sash glazed with three lights of glass, enclosing two separate air spaces. This can be accomplished by applying a storm panel to a sash that is glazed with insulating glass or on some units by applying inside and outside storm panels to a single-glazed sash.

Grade line.   The point at which the ground rests against the foundation wall.

Green lumber.   Lumber that has not been adequately dried out to remove moisture. As green lumber dries, it has a tendency to warp.

Grille.   A covering for any opening through which air can pass.

Gutter.   A channel at the eaves for carrying rain water away.

Gypsum.   A widely distributed mineral consisting of hydrous calcium sulfate that is used in making plaster of paris; plasterboard.

Hardboard.   Composition board made by compressing shredded wood chips often with a binder at high temperatures.

Hardwood.   The close-grained wood from broad-leaved trees such as oak or maple.

Headers.   Double wood pieces supporting joists in a floor or double wood members placed on edge over windows and doors to transfer the roof and floor weight to the studs.

Heat exchanger.   A device for transferring thermal energy from one fluid to another.

Heating capacity (Btu per hour).   The quantity of heat in Btu that a heater or a room air conditioner is capable of adding to a room in one hour.

Heat transfer liquid.   A liquid which is used to transport thermal energy.

Hermetic.   Airtight seal, impervious to external influence.

Humid.   Containing or characterized by perceptible moisture, especially to the point of being emotionally depressing.

Humidifier.   A device for supplying or maintaining humidity; a device for maintaining or putting moisture into the air.

Humidify.   To make humid; to add water into the air.

Humidistat.   An instrument for regulating or maintaining the degree of humidity.

Humidity.   A moderate degree of wetness especially of the atmosphere; dampness.

Hydronic.   Of, relating to, or being a system of heating or cooling that involves transfer of heat by a circulating fluid (as water or vapor in a closed system of pipes).

Hygroscopic.   Readily taking up and retaining moisture; taken up and retained under some conditions of humidity and temperature.

Indraft.   Air movement into a fireplace.

Insolate.   To expose to the sun's rays.

Insolation.   Solar radiation that has been received; the rate of delivery of all direct solar energy per unit of horizontal surface.

Insulation.   A material used to separate conducting bodies by means of nonconductors to prevent transfer of heat, electricity, or sound.

Jack stud.   A vertical support for a header placed above a window opening.

Jalousies.   Windows with movable, horizontal glass slats angled to admit ventilation and keep out rain. This term is also used for outside shutters of wood constructed in this way.

Jamb.   An upright surface that lines an opening for a door or window.

Joist.   A small rectangular sectional member arranged parallel from wall to wall in a building, or resting on beams or girders. Joists support a floor or the laths or furring strips of a ceiling.

K.   Thermal conductivity; k (written lower case) represents the amount of heat that passes through a uniform composition (homogenous) material one square foot, one inch thick, in one hour, with a temperature difference of one degree F between the inner and outer surfaces. The k value is expressed in Btu per hour. The lower the k value, the higher is its insulating ability.

Kiln dried.   Lumber that has had its moisture removed by placement in a heated enclosure (kiln); a method superior to air drying.

Kilowatt.   One thousand watts (about $1^1/_3$ horsepower).

Kilowatt-hour (kWh).   One thousand watt-hours. A unit of electrical energy equal to the energy delivered by the flow of one kilowatt of electrical power for one hour.

Lath.   One of a number of thin, narrow strips of wood nailed to rafters, ceiling joists, wall studs, etc., to make a groundwork or key for slates, tiles, or plastering.

Lee.   The side that is sheltered from the wind.

Leeward.   Being in or facing the direction toward which the wind is blowing; the side opposite the windward; the lee-side.

Light.   A medium (as a window or windowpane) through which light is admitted; an electromagnetic radiation in the wavelength range including infrared, visible, ultraviolet, and X rays and traveling in a vacuum with a speed of about 186,281 miles per second.

Lintel.   The top piece over a door or window which supports walls above the opening. The horizontal support member of a fireplace which is the top of the opening.

Load-bearing wall.   A strong wall capable of supporting weight.

Louver.   An opening with horizontal slats to permit passage of air, but excluding rain, sunlight, and view.

Masonry. Walls built by a mason using brick, stone, tile or similar materials.

Mcf. One thousand cubic feet (of natural gas).

Medium (media, pl.). A device for conveying or transferring something. In a humidifier, a medium is an absorbent material (as a wick) that transfers water into the path of an air stream for evaporation.

Mildew. A superficial, usually whitish growth produced on organic matter or living plants by fungi; a fungus-producing mildew. Mildew growth is increased by humid conditions and appears on paneling, draperies, carpeting, and books.

Moisture barrier. Treated paper, plastic or metal that retards or bars water vapor, used to keep moisture from passing into walls or floors.

Molding. A strip of decorative material having a plane or curved, narrow surface prepared for ornamental application. These strips are often used to hide gaps at wall junctions.

Mullion. The vertical or horizontal divisions or joints between single windows in a multiple window unit.

Mullion casing. An interior or exterior casing member used to cover the mullion joint between single windows.

Nuclear energy. Energy, largely in the form of heat, produced during nuclear chain reaction. This thermal energy can be transformed into electrical energy.

Opaque. Not pervious to radiant energy and especially light.

Passive solar system. An assembly of natural and architectural components including collectors, thermal storage device(s) and transfer fluid which converts solar energy into thermal energy in a controlled manner and in which no pumps are used to accomplish the transfer of thermal energy.

Perlite. Volcanic glass that has a concentric, shelly structure; appears as if composed of concretions, is usually grayish and sometimes spherulitic, and, when expanded by heat, forms a lightweight aggregate used especially in concrete and plaster.

Pitch. The angle of slope of a roof.

Plant. A young tree, vine, shrub, or herb planted or suitable for planting.

Plaster. A pasty composition (as of lime, water, and sand) that hardens on drying and is used for coating walls, ceilings, and partitions.

Plasterboard. A board used in large sheets as a backing or as a substitute for plaster in walls and consisting of several plies of fiberboard, paper, or felt, usually bonded to a hardened gypsum plaster core; gypsum board, used instead of plaster. See dry wall.

Plenum. A chamber which can serve as a distribution area for heating or cooling systems, generally between a false ceiling and the actual ceiling. Also, the air chamber above the burner of a hot air furnace.

Pointing. Treatment of joints in masonry by filling with mortar to improve appearance or protect against weather.

Post and beam construction. Wall construction in which beams are supported by heavy posts rather than many smaller studs.

R. The unit of thermal resistance. The R value represents the resistance of a material to the flow of heat. The higher the R value, the higher the insulating ability. Two different materials having the same R value, regardless of material or thickness, provide the same resistance to the flow of heat. R is the reciprocal of U which is the overall coefficient of heat transmission. Thus, $R = 1/U$.

Radiant heat. Coils of electric wire, hot water or steam pipes embedded in floors, ceilings, or walls to heat rooms; invisible heat rays that are emitted from fires or flow from hotter to colder objects through air; heat transmitted by radiation as contrasted to convection or conduction.

Rafter. One of a series of structural roof members spanning from an exterior wall to a center beam or ridge board.

Register. A grille equipped with a damper or volume control.

Relative humidity. The ratio of the amount of water vapor actually present in the air to the greatest amount possible at the same temperature.

Rock wool. Mineral wool made by blowing a jet of steam through molten rock (as limestone or siliceous rock) or through slag, and used chiefly for heat and sound insulation.

Roof sheathing. Sheets, usually plywood, which are nailed to the top edges of trusses or rafters to tie the roof together and support the roofing material.

Sash. The movable part of a window; the frame in which panes of glass are set in a window or door.

Sash balance. A system of weights, cords and pulleys, or coiled springs which assist in raising double-hung sashes and tend to keep the sash in any placed position by counterbalancing the weight of the sash.

Sash lock. Generally a cam-action type lock applied to the check rails of a sliding window or at the open edges of a projecting window to pull the check rails tightly together or to seal the sash tightly to the frame, both for security and weatherizing.

Shakes. Hand-cut wood shingles.

Sheathing. The first covering of boards or material on the outside wall or roof prior to installing the finished siding or roof covering.

Sheetrock. A trademark used for plasterboard.

Shingles. Pieces of wood, asbestos, or other material used as an overlapping outer covering on walls or roofs.

Siding. Boards of special design nailed horizontally to vertical studs with or without intervening sheathing to form the exposed surface of outside walls of frame buildings.

Sill. The horizontal member that forms the bottom of a window frame.

Sill plate. The lowest member of the house framing resting on top of the foundation wall. Also called the mud sill.

Slab. Concrete floor placed directly on earth or a gravel base and usually about four inches thick.

Soffit. The visible underside of structural members such as staircases, cornices, beams, a roof overhang, or eave.

Softwood. Easily worked wood or wood from a cone-bearing tree.

Soil pipe (stack). Vertical plumbing pipe for waste water.

Solar building. A building which utilizes solar energy by means of an active or passive solar system.

Solar domestic hot water system. The complete assembly of subsystems and components necessary to convert solar energy into thermal energy for domestic hot water in combination with auxiliary energy when required.

Solar energy. Energy radiated directly from the sun.

Stop.   A wood trim member nailed to the window frame to stop the sash of a projecting window when closed to prevent it from swinging through the opening. It also covers the perimeter crack between the sash and the window frame.

Storage device.   The device used to store thermal energy in a solar heating or cooling system. Usually an insulated tank holding a liquid, an insulated rock bin, or a combination of both.

Storage subsystem.   The assembly used for storing thermal energy so it can be used when required.

Studs.   The vertical members of a wall to which horizontal pieces are nailed. Studs are spaced either 16 inches or 24 inches center to center.

Subfloor.   Usually plywood sheets are nailed directly to the floor joists; receives the finish flooring.

Subsystem.   A major, separable, functional assembly of a system such as a complete collector or storage assembly.

Temperate.   Marked by moderation; keeping or held within limits; not extreme or excessive.

Therm.   A unit of heat equal to 100,000 Btu.

Thermal barrier.   A strip of nonconducting material, such as wood, vinyl, or foam rubber, which is used to separate the inside and outside surfaces of a metal window sash or frame, or a metal door or sill to stop the conduction of heat to the outside which results in a cold inside surface.

Thermal conductance.   Refer to C.

Thermal conductivity.   Refer to k.

Thermal energy.   A form of energy the effect of which (heat) is produced by accelerated vibration of molecules.

Thermal resistance.   Refer to R.

Thermostat.   An automatic device for regulating temperature.

Toenail.   Driving nails at an angle into corners or other joints.

Transfer fluid.   A liquid (usually water or water plus an additive) or air used to transfer thermal energy from the collector to the storage or distribution system.

Translucent.   Permitting the passage of light; clear.

U value.   A term used to describe the insulating characteristics of materials; the coefficient of transmission of heat. R value describes the resistance of a specified material where the U value includes all the components involved, such as the siding, sheathing, insulation, and wallboard of a wall. The lower the U value, the better the resistance to heat.

VAC.   Volts, alternating current.

Valley.   The depression at the meeting point of two roof slopes.

Vapor barrier.   Material such as paper, plastic, metal or paint which is used to prevent vapor from passing from rooms into the outside walls.

Vegetation.   Plant life or total plant cover.

Vent.   An opening for the escape of a gas or liquid, or for the relief of pressure.

Ventilation.   Circulation of air; a system or means of providing fresh air.

Ventilator.   A contrivance for introducing fresh air or expelling foul or stagnant air.

Vent pipe.   A pipe which allows gas to escape from plumbing systems.

Vermiculite. Any of various micaceous minerals that are hydrous silicates resulting usually from expansion of the granules of mica at high temperatures to give a lightweight, highly water-absorbent material.

Wall sheathing. Sheets of plywood, gypsum board, or other material nailed to the outside face of studs and used as a base for exterior siding.

Watt. The amount of power available from an electric current of one ampere at a potential of one volt.

Weather stripping. Metal, wood, plastic or other material installed around door and window openings to prevent heat escapement or cold infiltration.

Weep hole. A small hole at the bottom of a storm window which permits water to drain.

Wind cap. Raised metal plate on top of a chimney to change or increase draft.

Wind energy. Energy derived from the wind.

Window, awning. A projecting window, hinged at the top, opening up and out, like an awning.

Window, bay. Composed of three or more individual windows, generally with the side or flanker units at 45 degrees or 30 degrees to the wall. A bay projects from the wall of the structure.

Window, bow. Composed of three or more individual windows in a gently curved contour. Bow windows project from the wall of the structure.

Window, casement. A projecting window hinged at the sides and usually opening outward like a door.

Window, combination storm-screen. See *combination storm-screen.*

Window, double-hung. Two vertically sliding sash which bypass each other in a single frame. Sash may be counterbalanced by weights or springs.

Window, picture. A large, stationary (nonventilating) window which is designed for a maximum view without obstruction.

Window, sliding. A window with two or more sash that slide past each other within the frame (they may slide horizontally or vertically).

Windward. Being in or facing the direction from which the wind is blowing; the side or direction from which the wind is blowing.

Winter design temperature. The lowest temperature expected during a winter season (construction and insulation should be such as to provide efficient protection against this temperature).

INDEX

DATE DUE

30 505 JOSTEN'S